MANUAL ON THE
USE OF THERMOCOUPLES IN
TEMPERATURE MEASUREMENT

Sponsored by ASTM
Committee E-20 on
Temperature Measurement
and Subcommittee IV on
Thermocouples
AMERICAN SOCIETY FOR
TESTING AND MATERIALS

ASTM SPECIAL TECHNICAL PUBLICATION 470

List price $17.00

04–470000–40

 AMERICAN SOCIETY FOR TESTING AND MATERIALS
1916 Race Street, Philadelphia, Pa. 19103

NOTE

The Society is not responsible, as a body,
for the statements and opinions
advanced in this publication.

Printed in Lancaster, Pa.

Second Printing,
April, 1973

Foreword

The Manual on the Use of Thermocouples in Temperature Measurements was sponsored and compiled by Committee E-20 on Temperature Measurement and Subcommittee IV on Thermocouples of the American Society for Testing and Materials. The editorial work was co-ordinated by R. P. Benedict, Westinghouse Electric Corp.

**Related
ASTM PUBLICATIONS**

Thermal Conductivity Measurements of Insulating
Materials at Cryogenic Temperatures,
STP 411 (1967), $9.50

Contents

Acknowledgments

Editors for this Manual

R. P. Benedict (chairman), Westinghouse Electric Corp.
E. L. Lewis (secretary), Naval Ship Engineering Center, Philadelphia Div.
R. S. Flemons, Canadian General Electric Co., Ltd.
H. J. Greenberg, Engelhard Minerals and Chemical Corp.
J. L. Howard, The Boeing Co.
A. J. Otter, Atomic Energy of Canada, Ltd.
J. D. Sine, Honeywell, Inc.
R. C. Stroud, Leeds and Northrup Co.
J. F. Swindells, National Bureau of Standards, Washington, D.C.

Officers of Committee E-20

R. D. Thompson (chairman), Taylor Instrument Div., Sybron Corp.
D. I. Finch (vice chairman), Leeds and Northrup Co.
H. J. Greenberg (secretary), Engelhard Minerals and Chemical Corp.
R. F. Abrahamsen (membership), Combustion Engineering, Inc.

Officers of Subcommittee IV

D. I. Finch (chairman), Leeds and Northrup Co.
E. D. Zysk (secretary), Engelhard Minerals and Chemical Corp.
A. J. Gottlieb (membership), Wilbur B. Driver Co.

Those Primarily Responsible for Individual Sections of the Manual

Introduction–J. L. Howard, The Boeing Co.
Principles–R. P. Benedict, Westinghouse Electric Corp.
Common Thermocouples–J. D. Sine, Honeywell, Inc.
Extension Wires–F. S. Sibley, Hoskins Mfg. Co.
Nonstandard Thermocouples–H. W. Deem, Battelle Memorial Institute, J. A. Bard, Matthey Bishop, Inc.
Compatibility–D. L. Clark, Oak Ridge National Laboratory, J. L. Howard, The Boeing Co.
Fabrication–H. J. Greenberg, Engelhard Minerals and Chemical Co.
Sheathed Thermocouples–J. C. Faul, American Standard Aero Research Instrument Department
Potentiometers–R. C. Stroud, Leeds and Northrup Co.
Reference Junctions–R. S. Flemons, Canadian General Electric Co., Ltd.
Calibration–J. F. Swindells, National Bureau of Standards, Washington, D. C.
Single Element Calibration–C. L. Guettel, Driver-Harris Co.
Fluid Installations–R. P. Benedict, Westinghouse Electric Corp.
Solid Installations–A. J. Otter, Atomic Energy of Canada, Ltd.
Reference Tables–J. D. Sine, Honeywell, Inc.
Cryogenics–R. L. Powell, National Bureau of Standards, Boulder
Bibliography–E. L. Lewis, Naval Ship Engineering Center, Philadelphia Div.

In addition to those listed, many other members of Committee E-20 have made substantial contributions to this manual as authors and reviewers. Their help is gratefully acknowledged.

Manual on the Use of Thermocouples in Temperature Measurement

1. INTRODUCTION

This manual has been prepared by Subcommittee IV of ASTM Committee E-20 on Temperature Measurement. The responsibilities of ASTM Committee E-20 include "Assembling a consolidated source book covering all aspects relating to accuracy, application, and usefulness of thermometric methods." This manual is addressed to the thermocouple portion of this responsibility.

The contents include principles, circuits, standard electromotive force (emf) tables, stability and compatibility data, installation techniques, and other information required to aid both the beginner and the experienced user of thermocouples. While the manual is intended to be comprehensive, the material, however, will not be adequate to solve all the individual problems associated with many applications. To further aid the user in such instances, there are numerous references and an extensive bibliography. In addition to presenting technical information, an attempt is made to properly *orient* a potential user of thermocouples. Thus, it is hoped that the reader of this manual will make fewer mistakes than the nonreader.

Regardless of how many facts are presented herein and regardless of the percentage retained, all will be for naught unless one simple important fact is kept firmly in mind. The thermocouple reports only what it "feels." This may or may not be the temperature of interest. The thermocouple is influenced by its entire environment, and it will tend to attain thermal equilibrium with this environment, not merely part of it. Thus, the environment of each thermocouple installation should be considered unique until proven otherwise. Unless this is done, the designer will likely overlook some unusual, unexpected, influence.

Of all the available temperature transducers, why use a thermocouple in a particular application? There are numerous advantages to consider. Physically, the thermocouple is inherently simple, being only two wires joined together at the measuring end. The thermocouple can be made large or small depending on the life expectancy, drift, and response-time requirements. It may be flexible, rugged, and generally is easy to handle and install. A thermocouple normally covers a wide range of temperatures and its output is reasonably linear over portions of that range. Unlike many temperature transducers, the thermocouple is not subject to self-heating problems. In practice, thermocouples of the same type are interchangeable within specified limits of error. Also, thermocouple materials are readily available at reasonable cost, the expense in most cases being nominal.

The bulk of the manual is devoted to identifying material characteristics and discussing application techniques. Every section of the manual is essential to an understanding of thermocouple applications. Each section should be studied carefully. Information should not be used out of context. The general philosophy should be—let the user beware.

2. PRINCIPLES OF THERMOELECTRIC THERMOMETRY

The principles, or theory, underlying thermoelectric effects were not established by one man at one time, but by several scientists working over a span of many years beginning with Alessandro Volta, who concluded in 1800 that the electricity which caused Galvani's frog to twitch was due to a contact of two dissimilar metals. This conclusion was the forerunner of the principle of the thermocouple. Others built on this base; for example, Thomas Johann Seebeck (1821), Jean Charles Althanase Peltier (1834), and William Thomson—later Lord Kelvin—(1848–1854). During this same period, Jean Baptiste Joseph Fourier published his basic heat-conduction equation (1821), Georg Simon Ohm discovered his celebrated equation for electrical conduction (1826), James Prescott Joule found the principle of the first law of thermodynamics and the important I²R heating effect (1840–1848), and Rudolf Julius Emanuel Clausius announced the principle of the second law of thermodynamics and introduced the concept of entropy (1850).

2.1 Historical Development of Basic Relations

2.1.1 Seebeck

Seebeck discovered the existence of thermoelectric currents while observing electromagnetic effects associated with bismuth-copper and bismuth-antimony circuits. His experiments showed that, when the'junctions of two dissimilar metals forming a closed circuit are exposed to different temperatures, a net thermal electromotive force is generated which induces a continuous electric current.

The Seebeck effect concerns the net conversion of thermal energy into electrical energy with the appearance of an electric current. The Seebeck voltage refers to the net thermal electromotive force set up in a thermocouple under zero-current conditions. The direction and magnitude of the

Seebeck voltage, E_S, depend upon the temperature of the junctions and upon the materials making up the thermocouple. For a particular combination of materials, A and B, for a small temperature difference

$$dE_S = \alpha_{A,B} dT \qquad \text{.....................} \quad (1)$$

where $\alpha_{A,B}$ is a coefficient of proportionality called the Seebeck coefficient. (This commonly is called the thermoelectric power.) The Seebeck coefficient is obtained usually in one of two ways: (1) as an algebraic sum, $\alpha_{A,B}$, of relative Seebeck coefficients, α_{AR} and α_{BR}, where, for a given temperature difference and at given temperature levels, emf's of each of the substances, A and B, making up the thermocouple are obtained with respect to an arbitrary reference material, R; and (2) by numerically differentiating tabulated values of E_S versus T for a given reference temperature, T_R, according to the relation

$$E_S = \int_{T_R}^{T} \alpha_{A,B} dT \qquad \text{..................} \quad (2)$$

In either case, the Seebeck coefficient represents, for a given material combination, the net change in thermal emf caused by a unit temperature difference; that is,

$$\alpha_{A,B} = \lim_{\Delta T \to 0} \frac{\Delta E_S}{\Delta T} = \frac{dE_S}{dT} \qquad \text{..............} \quad (3)$$

Thus, if $E = aT + 1/2 bT^2$ is determined by calibration, then $\alpha = a + bT$. Note that, based on the validity of the experimental relation,

$$E_S = \int_{T_2}^{T} \alpha dT = \int_{T_1}^{T} \alpha dT - \int_{T_1}^{T_2} \alpha dT \qquad \text{.........} \quad (4)$$

where $T_1 < T_2 < T$, it follows that α is entirely independent of the reference temperature employed. In other words, for a given combination of materials, the Seebeck coefficient is a function of temperature level only.

2.1.2 Peltier

Peltier discovered peculiar thermal effects when he introduced small, external electric currents in Seebeck's bismuth-antimony thermocouple. His experiments show that, when a small electric current is passed across the junction of two dissimilar metals in one direction, the junction is cooled (that is, it acts as a heat sink) and thus absorbs heat from its surroundings. When the direction of the current is reversed, the junction is heated (that is, it acts as a heat source) and thus releases heat to its surroundings.

The Peltier effect concerns the reversible evolution, or absorption, of heat which usually takes place when an electric current crosses a junction between two dissimilar metals. (In certain combinations of metals, at certain temperatures, there are thermoelectric neutral points where no Peltier effect is apparent.) This Peltier effect takes place whether the current is introduced externally or is induced by the thermocouple itself. The Peltier heat was found early to be proportional to the current, and may be written

$$dQ_P = \pi I dt \quad \dots\dots\dots\dots\dots\dots \quad (5)$$

where π is a coefficient of proportionality known as the Peltier coefficient or the Peltier voltage. Note that π represents the reversible heat which is absorbed, or evolved, at the junction when unit current passes across the junction in unit time, and that it has the dimensions of voltage. The direction and magnitude of the Peltier voltage depend upon the temperature of the junction and upon the materials making up the junction; however, π at one junction is independent of the temperature of the other junction.

External heating, or cooling, of the junctions results in the converse of the Peltier effect. Even in the absence of all other thermoelectric effects, when the temperature of one junction (the reference junction) is held constant and when the temperature of the other junction is increased by external heating, a net electric current will be induced in one direction. If the temperature of the latter junction is reduced below the reference-junction temperature by external cooling, the direction of the electric current will be reversed. Thus, the Peltier effect is seen to be related closely to the Seebeck effect. Peltier himself observed that, for a given electric current, the rate of absorption, or liberation, of heat at a thermoelectric junction depends upon the Seebeck coefficient, α, of the two materials.

2.1.3 Thomson

It remained for Thomson (see the Kelvin relations discussed below) to show that α and π are related by the absolute temperature. (We might appropriately mention at this time that the Peltier thermal effects build up a potential difference opposing the thermoelectric current, thus negating the perpetual-motion question.) Thomson came to the remarkable conclusion that an electric current produces different thermal effects, depending upon the direction of its passage from hot to cold or from cold to hot, in the same metal. By applying the (then) new principles of thermodynamics to the thermocouple, and by disregarding (with tongue in cheek) the irreversible I^2R and conduction-heating processes, Thomson reasoned that, if an electric current produces only the reversible Peltier heating effects, then the net Peltier voltage will equal the Seebeck voltage and will be linearly proportional to the temperature difference at the junctions of the thermocouple.

This reasoning led to requirements at variance with observed characteristics (that is, $dE_S/dT \neq$ constant). Therefore, Thomson concluded that the net Peltier voltage is not the only source of emf in a thermocouple circuit, but that the single conductor itself, whenever it is exposed to a longitudinal temperature gradient, must be also a seat of emf. (Becquerel had at that time already discovered a thermoelectric neutral point, that is, $E_S = 0$, for an iron-copper couple at about 280 C. Thomson agreed with Becquerel's conclusion and started his thermodynamic reasoning from there.)

The Thomson effect concerns the reversible evolution, or absorption, of heat occurring whenever an electric current traverses a single homogeneous conductor, across which a temperature gradient is maintained, regardless of external introduction of the current or its induction by the thermocouple itself. The Thomson heat absorbed, or generated, in a unit volume of a conductor is proportional to the temperature difference and to the current, that is,

$$dQ_T = \pm \sigma dT I dt \qquad \ldots\ldots\ldots\ldots\ldots\ldots (6)$$

where σ is a coefficient of proportionality called the Thomson coefficient. Thomson refers to this as the specific heat of electricity because of an apparent analogy between σ and the usual specific heat, c, of thermodynamics. Note that σ represents the rate at which heat is absorbed, or evolved, per unit temperature difference per unit current, whereas c represents the heat transfer per unit temperature difference per unit mass. The Thomson coefficient is seen also to represent an emf-per-unit difference in temperature. Thus, the total Thomson voltage set up in a single conductor may be expressed as

$$E_T = \int_{T_1}^{T_2} \sigma dT \qquad \ldots\ldots\ldots\ldots\ldots\ldots (7)$$

where its direction and magnitude depend upon temperature level, temperature difference, and material considered. Note that the Thomson voltage alone cannot sustain a current in a single homogeneous conductor forming a closed circuit, since equal and opposite emf's will be set up in the two paths from heated to cooled parts.

Soon after his heuristic reasoning, Thomson succeeded in demonstrating indirectly the existence of the predicted Thomson emf's. He sent an external electric current through a closed circuit, formed of a single homogeneous conductor which was subjected to a temperature gradient, and found the I^2R heat to be augmented slightly, or diminished, by the reversible Thomson heat in the paths from cold to hot or from hot to cold, depending upon the direction of the current and the material under test.

2.14 *Interim Summary*

In summary, thermoelectric currents may exist whenever the junctions of a circuit formed of at least two dissimilar metals are exposed to different temperatures. This temperature difference always is accompanied by irreversible Fourier heat conduction, while the passage of electric currents always is accompanied by irreversible Joule heating effects. At the same time, the passage of electric currents always is accompanied by reversible Peltier heating or cooling effects at the junctions of the dissimilar metals, while the combined temperature difference and passage of electric current always is accompanied by reversible Thomson heating or cooling effects along the conductors. The two reversible heating-cooling effects are manifestations of four distinct emf's which make up the net Seebeck emf:

$$E_S = \pi_{A,B}|_{T_2} - \pi_{A,B}|_{T_1} + \int_{T_1}^{T_2} \sigma_A dT - \int_{T_1}^{T_2} \sigma_B dT = \int_{T_1}^{T_2} \alpha_{A,B} dT \dots (8)$$

where the three coefficients, α, π, σ, are related by the Kelvin relations.

2.1.5 *Kelvin Relations*

Assuming that the irreversible I^2R and heat-conduction effects can be disregarded completely (actually, they can be only minimized since, if thermal conductivity is decreased, electrical resistivity usually is increased, and vice versa), then the net rate of absorption of heat required by the thermocouple to maintain equilibrium in the presence of an electric current is

$$q = \frac{Q_{net}}{t} = \left[\pi_2 - \pi_1 + \int_1^2 (\sigma_A - \sigma_B) dT \right] I = E_S I \quad \dots \dots (9)$$

This is in accord with the first law of thermodynamics, according to which heat and work are mutually convertible. Thus, the net heat absorbed must equal the electric work accomplished or, in terms of a unit charge of electricity, the Seebeck emf, E_S, which may be expressed in the differential form

$$dE_S = d\pi + (\sigma_A - \sigma_B) dT \quad \dots \dots \dots \dots (10)$$

The second law of thermodynamics may be applied also to the thermocouple cycle, the unit charge of electricity again being considered, as

$$\Delta S_{rev} = \sum \frac{\Delta Q}{T_{abs}} = 0 \quad \dots \dots \dots \dots (11)$$

where ΔQ implies the various components of the net heat absorbed (that is, the components of E_S), and T_{abs} implies the temperature at which the heat is transferred across the system boundaries. Equation 11 can be expressed in the differential form

$$dS_{rev} = d\left(\frac{\pi}{T}\right) + \frac{(\sigma_A - \sigma_B)}{T} dT = 0 \quad \ldots\ldots\ldots\ldots \text{ (12)}$$

Combining the differential expressions for the first and second laws of thermodynamics, we obtain the Kelvin relations:

$$\pi_{A,B} = T_{abs}\left(\frac{dE_S}{dT}\right) = T_{abs}\alpha_{A,B} \quad \ldots\ldots\ldots\ldots \text{ (13)}$$

$$(\sigma_A - \sigma_B) = -T_{abs}\left(\frac{d^2E_S}{dT^2}\right) \quad \ldots\ldots\ldots\ldots\ldots \text{ (14)}$$

from which we can determine α, π, and $\Delta\sigma$, when E_S is obtained as a function of T. Thus, if

$$E_S = aT + \tfrac{1}{2}bT^2 + \cdots \quad \ldots\ldots\ldots\ldots\ldots \text{ (15)}$$

is taken to represent the thermoelectric characteristics of a thermocouple whose reference junction is maintained at 0 C, and where the coefficients, a and b, are obtained (for example) by the curve fitting of calibration data, then

$$\alpha = (a + bT + \cdots) \quad \ldots\ldots\ldots\ldots\ldots \text{ (16)}$$

$$\pi = T_{abs}(a + bT + \cdots) \quad \ldots\ldots\ldots\ldots \text{ (17)}$$

$$\Delta\sigma = -T_{abs}(b + \cdots) \quad \ldots\ldots\ldots\ldots \text{ (18)}$$

Examples of the use of these coefficients are given in Table 1.

TABLE 1—*Determination of various thermoelectric quantities applied to thermocouples.*

Given, the two constants, a and b, as determined with respect to platinum via Eq 15:

Metal	a, V/deg C	b, V/(deg C)²
Iron (Fe)	+16.7	−0.0297
Copper (Cu)...........................	+ 2.7	+0.0079
Constantan (Con)	−34.6	−0.0558

By way of illustration, consider the following combinations of materials: iron-copper and iron-constantan, with their measuring junctions at 200 C and their reference junctions at 0 C:

TABLE 1—*Determination of various thermoelectric quantities applied to thermocouples (cont'.).*

Iron-copper
$$a_{Fe-Cu} = a_{Fe} - a_{Cu} = 16.7 - 2.7 = 14 \ \mu V/\text{deg C}$$
$$b_{Fe-Cu} = b_{Fe} - b_{Cu} = -0.0297 - 0.0079$$
$$b_{Fe-Cu} = -0.0376 \ \mu V/(\text{deg C})^2$$

Iron-constantan
$$a_{Fe-Con} = a_{Fe} - a_{Con} = 16.7 - (-34.6) = 51.3 \ \mu V/\text{deg C}$$
$$b_{Fe-Con} = b_{Fe} - b_{Con} = -0.0297 - (-0.0558)$$
$$b_{Fe-Con} = 0.0261 \ \mu V/(\text{deg C})^2$$

Since Seebeck voltage $E_S = aT + 1/2bT^2$,

Iron-copper

$$E_S = 14(200) + \frac{1}{2}(-0.0376)(200)^2$$

$$E_S = 2048 \ \mu V$$

Iron-constantan

$$E_S = 51.3(200) + \frac{1}{2}(0.0261)(200)^2$$

$$E_S = 10,782 \ \mu V$$

Note how different combinations of materials give widely different thermal emf's.

Now we proceed to write expressions for α, π, and $\Delta\sigma$, to note how the separate emf's combine to give the (net) Seebeck emf: Since $\alpha_{A,B} = a_{A,B} + b_{A,B}T$ = Seebeck coefficient

Iron-copper
$$\alpha_0 = 14 + (-0.0376)(0) = 14 \ \mu V/\text{deg C}$$

$$\alpha_{200} = 14 + (-0.0376)(200) = 6.48 \ \mu V/\text{deg C}$$

Iron-constantan
$$\alpha_0 = 51.3 + 0.0261(0) = 51.3 \ \mu V/\text{deg C}$$

$$\alpha_{200} = 51.3 + 0.0261(200) = 56.52 \ \mu V/\text{deg C}$$

Note that it is the great difference in Seebeck coefficients (thermoelectric powers) for the two combinations which accounts for the difference in thermal emf's:

$$E_S = \int_{T_R}^{T} \alpha_{A,B} \, dT$$

Since $\pi_{A,B} = T_{abs}\alpha_{A,B}$ = Peltier coefficient = Peltier voltage

Iron-copper
$$\pi_0 = 273(14) = 3822 \ \mu V$$

$$\pi_{200} = 473(6.48) = 3065 \ \mu V$$

Iron-constantan
$$\pi_0 = 273(51.3) = 14.005 \ \mu V$$

$$\pi_{200} = 473(56.52) = 26.734 \ \mu V$$

TABLE 1—*Determination of various thermoelectric quantities applied to thermocouples (cont'.)*

Note that, in the case of the iron-copper (Fe-Cu) couple, $\pi_{cold} > \pi_{hot}$, whereas in the more usual Fe-Con couple, $\pi_{hot} > \pi_{cold}$.

Since $\Delta\sigma_{A,B} = -b_{A,B} T_{abs} = $ Thomson coefficient, and

$$E_T = \int_{T_{R\,abs}}^{T_{abs}} \Delta\sigma \, dT = \frac{1}{2} b_{A,B}(T_{R\,abs}^2 - T_{abs}^2) = \text{Thomson voltage}$$

Iron-copper

$$E_T = \frac{0.0376}{2}(273^2 - 473^2)$$

$$E_T = 2805 \; \mu V$$

Iron-constantan

$$E_T = \frac{0.0261}{2}(273^2 - 473^2)$$

$$E_T = -1947 \; \mu V$$

We sum the various components

$$E_S = \pi_2 - \pi_1 + \int_1^2 \Delta\sigma \, dT = \text{Seebeck voltage}$$

Iron-copper
$$E_S = 3065 - 3822 + 2805$$

$$E_S = 2048 \; \mu V$$

Iron-constantan
$$E_S = 26.734 - 14.005 - 1947$$

$$E_S = 10.782 \; \mu V$$

These figures of course, check with the original calculations. Note that, in the Fe-Cu case, the net Thomson emf far outweighs in importance the net Peltier emf, whereas in the Fe-Con case, the converse is true.

2.1.6 *Onsager Relations*

The historical viewpoint presented thus far has avoided the very real irreversible I^2R and heat conduction in order to arrive at the useful and experimentally confirmed Kelvin relations. We shall now discuss how the present-day, irreversible thermodynamic viewpoint removes this flaw in our reasoning.

Basically, we judge whether a given process is reversible or irreversible by noting the change in entropy accompanying a given change in the thermodynamic state. Thus, if $dS > \delta Q_q/T_{abs}$, we say the process is irreversible; or, stated in a more useful manner,

$$dS_{system} = dS_{\substack{across \\ boundary}} + dS_{\substack{produced \\ inside}} \qquad \ldots\ldots\ldots\ldots (19)$$

or

$$dS_s = dS_0 + dS_i = \frac{\delta Q_q}{T_{abs}} + \frac{\delta F}{T_{abs}} \qquad \dots \dots \dots \dots (20)$$

Hence, only in the absence of entropy within the system boundaries do we have the reversible case, $dS_{rev} = \delta Q_q/T_{abs}$, which may be handled adequately by classical thermodynamics in the steady and quasi-steady states. Evidently, the rate of production of entropy per unit volume, ξ, is an important quantity in irreversible thermodynamics, which may be expressed as

$$\xi = \left(\frac{1}{Adx}\right)\frac{dS_i}{dt} = \left(\frac{1}{Adx}\right)\frac{\delta F}{T_{abs}dt} \qquad \dots \dots \dots \dots (21)$$

where Adx is the area times the differential length.

Another significant quantity, the product $T_{abs}\xi$ (called the dissipation), always can be split into two terms or a sum of two terms; one associated with a flow, J, and the other associated with a force, X. Furthermore, in many simple cases a linear relation is found (by experiment) to exist between the flow and force terms so defined. For example, in the one-dimensional, isothermal, steady flow of electric charges, $\delta Q_e/dt$ across a potential gradient, $-dE/dx$, it may be shown that

$$T_{abs}\xi = \left(\frac{I}{A}\right)\left(-\frac{dE}{dx}\right) = J_e X_e \qquad \dots \dots \dots \dots (22)$$

where J_e and X_e represent the electric flow and force terms, respectively, as defined by the entropy production method. The term J_e represents the electric-current density and the term X_e the electric-field strength or the electromotive force, which of course are related by the linear Ohm's law (that is, $J_e = L_e X_e$, where L_e represents the electrical conductivity). Again, in the one-dimensional, steady flow of thermal charges, dQ_q/dt, across a temperature gradient, $-dT/dx$, it may be shown that

$$T_{abs}\xi = \left(\frac{Q}{A}\right)\left(\frac{1}{T_{abs}}\frac{dT}{dx}\right) = J_q X_q \qquad \dots \dots \dots \dots (23)$$

where J_q and X_q represent the thermal flow and force, respectively, as defined by the entropy production method. The term J_q represents the thermal current density, and the term X_q represents the thermomotive force, which are, of course, related by the linear Fourier's law (that is, $J_q = L_q X_q$ where L_q represents the product of the thermal conductivity and the absolute temperature). It has been found that, even in complex situations, it always may be stated that

$$T_{abs}\xi = \sum J_K X_K \quad \dots\dots\dots\dots\dots \quad (24)$$

When several irreversible transport processes occur simultaneously (as, for example, the electric and thermal conduction in a thermocouple), they usually will interfere with each other; therefore, the linear relations must be generalized to include the various possible interaction terms. Thus, for the combined electric and thermal effects we would write

$$J_e = L_{ee}X_e + L_{eq}X_q \quad \dots\dots\dots\dots\dots \quad (25)$$

$$J_q = L_{qe}X_e + L_{qq}X_q \quad \dots\dots\dots\dots\dots \quad (26)$$

or, in general

$$J_i = \sum L_{ij}X_j \quad \dots\dots\dots\dots\dots \quad (27)$$

We have just seen that an entropy production necessarily accompanies both the I^2R and heat conduction effects (that is, they are irreversible); therefore, the Kelvin relations could not follow from reversible thermodynamic theory without certain intuitive assumptions. By reasoning that the electric and thermal currents were independent, Thomson tacitly assumed that $L_{eq} = L_{qe}$ as we shall subsequently show. Experimentally, this reciprocal relationship often was found to be true. The American chemist, Lars Onsager, proved in 1931 from a statistical-mechanics viewpoint that the assumption

$$L_{ij} = L_{ji} \quad \dots\dots\dots\dots\dots \quad (28)$$

is always true when the linear relations between flows, J_k, and forces, X_k, are valid. The Onsager reciprocal relation forms the basis of irreversible thermodynamics. By applying these concepts to the processes involved in the thermocouple, we are led rationally and unambiguously to the Kelvin relations. Thus, whenever the junctions of a thermocouple are maintained at different temperatures, we expect that an electric potential difference, an electric current, and a thermal current will be present. The dissipation for this thermoelectric process is simply the sum of the electric and thermal terms previously given. That is

$$T_{abs}\xi = \frac{I}{A}\left(-\frac{dE}{dT}\right) + \frac{Q}{A}\left(\frac{1}{T_{abs}}\frac{dT}{dx}\right) \quad \dots\dots\dots\dots \quad (29)$$

The generalized linear laws for this case also have been given as

$$J_e = L_{ee}\left(-\frac{dE}{dT}\right) + \frac{L_{eq}}{T_{abs}}\left(\frac{dT}{dx}\right) \quad \dots\dots\dots\dots (30)$$

$$J_q = L_{qe}\left(-\frac{dE}{dx}\right) + \frac{L_{qq}}{T_{abs}}\left(\frac{dT}{dx}\right) \quad \dots\dots\dots\dots (31)$$

Recalling that the Seebeck emf is determined under conditions of zero electric current, the Seebeck coefficient, α, may be expressed in terms of the Onsager coefficients as

$$\alpha = \left(\frac{dE_S}{dT}\right)_{I=0} = \frac{L_{eq}}{L_{ee}T_{abs}} \quad \dots\dots\dots\dots (32)$$

Recalling that the Peltier coefficient, π, represents the heat absorbed, or evolved, with the passage of an electric current across an isothermal junction, this too may be expressed in terms of the Onsager coefficients as

$$\pi = \left(\frac{J_q}{J_e}\right)_{dT=0} = \frac{L_{qe}}{L_{ee}} \quad \dots\dots\dots\dots (33)$$

Finally, we recall that Thomson found experimentally (and expressed in the Kelvin relations) that the Seebeck and Peltier coefficients are related, as shown in Eq 13.

$$\pi = T_{abs}\left(\frac{dE_S}{dT}\right) \quad \dots\dots\dots\dots (34)$$

In terms of the Onsager coefficients, this requires that

$$\frac{L_{qe}}{L_{ee}} = T_{abs}\left(\frac{L_{eq}}{L_{ee}T_{abs}}\right) \quad \dots\dots\dots\dots (35)$$

or

$$L_{qe} = L_{eq} \quad \dots\dots\dots\dots (36)$$

which indicates that the experimental results agree with those which are predicted by the entropy production-linear law-Onsager reciprocal relation approach; in other words, by irreversible thermodynamics, without using any intuitive assumption. The Kelvin relations, also in accord with experiment, must follow.

2.2 Laws of Thermoelectric Circuits

Numerous investigations of thermoelectric circuits in which accurate measurements were made of the current, resistance, and electromotive force have resulted in the establishment of several basic laws. These laws have been established experimentally beyond a reasonable doubt and may be accepted in spite of any lack of a theoretical development.

2.2.1 Law of Homogeneous Metals

A thermoelectric current cannot be sustained in a circuit of a single homogeneous material, however varying in cross section, by the application of heat alone.

A consequence of this law is that two different materials are required for any thermocouple circuit. Experiments have been reported suggesting that a nonsymmetrical temperature gradient in a homogeneous wire gives rise to a measurable thermoelectric emf. A preponderance of evidence indicates, however, that any emf observed in such a circuit arises from the effects of local inhomogeneities. Furthermore, any current detected in such a circuit when the wire is heated in any way whatever is taken as evidence that the wire is inhomogeneous.

2.2.2 Law of Intermediate Metals

The algebraic sum of the thermoelectromotive forces in a circuit composed of any number of dissimilar materials is zero if all of the circuit is at a uniform temperature.

A consequence of this law is that a third homogeneous material always can be added in a circuit with no effect on the net emf of the circuit so long as its extremities are at the same temperature. Therefore, it is evident that a device for measuring the thermoelectromotive force may be introduced into a circuit at any point without affecting the resultant emf, provided all of the junctions which are added to the circuit by introducing the device are all at the same temperature. It also follows that any junction whose temperature is uniform and which makes a good electrical contact does not affect the emf of the thermoelectric circuit regardless of the method employed in forming the junction (Fig. 1).

Another consequence of this law may be stated as follows. If the thermal emfs of any two metals with respect to a reference metal (such as C) are known, then the emf of the combination of the two metals is the algebraic sum of their emfs against the reference metal (Fig. 2).

2.2.3 Law of Successive or Intermediate Temperatures

If two dissimilar homogeneous metals produce a thermal emf of E_1, when the junctions are at temperatures T_1 and T_2, and a thermal emf of E_2, when the junctions are at T_2 and T_3, the emf generated when the junctions are at T_1 and T_3, will be $E_1 + E_2$.

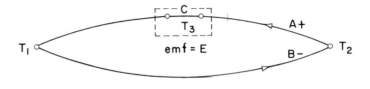

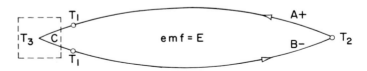

FIG. 1—*E unaffected by third material, C.*

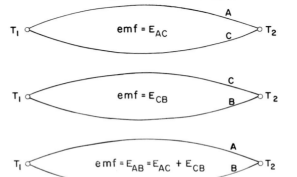

FIG. 2—*Emf's are additive for materials.*

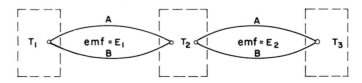

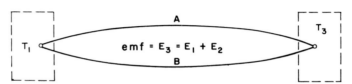

FIG. 3—*Emf's are additive for temperature intervals.*

One consequence of this law permits a thermocouple, calibrated for a given reference temperature, to be used with any other reference temperature through the use of a suitable correction (see Fig. 3 for a schematic example).

Another consequence of this law is that extension wires, having the same thermoelectric characteristics as those of the thermocouple wires, can be introduced in the thermocouple circuit (say from region T_2 to region T_3 in Fig. 3) without affecting the net emf of the thermocouple.

2.3 Elementary Thermoelectric Circuits

Two continuous, dissimilar thermocouple wires extending from the measuring junction to the reference junction, when used together with copper connecting wires and a potentiometer, connected as shown in Fig. 4, make up the basic thermocouple circuit.

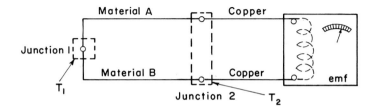

(a) For temperature level (Junction 2 is held at a constant, known reference temperature)

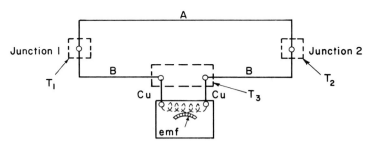

(b) For temperature difference (Junctions I and 2 are each exposed to unknown environment temperatures)

FIG. 4—*Several methods for introducing copper extension wires in elementary thermocouple circuits.*

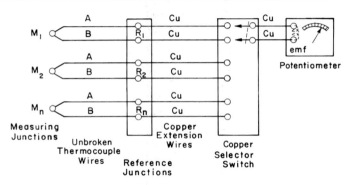

FIG. 5—*Basic thermocouple circuit.*

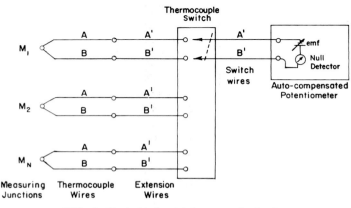

FIG. 6—*Typical industrial thermocouple circuits.*

An ideal circuit is given in Fig. 5 for use when more than one thermocouple is involved. The usual thermocouple circuit, however, includes: measuring junctions, thermocouple extension wires, reference junctions, copper connecting wires, a selector switch, and potentiometer, as indicated in Fig. 6. Many different circuit arrangements of the above components are also acceptable, depending on given circumstances, and these are discussed in the appropriate sections which follow.

2.4 Bibliography

2.4.1 *Early Historical References*

Volta, A., "On the Electricity Excited by Mere Contact of Conducting Substances of Different Kinds," *Philosophical Transactions*, 1800, p. 403.

Seebeck, T. J., "Evidence of the Thermal Current of the Combination Bi-Cu by its Action on Magnetic Needle, *Royal Academy of Science*, Berlin, 1822–1823, p. 265.

Fourier, J. B. J., *Analytical Theory of Heat*, Gautheir-Villars et Cie., Paris, 1822; English translation by Freeman, A., Cambridge University Press, Cambridge, 1878.

Ohm, G. S., "Determination of the Laws by Which Metals Conduct the Contact Electricity, and also a Draft for a Theory of the Voltage Apparatus," *Journal for Chemie and Physik (Schweigger's Journal)*, Vol. 46, 1826, p. 137.

Peltier, J. C. A., "Investigation of the Heat Developed by Electric Currents in Homogeneous Materials and at the Junction of Two Different Conductors," *Annalis de Chimie et de Physique*, Vol. 56 (2nd Series), 1834, p. 371.

Joule, J. P., "On the Production of Heat by Voltaic Electricity," *Proceedings of the Royal Society*, Dec. 1840.

Thomson, W., "On an Absolute Thermometric Scale Founded on Carnot's Theory of the Motive Power of Heat, and Calculated from Regnault's Observations," *Proceedings of the Cambridge Philosophical Society*, June 1848.

Clausius, R. J. E., "About the Motive Force of Heat," *Annalen der Physik und Chemie*, Vol. 79, 1850, pp. 368 and 500.

Thomson, W., "On a Mechanical Theory of Thermo-Electric Currents," *Proceedings of the Royal Society of Edinburgh*, Dec. 1851.

Thomson, W., "On the Thermal Effects of Electric Currents in Unequal Heated Conductors," *Proceedings of the Royal Society*, Vol. VII, May 1854.

2.4.2 *Recent References*

Benedict, R. P., "Thermoelectric Effects," *Electrical Manufacturing*, Feb. 1960, p. 103.

Finch, D. I., "General Principles of Thermoelectric Thermometry," *Temperature*, Vol. 3, Part 2, Reinhold, New York, 1962.

Roeser, W. F., "Thermoelectric Circuitry," *Journal of Applied Physics*, Vol. 11, 1940, p. 388.

Dike, P. H., "Thermoelectric Thermometry," *Leeds and Northrup Technical Publication EN-33A (1a)*, 3rd ed., April 1955.

Stratton, R., "On the Elementary Theory of Thermoelectric Phenomena," *British Journal of Applied Physics*, Vol. 8, Aug. 1949, p. 33.

Miller, D. G., "Thermodynamic Theory of Irreversible Processes," *American Journal of Physics*, May 1956, p. 433.

Onsager, L., "Reciprocal Relations in Irreversible Processes," *Physical Review*, Vol. 37, Feb. 1931, p. 405.

Callen, H. B., "The Application of Onsager's Reciprocal Relations to Thermoelectric, Thermomagnetic, and Galvanomagnetic Effects," *Physical Review*, Vol. 73, June 1948, p. 3149.

Hatsopoulos, G. N. and Keenan, J. H., "Analyses of the Thermoelectric Effects of Methods of Irreversible Thermodynamics," *Journal of Applied Mechanics*, Vol. 25., Dec. 1958, p. 428.

Benedict, R. P., *Fundamentals of Temperature, Pressure, and Flow Measurements*, Wiley, New York, 1969.

2.5 **Nomenclature**

Roman

a,b	Coefficients
A	Area
E	Electrical potential
F	Frictional loss
I	Electric current
J	Flow
k	Thermal conductivity
L	Constant of proportionality
Q	Heat
S	Entropy
t	Time
T	Temperature

U	Internal energy
W	Work
x	Length or thickness
X	Force
Subscripts	
$A,B,C,$	Thermocouple materials
e	Electrical
i,j,k	General subscripts
i	Internal
P	Peltier
q	Thermal
R	Reference
S	Seebeck
T	Thomson
abs	Absolute
rev	Reversible
1,2	States
Greek	
α	Seebeck coefficient
Δ	Finite difference
π	Peltier coefficient
σ	Thomson coefficient
$\sum$	Sum
ζ	Entropy production/volume

3. THERMOCOUPLE MATERIALS

3.1 Common Thermocouple Types

The commonly used thermocouple types are identified by letter designations originally assigned by the Instrument Society of America (ISA) and adopted as an American Standard in ASA C96.1-1964. This section covers general application data on the atmospheres in which each thermocouple type can be used, recommended temperature ranges, limitations, etc. Physical and thermoelectric properties of the thermoelement materials used in each of these thermocouple types are tabulated in Section 3.1.2.

The following thermocouple types are included (these are defined as having the emf-temperature relationship given in the corresponding letter-designated Table in Section 10. within the limits of error specified in Table 41 of that section):

Type T—Copper (+) Constantan (−)
Type J—Iron (+) Constantan (−)
Type K—Originally Chromel[1] (+) Alumel[1] (−)
Type E—Originally Chromel (+) Constantan (−)
Type R—Platinum/13% Rhodium (+) versus Platinum (−)
Type S—Platinum/10% Rhodium (+) versus Platinum (−)
Type B—Platinum/30% Rhodium (+) versus Platinum/6% Rhodium (−)

[1] Trademark–Hoskins Manufacturing Co.

Temperature limits stated in the text are maximum values. Table 2 gives recommended maximum temperature limits for various gage sizes of wire. Table 3 gives nominal Seebeck coefficients for the various types. Temperature-emf equivalents and commercial limits of error for these common thermocouple types are given in Section 10.

TABLE 2—*Recommended upper temperature limits for protected thermocouples.*

	Upper Temperature Limits for Various Wire Sizes (Awg)				
Thermocouple	No. 8 (0.128 in.)	No. 14 (0.064 in.)	No. 20 (0.032 in.)	No. 24 (0.020 in.)	No. 28 (0.013 in.)
J	760 C (1400 F)	593 C (1100 F)	482 C (900 F)	371 C (700 F)	371 C (700 F)
E	871 C (1600 F)	649 C (1200 F)	538 C (1000 F)	427 C (800 F)	427 C (800 F)
T	...	371 C (700 F)	260 C (500 F)	204 C (400 F)	204 C (400 F)
K	1260 C (2300 F)	1093 C (2000 F)	982 C (1800 F)	871 C (1600 F)	871 C (1600 F)
R and S	...	...	...	1482 C (2700 F)	...
B	...	...	...	1704 C (3100 F)	...

NOTE—This table gives the recommended upper temperature limits for the various thermocouples and wire sizes. These limits apply to protected thermocouples, that is, thermocouples in conventional closed-end protecting tubes. They do not apply to sheathed thermocouples having compacted mineral oxide insulation. In any general recommendation of thermocouple temperature limits, it is not practicable to take into account special cases. In actual operation, there may be instances where the temperature limits recommended can be exceeded. Likewise, there may be applications where satisfactory life will not be obtained at the recommended temperature limits. However, in general, the temperature limits listed are such as to provide satisfactory thermocouple life when the wires are operated continuously at these temperatures.

3.1.1 *General Application Data*

Type T—These thermocouples are resistant to corrosion in moist atmospheres and are excellent for subzero temperature measurements. They have an upper temperature limit of 700 F (371 C) and can be used in a vacuum and in oxidizing, reducing, or inert atmospheres. This is the only thermocouple type for which limits of error are guaranteed in the subzero temperature range.

Type J—These thermocouples are suitable for use in vacuum and in oxidizing, reducing, or inert atmospheres, at temperatures up to 1400 F (760 C). The rate of oxidation of the iron thermoelement is rapid above 1000 F (538 C), however, and the use of heavy-gage wires is recommended when long life is required at the higher temperatures.

Bare thermocouples should not be used in sulfurous atmospheres above 1000 F (538 C).

TABLE 3—*Nominal Seebeck coefficients (thermoelectric power).*

Temperature	Thermocouple Type						
	E	J	K	R	S	T	B
Deg C	Seebeck Coefficient–Microvolts/Deg C						
−190	27.3	24.2	17.1	...	...	17.1	
−100	44.8	41.4	30.6	...	...	28.4	
0	58.5	50.2	39.4	...	...	38.0	...
200	74.5	55.8	40.0	8.8	8.5	53.0	2.0
400	80.0	55.3	42.3	10.5	9.5	...	4.0
600	81.0	58.5	42.6	11.5	10.3	...	6.0
800	78.5	64.3	41.0	12.3	11.0	...	7.7
1000	...	...	39.0	13.0	11.5	...	9.2
1200	...	...	36.5	13.8	12.0	...	10.3
1400	...	...	...	13.8	12.0	...	11.3
1600	...	...	...	...	11.8	...	11.6
Deg F	Seebeck Coefficient–Microvolts/Deg F						
−300	15.5	14.4	...	...	...	9.7	
−200	22.0	20.6	...	...	...	13.7	
−100	27.0	24.6	...	...	...	17.3	
32	32.5	28.0	21.7	3.0	3.0	21.3	...
200	37.5	30.1	23.2	4.1	4.0	25.7	0.5
400	41.5	30.9	22.3	4.9	4.8	29.8	1.1
600	43.5	30.7	23.1	5.5	5.1	32.7	1.8
800	45.0	30.6	23.5	5.9	5.3	...	2.4
1000	45.0	31.7	23.7	6.2	5.5	...	3.0
1500	44.0	35.7	22.8	6.9	6.1	...	4.4
2000	...	...	21.1	7.6	6.6	...	5.4
2500	...	...	...	7.6	6.6	...	6.2
3000	...	...	...	7.6	6.5	...	6.5

This thermocouple is sometimes used for subzero temperatures, but the possible rusting and embrittlement of the iron wire under these conditions makes its use less desirable than Type T for low temperature measurements. Limits of error have not been established for Type J thermocouples at subzero temperatures (see Section 10.).

Type K—Type K thermocouples are recommended for continuous use in oxidizing or inert atmospheres at temperatures up to 2300 F (1260 C). Because their oxidation resistance characteristics are better than those of other base metal thermocouples, they find widest use at temperatures above 1000 F (538 C). However, this thermocouple is suitable for temperature measurements as low as -420 F (-250 C), although limits of error have been established only for the temperature range 0 to 2300 F (-18 to 1260 C) (see Section 10.).

The Type K thermocouple may be used in hydrogen or cracked ammonia atmospheres if the dewpoint is below -40 F (-40 C). However, they should not be used in:

1. Atmospheres that are reducing or alternately oxidizing and reducing unless suitably protected with protection tubes.

2. Sulfurous atmospheres unless properly protected. Sulfur will attack both thermoelements and will cause rapid embrittlement and breakage of the negative thermoelement wire through interangular corrosion.

3. Vacuum except for short time periods (preferential vaporization of chromium from the positive element will alter calibration).

4. Atmospheres that promote "green-rot" corrosion of the positive thermoelement. Such corrosion results from preferential oxidation of chromium when the oxygen content of the atmosphere surrounding the thermocouple is low and in a certain range. It can cause large negative errors in calibration and is most serious in the temperature range 1500 to 1900 F (816 to 1038 C).

Green-rot corrosion frequently occurs when thermocouples are used in long unventilated protecting tubes of small diameter. It can be minimized by increasing the oxygen supply through the use of large diameter protecting tubes or ventilated protecting tubes. Another approach is to decrease the oxygen content below that which will promote preferential oxidation by inserting a "getter" to absorb the oxygen in a sealed protection tube.

Type E—Type E thermocouples are recommended for use over the temperature range of -420 to $+1600$ F (-250 to 871 C) in oxidizing or inert atmospheres. In reducing atmospheres, alternately oxidizing and reducing atmospheres, marginally oxidizing atmospheres, and in vacuum they are subject to the same limitations as Type K thermocouples.

These thermocouples are suitable for subzero temperature measurements since they are not subject to corrosion in atmospheres with high moisture content. However, limits of error for the subzero range have not been established.

Type E thermocouples develop the highest emf per degree of all the commonly used types and are often used primarily because of this feature.

Types R and S—Type R and S thermocouples are recommended for continuous use in oxidizing or inert atmospheres at temperatures up to 2550 F (1399 C); intermittently up to 2700 F (1482 C).

They should not be used in reducing atmospheres, nor those containing metallic or nonmetallic vapors, unless suitably protected with nonmetallic protecting tubes. They never should be inserted directly into a metallic primary protecting tube.

Types R and S thermocouples may be used in a vacuum for short periods of time, but greater stability will be obtained by using Type B thermo-couples for such applications.

Continued use of Types R and S thermocouples at high temperatures causes excessive grain growth which can result in mechanical failure of the platinum element. It also renders the platinum susceptible to contamination which causes negative drifts in calibration, that is, a reduction in the emf output of the thermocouple.

Calibration changes also are caused by diffusion of rhodium from the alloy wire into the platinum, or by volatilization of rhodium from the alloy. All of these effects tend to produce negative calibration shifts.

Type B—Type B thermocouples are recommended for continuous use in oxidizing or inert atmospheres at temperatures up to 3100 F (1704 C). They are also suitable for short term use in vacuum to this temperature.

They should not be used in reducing atmospheres, nor those containing metallic or nonmetallic vapors, unless suitably protected with nonmetallic protecting tubes. They should never be inserted directly into a metallic primary protecting tube.

Under corresponding conditions of temperature and environment Type B thermocouples will show less grain growth and less drift in calibration than Type R or S thermocouples.

3.1.2 *Properties of Thermoelement Materials*

This section indicates in Tables 4 to 10 and in Fig. 7 the physical and electrical properties of thermoelement materials as used for the common letter-designated thermocouple types (Types E, J, K, R, S, and T). These are typical data and are listed for information only. They are not intended for use as specifications for ordering thermocouple materials.

Thermoelement materials are designated in the tables by the established American Standard letter symbols JP, JN, etc. The first letter of the symbol designates the type of thermocouple. The second letter, P or N, denotes the positive or negative thermoelement, respectively. Typical materials to which these letter designations apply are listed below:

JP	Iron, ThermoKanthal[2] JP
JN, EN, or TN	Constantan, Cupron,[3] Advance,[3] ThermoKanthal JN
TP	Copper
KP or EP	Chromel, Tophel,[3] T–1,[4] ThermoKanthal KP
KN	Alumel, Nial,[3] T–2,[4] ThermoKanthal KN
RP	Platinum 13 percent rhodium
SP	Platinum 10 percent rhodium
RN or SN	Platinum

[2] Trademark–Kanthal Corp.
[3] Trademark–Wilbur B. Driver Co.
[4] Trademark–Driver-Harris Co.

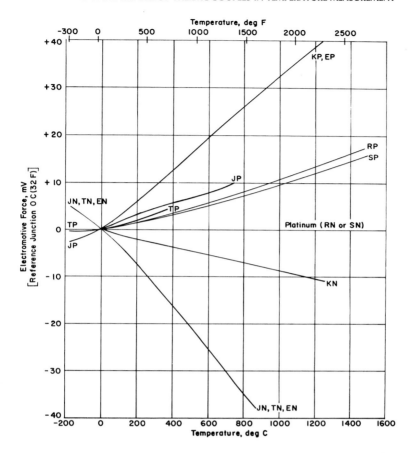

FIG. 7—*Thermal emf of thermoelements relative to platinum.*

Note that JN, EN, and TN thermoelements, as listed above, are composed of the same basic types of material. The typical data contained in the following pages are applicable to any of these thermoelements, but the thermal emf (versus platinum) of the three types may differ significantly depending on the type of thermocouple for which each is intended.

It also should be noted that positive and negative thermoelements for a given type of thermocouple, as supplied by any one manufacturer, will conform to the calibration curve for that thermocouple within specified limits of error. However, because materials used for a given thermoelement by various manufacturers may differ slightly in thermal emf, larger errors may occur if positive and negative thermoelements from different sources are combined. This is particularly true of thermoelements for Types J, R, and S thermocouples.

TABLE 4—*Typical physical properies of a variety of thermoelement materials.*

Property	JP	JN, EN, TN	TP	KP, EP	KN	RP	SP	RN, SN
Melting point (solidus temperature):								
deg C	1490	1220	1083	1427	1399	1860	1850	1769
deg F	2715	2228	1981	2600	2550	3380	3362	3216
Resistivity:								
microhm-cm:								
at 0 C	8.57	48.9	1.56	70	28.1	19.0	18.4	9.83
at 20 C	9.67	48.9	1.724	70.6	29.4	19.6	18.9	10.4
ohms/cir mil ft:								
at 0 C	51.5	294.2	9.38	421	169	114.3	110.7	59.1
at 20 C	58.2	294	10.37	425	177	117.7	114.0	62.4
Temperature coefficient of resistance, ohms/ohm deg C (0 to 100 C)	65×10^{-4}	-0.1×10^{-4}	43×10^{-4}	4.1×10^{-4}	23.9×10^{-4}	15.6×10^{-4}	16.6×10^{-4}	39.2×10^{-4}
Coefficient of thermal expansion, in./in. deg C (20 to 100 C)	11.7×10^{-6}	14.9×10^{-6}	16.6×10^{-6}	13.1×10^{-6}	12.0×10^{-6}	9.0×10^{-6}	9.0×10^{-6}	9.0×10^{-6}
Thermal conductivity at 100 C:								
Cal cm/sec cm² deg C	0.162	0.0506	0.901	0.046	0.071	0.088	0.090	0.171
Btu ft/h ft² deg F	39.2	12.2	218	11.1	17.2	21.3	21.8	41.4
Specific heat at 20 C, cal/g deg C	0.107	0.094	0.092	0.107	0.125	...	...	0.032
Density:								
g/cm³	7.86	8.92	8.92	8.73	8.60	19.61	19.97	21.45
lb/in.³	0.284	0.322	0.322	0.315	0.311	0.708	0.721	0.775
Tensile strength (annealed):								
kg/cm²	3 500	5 600	2 500	6 700	6 000	3 200	3 200	1 400
psi	50 000	80 000	35 000	95 000	85 000	46 000	45 000	20 000
Magnetic attraction	strong	none	none	none	moderate	none	none	none

TABLE 5—*Nominal chemical composition of thermoelements.*

Element	JP	JN, TN, EN[a]	TP	KP, EP	KN	RP	SP	RN, SN
				Nominal Chemical Composition, percent				
Iron	99.5	...	...	...	...	...	...	...
Carbon	...[b]	...	...	...	...	...	...	...
Manganese	...[b]	...	...	...	2	...	...	...
Sulfur	...[b]	...	...	...	...	...	...	...
Phosphorus =	...[b]	...	...	...	...	...	...	...
Silicon	...[b]	...	...	...	1	...	...	...
Nickel	...[b]	45	...	90	95	...	...	...
Copper	...[b]	55	100	...	...	...	...	...
Chromium	...[b]	...	...	10	...	...	...	...
Aluminum	...	...	...	...	2	...	...	...
Platinum	...	...	...	...	...	87	90	100
Rhodium	...	...	...	...	...	13	10	...

[a] Types JN, TN, and EN thermoelements usually contain small amounts of various elements for control of thermal emf, with corresponding reductions in the nickel or copper content, or, both.

[b] Thermoelectric iron (JP) contains small but varying amounts of these elements.

TABLE 6—*Environmental limitations of thermoelements.*

Thermoelement	Environmental Recommendations and Limitations (see Notes)
JP	For use in oxidizing, reducing, or inert atmospheres or in vacuum. Oxidizes rapidly above 540 C (1000 F). Will rust in moist atmospheres as in subzero applications.
	Stable to neutron radiation transmutation. Change in composition is only 0.5 percent (increase in manganese) in a 20-year period.
JN, TN, EN	Suitable for use in oxidizing, reducing, and inert atmospheres or in vacuum. Should not be used unprotected in sulfurous atmospheres above 540 C (1000 F).
	Composition changes under neutron radiation since copper content is converted to nickel and zinc. Nickel content increases 5 percent in a 20-year period.
TP	Can be used in vacuum or in oxidizing, reducing or inert atmospheres. Oxidizes rapidly above 370 C (700 F). Preferred to Type JP element for subzero use because of its superior corrosion resistance in moist atmospheres.
	Radiation transmutation causes significant changes in composition. Nickel and zinc grow into the material in amounts of 10 percent each in a 20-year period.
KP, EP	For use in oxidizing or inert atmospheres. Can be used in hydrogen or cracked ammonia atmospheres if dew point is below −40 C (−40 F). Do not use unprotected in sulfurous atmospheres above 540 C (1000 F).
	Not recommended for service in vacuum at high temperatures except for short time periods because preferential vaporization of chromium will alter calibration. Large negative calibration shifts will occur if exposed to marginally oxidizing atmospheres in temperature range 815 to 1040 C (1500 to 1900 F).
	Quite stable to radiation transmutation. Composition change is less than 1 percent in a 20-year period.
KN	Can be used in oxidizing or inert atmospheres. Do not use unprotected in sulfurous atmospheres as intergranular corrosion will cause severe embrittlement.
	Relatively stable to radiation transmutation. In a 20-year period, iron content will increase approximately 2 percent. The manganese and cobalt contents will decrease slightly.
RP, SP, SN, RN	For use in oxidizing or inert atmospheres. Do not use unprotected in reducing atmospheres in the presence of easily reduced oxides, atmospheres containing metallic vapors such as lead or zinc, or those containing nonmetallic vapors such as arsenic, phosphorus, or sulfur. Do not insert directly into metallic protecting tubes. Not recommended for service in vacuum at high temperatures except for short time periods.
	Type SN elements are relatively stable to radiation transmutation. Types RP and SP elements are unstable because of the rapid depletion of rhodium. Essentially, all the rhodium will be converted to palladium in a 10-year period.

NOTE 1—Refer to Table 7 for recommended upper temperature limits.

NOTE 2—Stability under neutron radiation refers to chemical composition of thermoelement, not to stability of thermal emf.

NOTE 3—Radiation transmutation rates are based on exposure to a thermal neutron flux of 1×10^{14} neutrons/cm^{2}s.

TABLE 7—*Recommended upper temperature limits for protected thermoelements.*

Thermoelement	Upper Temperature Limits for Various Wire Sizes (awg)				
	No. 8 (0.128 in.)	No. 14 (0.064 in.)	No. 20 (0.032 in.)	No. 24 (0.020 in.)	No. 28 (0.013 in.)
JP	760 C (1400 F)	593 C (1100 F)	482 C (900 F)	371 C (700 F)	371 C (700 F)
JN, TN, EN	871 C (1600 F)	649 C (1200 F)	538 C (1000 F)	427 C (800 F)	427 C (800 F)
TP	...	371 C (700 F)	260 C (500 F)	204 C (400 F)	204 C (400 F)
KP, EP, KN......................	1260 C (2300 F)	1093 C (2000 F)	982 C (1800 F)	871 C (1600 F)	871 C (1600 F)
RP, SP, RN, SN	...	...	...	1482 C (2700 F)	...

NOTE—This table gives the recommended upper temperature limits for the various thermo-elements and wire sizes. These limits apply to protected thermoelements, that is, thermo-elements in conventional closed-end protecting tubes. They do not apply to sheathed thermo-elements having compacted mineral oxide insulation. In any general recommendation of thermoelement temperature limits, it is not practicable to take into account special cases. In actual operation, there may be instances where the temperature limits recommended can be exceeded. Likewise, there may be applications where satisfactory life will not be obtained at the recommended temperature limits. However, in general, the temperature limits listed are such as to provide satisfactory thermoelement life when the wires are operated continuously at these temperatures.

TABLE 8—*Seebeck coefficient (thermoelectric power) of thermoelements with respect to platinum 27 (typical values).*

Temperature, deg C	Thermoelement	Seebeck Coefficient, µV/deg C						
		JP	JN, TN, EN	TP	KP, EP	KN	RP	SP
−190		+6.3	−20.9	−4.1	...	...	...	...
−100		14.4	27.0	+1.1	...	...	...	...
0		17.8	32.2	5.9	+25.7	−13.5	+5.5	+5.5
200		14.6	41.0	12.0	32.7	7.4	8.5	8.5
400		9.7	45.5	16.2	34.6	7.7	10.5	9.5
600		11.7	46.8	...	33.8	8.8	11.5	10.0
800		17.8	46.4	...	32.2	8.8	12.5	11.0
1000		...	...	...	30.8	8.3	13.0	11.5
1200		...	...	...	29.1	7.4	14.0	12.0
1400		...	...	...	...	...	14.0	12.0
1600		...	...	...	...	...	13.5	12.0

Temperature, deg F	Seebeck Coefficient, µV/deg F						
	JP	JN, TN, EN	TP	KP, EP	KN	RP	SP
−300	+2.5	−11.9	−2.1	...	...	...	...
−200	6.7	14.0	−0.2	...	...	...	...
−100	8.8	15.8	+1.5	...	...	...	...
32	9.9	17.9	3.3	+14.3	−7.5	+3.0	+3.0
200	9.6	20.5	5.0	16.7	6.5	4.1	4.0
400	8.0	22.9	6.7	18.3	4.0	4.9	4.7
600	6.2	24.5	8.2	19.0	4.1	5.5	5.2
800	5.3	25.3	...	19.1	4.4	5.8	5.4
1000	5.7	26.0	...	18.9	4.8	6.2	5.5
1500	9.9	25.8	...	17.8	4.9	6.8	6.1
2000	...	...	...	16.7	4.3	7.6	6.6
2500	...	...	...	14.9	4.0	7.7	6.7
3000	...	...	...	...	...	7.6	6.5

TABLE 9 — *Thermoelements–resistance change with increasing temperature.*

Thermoelement	Ratio of Resistance at Temperature Indicated to Resistance at 0 C (32 F)									
	0 C (32 F)	20 C (68 F)	200 C (392 F)	400 C (752 F)	600 C (1112 F)	800 C (1472 F)	1000 C (1852 F)	1200 C (2192 F)	1400 C (2552 F)	1500 C (2732 F)
JP	1.00	1.13	2.46	4.72	7.84	12.00	13.07	...	...	...
JN, TN, EN	1.00	0.999	0.996	0.994	1.024	1.056	1.092	...	...	...
TP	1.00	1.11	1.86	2.75	3.70	4.75	5.96	...	...	...
KP	1.00	1.01	1.09	1.19	1.25	1.30	1.37	1.43	...	...
KN	1.00	1.05	1.43	1.64	1.82	1.98	2.15	2.32	...	...
RP	1.00	1.03	1.31	1.60	1.89	2.16	2.41	2.66	2.90	3.01
SP	1.00	1.03	1.33	1.65	1.95	2.23	2.50	2.76	3.01	3.13
RN, SN	1.00	1.06	1.77	2.50	3.18	3.81	4.40	4.94	5.42	5.66

TABLE 10—*Nominal resistance of thermoelements.*

Awg No.	Diameter Inches	Nominal Resistance, ohms per foot at 20 C (68 F)						
		KN	KP, EP	TN, JN, EN	TP	JP	RN, SN	SP
6......	0.1620	0.0067	0.0162	0.0112	0.000395	0.0027	0.00243	0.00438
9ª.....	0.148	...	...	...	...	0.0033	...	...
8......	0.1285	0.0107	0.0257	0.0179	0.000628	0.0043	0.00386	0.00697
10......	0.1019	0.0170	0.041	0.0283	0.000999	0.0069	0.00614	0.01108
12......	0.0808	0.0270	0.065	0.0448	0.00159	0.0109	0.00976	0.01761
14......	0.0641	0.0432	0.104	0.0718	0.00253	0.0174	0.0155	0.0280
16......	0.0508	0.0683	0.164	0.113	0.00402	0.0276	0.0247	0.0445
17......	0.0453	0.0874	0.209	0.145	0.00506	0.0349	0.0311	0.0562
18......	0.0403	0.111	0.266	0.184	0.00648	0.0446	0.0399	0.0719
20......	0.0320	0.173	0.415	0.287	0.0102	0.0699	0.0624	0.1125
22......	0.0253	0.276	0.663	0.456	0.0161	0.1111	0.0993	0.1790
23......	0.0226	0.347	0.833	0.576	0.0204	0.1401	0.1251	0.2257
24......	0.0201	0.438	1.05	0.728	0.0257	0.1767	0.1578	0.2847
25......	0.0179	0.553	1.33	0.918	0.0324	0.2228	0.1990	0.3589
26......	0.0159	0.700	1.68	1.16	0.0408	0.281	0.2509	0.4526
28......	0.0126	1.11	2.68	1.85	0.0649	0.447	0.3989	0.7197
30......	0.0100	1.77	4.25	2.94	0.1032	0.710	0.6344	1.144
32......	0.0080	2.76	6.65	4.59	0.1641	1.13	1.009	1.819
34......	0.0063	4.45	10.7	7.41	0.2609	1.80	1.604	2.893
36......	0.0050	7.08	17.0	11.8	0.4148	2.86	2.550	4.600
38......	0.0040	11.1	26.6	18.4	0.6597	4.54	4.056	7.316
40......	0.0031	18.4	44.2	30.6	1.049	7.22	6.448	11.63

ª #9 Birmingham wire gage.

3.2 Extension Wires

3.2.1 *General Information*

Extension wires are inserted between the measuring junction and the reference junction and have approximately the same thermoelectric properties as the thermocouple wires with which they are used. Table 11 gives comparative data on extension wires available for thermocouples in common use. Extension wires are normally available as single or duplex, solid or stranded, insulated wires in sizes ranging from 14 to 20 B&S gage. A variety of insulations and protective coverings is available in several combinations to suit the many types of environments encountered in industrial service (see Section 4.).

Some advantages of using extension wires are:

1. Improvement in mechanical or physical properties of the thermo-

TABLE 11—*Extension wires for common thermocouples.* [a]

Thermocouple Type	Extension Type	Alloy Type		Temperature Range, deg F	Limits of Error, deg F		Magnetic[b] Response	
		Positive Element	Negative Element		Standard	Special	P	N
Base Metal	*Category 1*							
E	EX	NiCr (Chromel)[c]	constantan	32 to 400	±3	...	O	O
J	JX	iron	constantan	32 to 400	±4	±2	M	O
K	KX	NiCr (Chromel)[c]	NiAl (Alumel)[c]	32 to 400	±4	...	O	M
T	TX	copper	constantan	−75 to 200	±1½	±¾	O	O
	Category 2							
K	WX	iron	copper-nickel alloy	75 to 400	±6	...	M	O
Noble Metal								
R	SX	copper	copper alloy	75 to 400	±12	...	O	O
S	SX	copper	copper alloy	75 to 400	±12	...	O	O

[a] Extension wires for other thermocouple types will be found in Section 3.3.
[b] M denotes ferro-magnetic alloy:
  O denotes nonferro-magnetic alloy.
[c] Former (proprietary) designation.

electric circuit. For example, the use of stranded construction or smaller diameter solid wire may increase the flexibility of a portion of the circuit. Extension wires also may be selected to adjust the electrical resistance of the circuit.

2. Cost improvement in thermoelectric circuitry. For example, certain base metal extension wires may be substituted for noble metal wires when the reference junction is situated at a distance from a noble metal thermocouple.

Extension wires may be separated into two categories having the following characteristics:

Category 1—Alloys substantially the same as used in the thermocouple. This type of extension wire normally is used with base metal thermocouples.

Category 2—Alloys differing from those used in the thermocouple. This type of extension wire normally is used with noble metal thermocouples and with several of the nonstandardized thermocouples (see Section 3.3).

3.2.2 *Sources of Error*

Several possible sources of error in temperature measurement accompany the use of extension wires in thermocouple circuits. Most of the errors can be avoided, however, by exercising proper precautions.

One type of error arises from the disparity in thermal emf between thermocouples and nominally identical extension wire components of Category 1. The disparity results from the variations occurring among thermoelements lying within the standard limits of error for each type of thermocouple and extension wire. Thus, for example, it is possible that an error as great as ± 8 F could occur in the Type K/KX and J/JX thermocouple-extension wire combinations, where the standard limits of error are ± 4 F for the thermocouple and the extension wires treated as separate combinations. Such errors can be eliminated substantially by selecting extension wires whose emf closely matches that of the specific thermocouple, up to the maximum temperature of the thermocouple-extension wire junction.

A second source of error can arise if a temperature difference exists between the two thermoelement-extension wire junctions. Errors of this type are potentially greater in circuits employing extension wires of Category 2, where each extension element may differ significantly in emf from the corresponding thermoelement. Such errors may occur even though the extension pair emf exactly matches the thermocouple emf at each temperature. Referring to Fig. 8, schematically representing emf versus temperature curves for positive and negative thermoelements P and N, and corresponding extension wire elements PX and NX, the following relationships apply at any temperature T within the operating range of the extension wires:

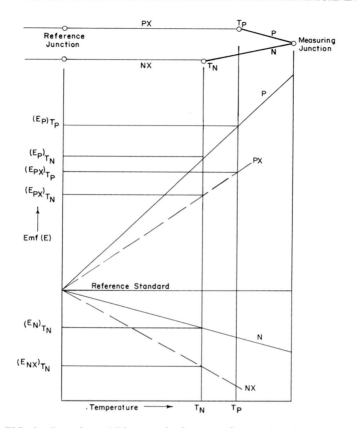

FIG. 8—*Error due to ΔT between the thermocouple-extension wire junctions.*

Thermocouple output = extension pair output
that is,

$$E_P - E_N = E_{PX} - E_{NX} \qquad \dots \dots \dots \dots \dots \quad (37)$$

Rearranging to

$$E_P - E_{PX} = E_N - E_{NX} \qquad \dots \dots \dots \dots \dots \quad (38)$$

If a temperature difference exists between the two junctions such that P joins PX at T_P, and N joins NX at T_N, an unwanted emf will exist across the two junctions, of magnitude

$$\Delta E = (E_P - E_{PX})_{T_P} - (E_N - E_{NX})_{T_N} \qquad \dots \dots \dots \dots \quad (39)$$

Rearranging Eq 39 according to Eq 38:

$$\Delta E = (E_P - E_{PX})_{T_P} - (E_P - E_{PX})_{T_N} \quad \dots\dots\dots (40)$$

The sign of ΔE will depend on the relationship of temperature T_P to T_N and the relationship of PX and NX to P and N.

This ΔE will be interpreted as an error in the output of the measuring thermocouple. Such errors do not exceed about one degree at the measuring junction, per degree of ΔT between the thermocouple-extension wire junctions, for any of the thermocouples and extension wires listed in Table 2. These errors can be essentially eliminated by taking steps to equalize the temperatures of the two junctions.

A third source of error lies in the presence of reversed polarity at the thermocouple-extension wire junctions, or at the extension wire-instrument junctions. Although a single reversal of polarity in the assembly would be noticeable, an inadvertent double reversal likewise may produce measurement errors, but could escape immediate detection.

A fourth source of error concerns the use of connectors in the thermocouple assembly. If the connector material has thermal emf characteristics which differ appreciably from those of the thermocouple extension wires, then it is important that a negligible temperature difference be maintained across the connector. This follows directly from the Law of Intermediate Metals (see Section 2.2.2). Thus, in situations where a connector made of a third metal spans a substantial temperature gradient, unwanted emfs are generated between the thermoelectric materials and the extremities of the connector, and they appear as errors in the output of the thermocouple. The magnitude of errors of this type can vary over a wide range depending on the materials involved and the temperature difference spanned by the connector.

If the emf errors arising from the use of extension wires or from other sources are to be expressed as temperature errors, the Seebeck coefficient at the measuring junction temperature must be used.

A useful graphical method of evaluating error sources in thermoelectric circuits is detailed in a paper by Moffat (see Ref *34* on p. 103).

3.3 Nonstandardized Thermocouple Types

Newer thermocouple materials are being evaluated constantly to find combinations which perform special functions more reliably than the common thermocouples. The special functions for which these newer combinations are required frequently involve very high temperatures, but also may include unusual environments such as special atmospheres or areas susceptible to vibration.

Each of the combinations described in this section has been designed to measure temperatures under specific conditions and to perform with a degree of reliability superior to other combinations under these same conditions. The properties of each combination are detailed to allow a quick selection of a combination which is most likely to be suitable for a special condition. Thermocouple compositions are given in weight percent with the positive thermoelement of the thermocouple named first.

The information on newer thermocouple materials is presented using comments, tables, and curves. The comments made for the various thermocouple systems are intended to convey information not easily shown by tables or curves. The information contained in the tables is intended to help the reader to quickly decide if a certain thermocouple system, or a specific thermocouple, is suited to his particular needs. The information given is general and nominal, and cannot be used too literally. For example, the useful maximum temperature of a thermocouple depends in part on wire size, insulation used, method of installation, atmosphere conditions, vibration present, etc. The evaluation of certain properties as good, fair, or poor is subject to wide ranges of interpretation in terms of a particular application; hence, no attempt is made to define these terms. Approximate millivolt-versus-temperature relations for the various thermocouples are shown by curves. The curves are presented to show general temperature ranges for the various thermocouples but are not intended for use in converting emf to temperature. The reader should contact the wire manufacturer for temperature-emf tables.

The thermocouples described here are in use, and sufficient data and experience are available to warrant their inclusion. No attempt is made to include the many other thermocouple materials described in the literature which may have limited uses, or for which there are limited data, or for which there are serious problems of stability, emf reversibility, structural strength, etc.

The best source of information for a specific thermocouple in the "newer material" classification is considered to be the manufacturer of the particular thermocouple under consideration. Other useful information can be found in Ref *1*.[5]

3.3.1 *Platinum Types*

3.3.1.1 *Platinum-Rhodium Versus Platinum-Rhodium Thermocouples—* The standard Type R and Type S thermocouples can be used for temperature measurement to the melting point of platinum, 1769 C (3216 F) on a short-term basis, but, for improved service life at temperatures over 1200 C (2192 F), special platinum/rhodium thermocouples are recommended.

The platinum-40 percent rhodium versus platinum-20 percent rhodium thermocouple, called the "Land-Jewell" thermocouple, is especially useful

[5] The italic numbers in brackets refer to the list of references in this manual.

for continuous use to 1800 C (3272 F) or occasional use to 1850 C (3362 F). However, it is seldom used where the platinum-30 percent rhodium versus platinum-6 percent rhodium thermocouple will suffice because of lower output and greater cost.

Other thermocouples suggested for high-temperature measurement have been a platinum-13 percent rhodium versus platinum-1 percent rhodium combination and a platinum-20 percent rhodium versus platinum-5 percent rhodium combination. The former shows slightly less tendency toward mechanical failure or contamination at high temperatures than the standard Type R and Type S thermocouples, while the latter has properties very similar to those of the platinum-30 percent rhodium versus platinum-6 percent rhodium thermocouple.

Figure 9 and Table 12 show the characteristics of these alloys.

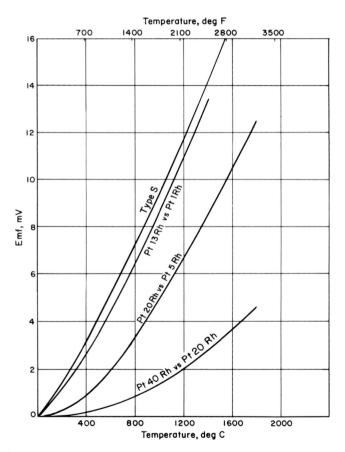

FIG. 9—*Thermal emf of platinum-rhodium versus platinum-rhodium thermocouples.*

TABLE 12—*Platinum-rhodium versus platinum-rhodium thermocouples.*

	Pt-20Rh versus Pt-5Rh	Pt-40Rh versus Pt-20Rh	Pt-13Rh versus Pt-1Rh
Nominal operating temperature range, in:			
Reducing atmosphere (nonhydrogen)	NR[a]	NR	NR
Wet hydrogen	NR	NR	NR
Dry hydrogen	NR	NR	NR
Inert atmosphere	1700 C (3092 F)	1800 C (3272 F)	1600 C (2912 F)
Oxidizing atmosphere	1700 C (3092 F)	1800 C (3272 F)	1600 C (2912 F)
Vacuum (short-time use)	1700 C (3092 F)	1800 C (3272 F)	1600 C (2912 F)
Maximum short-time temperature	1770 C (3218 F)	1850 C (3362 F)	1770 C (3218 F)
Approximate microvolts per degree:			
Mean, over nominal operating range..............	6.8/C (12.2/F)	2.5/C (4.5/F)	9.9/C (17.8/F)
At top temperature of normal range..................	9.9/C (17.8/F)	4.7/C (8.45/F)	12.2/C (22.0/F)
Melting temperature, nominal:			
Positive thermoelement	1900 C (3452 F)	1930 C (3520 F)	1865 C (3389 F)
Negative thermoelement	1820 C (3308 F)	1900 C (3452 F)	1771 C (3220 F)
Stability with thermal cycling ...	good	good	good
High-temperature tensile properties	good	good	good
Stability under mechanical working	good	fair	good
Ductility (of most brittle thermoelement) after use ...	good	fair	good
Resistance to handling contamination	fair	fair	fair
Recommended extension wire, 175 C (347 F) max:			
Positive conductor	Cu	Cu	Cu
Negative conductor	Cu	Cu	Cu

[a] NR = not recommended.

All special platinum-rhodium versus platinum-rhodium thermocouples, like the standard Type R and Type S thermocouples, show improved life at high temperatures when protected by double-bore, full-length insulators of high-purity alumina.

3.3.1.2 *Platinum-15 Percent Iridium Versus Palladium Thermocouples*— The platinum-15 percent iridium versus palladium combination was developed as a high-output noble-metal thermocouple. It combines the desirable attributes of noble metals with a high emf output at a lower cost than other noble-metal thermocouples.

The output becomes more linear and the Seebeck coefficient (thermoelectric power) increases with increasing temperature. In the absence of of vibration, the useful range can probably be extended closer to the melting point of palladium, 1550 C (2826 F).

Figure 10 and Table 13 show the characteristics of these alloys.

Extension wires of base metals have been developed to provide a reasonable match with the thermocouple to about 700 C (1292 F).

Resistance to corrosion of the platinum-15 percent iridium alloy is better than that of the platinum-rhodium alloys in current use. Palladium is slightly less resistant to corrosion than the platinum alloy group. It will superficially oxidize at 700 C (1292 F). The oxide decomposes at about 875 C (1607 F) leaving a bright metal. When subjected to alternating oxidizing and reducing atmospheres, surface blistering may result. As with all noble metals, the catalytic effect of the wires must be considered in combustible atmospheres. Its use may be preferred to base metals, however, for many applications. Both wires are ductile and may be reduced to very small sizes and still be handled with relative ease.

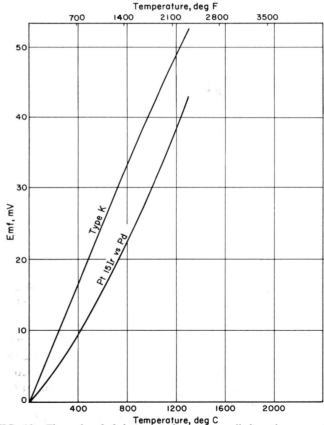

FIG. 10—*Thermal emf of platinum-iridium versus palladium thermocouple.*

TABLE 13—*Platinum-iridium versus palladium thermocouple.*

	Pt-15Ir versus Pd
Nominal operating temperature range, in:	
Reducing atmosphere (nonhydrogen)	NR[a]
Wet hydrogen	NR
Dry hydrogen	NR
Inert atmosphere	1370 C (2500 F)
Oxidizing atmosphere	1370 C (2500 F)
Vacuum	NR
Maximum short-time temperature	1550 C (2826 F)
Approximate microvolts per degree:	
Mean, over hominal operating range	12/C (22/F)
At top temperature of normal range	13.6/C (24.6/F)
Melting temperature, nominal:	
Positive thermoelement	1785 C (3245 F)
Negative thermoelement	1550 C (2826 F)
Stability with thermal cycling	good
High-temperature tensile properties	fair
Stability under mechanical working	good
Ductility (of most brittle thermoelement) after use	good
Resistance to handling contamination	fair
Recommended extension wire:	
Positive conductor	base metal alloys[b]
Negative conductor	base metal alloys[b]

[a] NR = not recommended.
[b] General Electric Company.

3.3.1.3 *Platinum-5 Percent Molybdenum Versus Platinum-0.1 Percent Molybdenum Thermocouple*—Platinum alloys containing rhodium are not suitable for use under neutron irradiation since the rhodium changes slowly to palladium. This causes a drift in the calibration of thermocouples containing rhodium. However, a thermocouple of platinum-5 percent molybdenum versus platinum-0.1 percent molybdenum is suitable for use in the helium atmosphere of a gas-cooled atomic reactor. Good stability at temperatures up to 1400 C (2552 F) has been reported. The output of the thermocouple is high and increases in a fairly uniform manner with increasing temperature.

Figure 11 and Table 14 show the characteristics of these alloys.

The thermocouple usually is used in an insulated metallic sheath of platinum-5 percent molybdenum alloy. The sheath may be joined to a Type 321 stainless steel sheath beyond the area of the helium atmosphere. Both the platinum-molybdenum alloy and the Type 321 stainless steel behave well under neutron irradiation and are compatible with graphite which normally is used in the reactor.

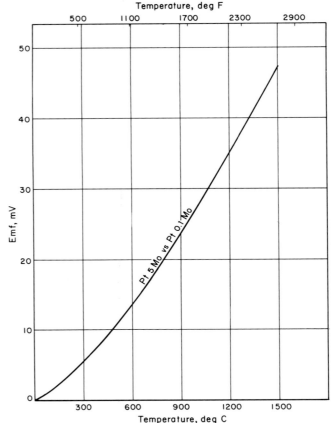

FIG. 11—*Thermal emf of platinum-molybdenum versus platinum-molybdenum thermocouple.*

Extension wires for this thermocouple can be copper for the positive conductor and copper-1.6 percent nickel for the negative conductor. Using these materials the junctions between the thermocouple and the extension wires should be maintained below 70 C (158 F).

3.3.2 *Iridium-Rhodium Types*

3.3.2.1 *Iridium-Rhodium Versus Iridium Thermocouples*—Iridium-rhodium versus iridium thermocouples are suitable for measuring temperature to approximately 2000 C (3632 F), and generally are used above the range served by platinum-rhodium versus platinum thermocouples. They can be used in inert atmospheres and in vacuum, but not in reducing atmospheres, and they may be used in oxidizing atmospheres with shortened life.

The alloys of principal interest are those containing 40, 50, and 60 percent rhodium. They may be used for short times at maximum temperatures

TABLE 14—*Platinum molybdenum versus platinum molybdenum.*

	Pt-5Mo versus Pt-0.1Mo
Nominal operating temperature range, in:	
Reducing atmosphere (nonhydrogen)	NR[a]
Wet hydrogen	NR
Dry hydrogen	NR
Inert atmosphere (helium)	1400 C (2552 F)
Oxidizing atmosphere	NR
Vacuum	NR
Maximum short-time temperature	1550 C (2822 F)
Approximate microvolts per degree:	
Mean, over nominal operating range	29/C (51.2/F)
At top temperature of normal range	30/C (54/F)
Melting temperature, nominal:	
Positive thermoelement	1788 C (3250 F)
Negative thermoelement	1770 C (3218 F)
Stability with thermal cycling	good
High-temperature tensile properties	fair
Stability under mechanical working	good
Ductility (of most brittle thermoelement) after use	good
Resistance to handling contamination	fair
Recommended extension wire 70 C (158 F) max:	
Positive conductor	Cu
Negative conductor	Cu-1.6Ni

[a] NR = not recommended.

2180, 2140, and 2090 C (3956, 3884, and 3794 F, respectively), these temperatures being 60 C (140 F) or more below the respective melting points.

Figure 12 and Table 15 show the characteristics of these alloys.

The wires must be handled carefully. They are flexible in the fibrous (as drawn) state, but when annealed are broken easily by repeated bending.

Metals said to be suitable for extension wires are copper for the positive conductor and stainless steel or an aluminum alloy for the negative conductor.

3.3.3 *Platinel Types*

3.3.3.1 *Platinel Thermocouples*—Platinel,[6] a noble-metal thermocouple combination, was metallurgically designed for high-temperature indication and control in turbo-prop engines. This combination approximates within reasonable tolerances the Type K thermocouple curve.

Actually, two combinations have been produced and are called Platinel I and Platinel II. The negative thermoelement in both thermocouples is a

[6] Trademark–Engelhard Industries, Inc.

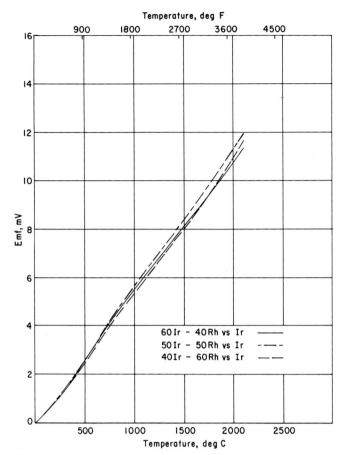

FIG. 12—*Thermal emf of iridium-rhodium versus iridium thermocouples.*

65 percent gold-35 percent palladium alloy (Platinel 1503), but the positive one in Platinel I is composed of 85 percent palladium, 14 percent platinum, and 3 percent gold (Platinel 1786), while that used in Platinel II contains 55 percent palladium, 31 percent platinum, and 14 percent gold (Platinel 1813). Platinel II is the preferred type and has superior mechanical fatigue properties. The thermal emf's of these combinations differ little, as shown in Fig. 13. Other properties are given in Table 16.

From Fig. 13 it is apparent that the emf match with the Type K thermocouple is excellent at high temperatures, but some departure occurs at low temperatures. Generally, the user of Platinel makes the connection between the thermocouple and the extension wire (Type K thermocouple wire) at an elevated temperature (800 C) where the match is good. However, if this is done, care should be taken to ensure that the junctions of both conductors are at the same temperature. If the junction is made at a temperature where

TABLE 15—*Iridium-rhodium versus iridium thermocouples.*

	60Ir-40Rh versus Ir	50Ir-50Rh versus Ir	40Ir-60Rh versus Ir
Nominal operating temperature range in:			
Reducing atmosphere (nonhydrogen)			
Wet hydrogen	NR[a]	NR	NR
Dry hydrogen	NR	NR	NR
Inert atmosphere	2100 C (3812 F)	2050 C (3722 F)	2000 C (3632 F)
Oxidizing atmosphere	NR	NR	NR
Vacuum	2100 C (3812 F)	2050 C (3722 F)	2000 C (3632 F)
Maximum short-time temperature	2190 C (3974 F)	2140 C (3884 F)	2090 C (3794 F)
Approximate microvolts per degree:			
Mean over nominal operating range	5.3 C (2.9 F)	5.7 C (3.2 F)	5.2 C (2.9 F)
At top temperature of normal range	5.6 C (3.1 F)	6.2 C (3.5 F)	5.0 C (2.8 F)
Melting temperature, nominal:			
Positive thermoelement	2250 C (4082 F)	2202 C (3996 F)	2153 C (3907 F)
Negative thermoelement	2443 C (4429 F)	2443 C (4429 F)	2443 C (4429 F)
Stability with thermal cycling	fair	fair	fair
High-temperature tensile properties ...	...	...	...
Stability under mechanical working ...	...	...	...
Ductility (of more brittle thermoelement) after use	poor	poor	poor
Resistance to handling contamination:			
Recommended extension wire	...	...	...
Positive conductor	...	...	...
Negative conductor	...	...	...

[a] NR = not recommended.

the extension wire/thermocouple emf match is not too close, then corrections should be made. Other base-metal extension wires capable of matching the emf of the Platinels very closely at low temperatures [to 160 C (320 F)] are also available.

It is recommended that precautions usually followed with the use of platinum-rhodium versus platinum thermocouples be observed when the Platinels are employed. Tests have shown that phosphorus, sulfur, and silicon have a deleterious effect on the life of the thermocouples.

3.3.4 *Nickel-Chromium Types*

3.3.4.1 *Nickel-Chromium Alloy Thermocouples*—Special nickel-chromium alloys are supplied by various manufacturers as detailed in the following paragraphs. Figure 14 and Table 17 give characteristics of these alloys.

3.3.4.1.1 *Geminol*—The Geminol[7] thermocouple was developed primarily for improved resistance to deterioration in reducing atmospheres.

[7] Trademark—Driver Harris Co.

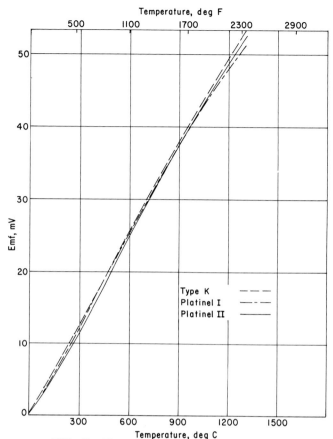

FIG. 13—*Thermal emf of platinel thermocouples.*

The composition of the positive thermoelement has been adjusted specifically to combat in reducing atmospheres the destructive corrosion known as "green rot."

The substitution of an 80 percent nickel-20 percent chromium type alloy for the conventional (Type KP) 90 percent nickel-10 percent chromium alloy positive thermoelement, and a 3 percent silicon in nickel alloy for the conventional (Type KN) manganese-aluminum-silicon in nickel alloy negative thermoelement, results in a more oxidation-resisting thermocouple.

The temperature-emf curve is practically parallel to that of the conventional Type K thermocouple above 760 C (1400 F).

3.3.4.1.2 *Thermo-Kanthal special*—The Thermo-Kanthal special thermocouple was developed to give improved stability at temperatures between 982 C (1800 F) and 1260 C (2300 F) over that obtained with conventional base-metal thermocouple materials.

TABLE 16—*Platinel thermocouples.*

	Platinel II	Platinel I
Nominal operating temperature range, in:		
Reducing atmosphere (nonhydrogen)	NR[b]	NR
Wet hydrogen	NR	NR
Dry hydrogen[a]	1010 C (1850 F)	1010 C (1850 F)
Inert atmosphere	1260 C (2300 F)	1260 C (2300 F)
Oxidizing atmosphere	1260 C (2300 F)	1260 C (2300 F)
Vacuum..	NR	NR
Maximum short-time temperature (< 1 h)	1360 C (2480 F)	1360 C (2480 F)
Approximate microvolts per degree:		
Mean, over nominal operating range		
(100 C to 1000 C).....................................	42.5/C (23.5/F)	41.9/C (23.3/F)
At top temperature of normal range		
(1000 C to 1300 C)	35.5/C (19.6/F)	33.1/C (18.4/F)
Melting temperature, nominal:		
Positive thermoelement—solidus	1500 C (2732 F)	1580 C (2876 F)
Negative thermoelement—solidus	1426 C (2599 F)	1426 C (2599 F)
Stability with thermal cycling............................	good	good
High-temperature tensile properties	fair	fair
Stability under mechanical working	?	?
Ductility (of most brittle thermoelement) after use ...	good	good
Resistance to handling contamination	?	?
Recommended extension wire at approximately		
800 C (1472 F):		
Positive conductor	Type KP	Type KP
Negative conductor	Type KN	Type KN

[a] High-purity alumina insulators are recommended.
[b] NR = not recommended.

3.3.4.1.3 *Tophel II-Nial II*—1. The Tophel II-Nial II thermocouple was developed for improved oxidation resistance and emf stability over the conventional Type K thermocouple alloys in both oxidizing and reducing atmospheres at elevated temperatures.

2. Tophel II, which is the positive thermoelement, is a Ni-10Cr base alloy with additions to resist "green rot" attack in reducing atmospheres at elevated temperatures. The emf of Tophel II is within the standard tolerance of the conventional Type K positive thermoelement over the entire temperature range of 32 to 2300 F (0 to 1260 C). Tophel II can be matched with any acceptable Type K negative thermoelement to form a couple which is within the standard tolerance for the Type K thermocouple.

3. Nial II, which is the negative thermoelement, is a Ni-2.5Si base alloy with additions to improve the oxidation resistance and emf stability in an oxidizing atmosphere at elevated temperatures. The emf of Nial II is within the standard tolerance of the conventional Type K thermoelement between the range of 300 to 2300 F (149 to 1260 C). From 32 to 300 F (0 to 149 C),

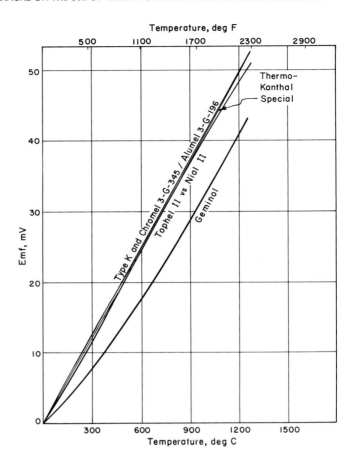

FIG. 14—*Thermal emf of nickel-chromium alloy thermocouples.*

the Nial II is about 0.1 mV less negative than the conventional Type K negative thermoelement with reference to platinum.

4. The Tophel II-Nial II thermocouple meets the emf tolerances designated in ASTM Specification E 230 for the Type K thermocouple from 300 to 2000 F (149 to 1093 C). From 32 to 300 F (0 to 149 C), the Tophel II-Nial II thermocouple generates 0.1 mV (or 5 deg equivalent) less than the standard Type K thermocouple at the same temperature.

5. Tophel II-Nial II thermocouples can be used on existing instruments designed for the Type K thermocouples for temperatures sensing and control within the range of 300 to 2300 F (149 to 1260 C). If extension wire is needed, the negative extension wire should be Nial II, while the positive extension wire could either be Tophel II or any acceptable Type K (+) extension wire.

TABLE 17—*Nickel-chromium alloy thermocouples.*

	Geminol	Thermo-Kanthal Special	Tophel II-Nial II	Chromel 3-G-345-Alumel 3-G-196
Nominal operating temperature range, in:				
Reducing atmosphere (nonhydrogen)	1205 C (2200 F)		1205 C (2200 F)	1205 C (2200 F)
Wet hydrogen	1205 C (2200 F)	1205 C (2200 F)	1205 C (2200 F)	1205 C (2200 F)
Dry hydrogen	1205 C (2200 F)	1205 C (2200 F)	1205 C (2200 F)	1205 C (2250 F)
Inert atmosphere	1205 C (2200 F)	1205 C (2200 F)	1205 C (2200 F)	1205 C (2200 F)
Oxidizing atmosphere	1205 C (2200 F)	1205 C (2200 F)	1205 C (2200 F)	1205 C (2200 F)
Vacuum	1040 C (1904 F)		1040 C (2000 F)	1040 C (2000 F)
Maximum short-time temperature	1260 C (2300 F)	1260 C (2300 F)	1260 C (2300 F)	1260 C (2300 F)
Approximate microvolts per degree:				
Mean, over nominal operating range	18.7/C (10.4/F)	22.6/C (12.6/F)	40 µV/C (22.5 µV/F)	40.7 µV/C (22.6 µV/F)
At top temperature of normal range	22.2/C (12.3/F)	20.0/C (11.1/F)	36 µV/C (20 µV/F)	36 µV/C (20 µV/F)
Melting temperature, nominal:				
Positive thermoelement	1400 C (2550 F)	1432 C (2610 F)	1430 C (2600 F)	1430 C (2600 F)
Negative thermoelement	1430 C (2600 F)	1410 C (2570 F)	1400 C (2550 F)	1400 C (2550 F)
Stability with thermal cycling	good	good	good	good
High-temperature tensile properties	good	good	good	good
Stability under mechanical working	intermediate	intermediate	fair	fair
Ductility (of most brittle thermoelement) after use	good	good	good	good
Resistance to handling contamination	good	fair	good	good
Recommended extension wire:				
Positive conductor	Geminol P	Thermo-Kanthal P	Tophel II or any Type K(+)	Chromel 3-G-345 or any Type K(+)
Negative conductor	Geminol N	Thermo-Kanthal N	Nial II	Alumel 3-G-196 any Type K(−)

6. Through the improvements in both oxidation resistance and emf stability, Tophel II-Nial II thermocouples offers longer useful and total service life than conventional Type K thermocouples of the same size. As a corollary benefit, finer size Tophel II-Nial II thermocouples can be used to achieve equivalent or better stability than conventional Type K couples of larger sizes.

3.3.4.1.4 *Chromel 3-G-345-Alumel 3-G-196*—The Chromel 3-G-345-Alumel 3-G-196 thermocouple is designed to provide improved performance under extreme environmental conditions where the conventional Type K thermocouple is subject to accelerated loss of stability.

More specifically, Chromel 3-G-345 is a Type K positive thermoelement in which the basic 10 percent chromium-nickel alloy is modified to give improved resistance to preferential chromium oxidation ("green rot"). At temperatures from 1600 to 1900 F (871 to 1038 C), conventional Type K positive thermoelements operating in marginally oxidizing environments are subject to embrittlement and loss of output as a result of such attack.

Under those conditions, Type K thermocouples employing Chromel 3-G-345 positive thermoelements offer greater stability than conventional Type K thermocouples. The usual precautions regarding protection of Type K thermocouples in corrosive environments apply to the special thermocouple as well.

The modified Chromel thermoelement meets the accepted curve of emf versus platinum for Type K positive thermoelements, within standard tolerances. It can be combined with either Alumel 3-G-196 or conventional Alumel to form Type K thermocouples meeting standard emf tolerances.

Alumel 3-G-196 is a Type K negative thermoelement of greatly improved oxidation resistance. It is suited to use in both reducing and oxidizing atmospheres, where its stability of output is especially advantageous in fine wire applications at high temperatures. It is nominally 2.5Si-Ni.

Alumel 3-G-196 meets the accepted curve of emf versus platinum for Type K negative thermoelements at all temperatures from 32 to 2300 F (0 to 1260 C). It can be combined with either Chromel 3-G-345 or regular Chromel to form thermocouples meeting standard Type K thermocouple tolerances over the entire range from 32 to 2300 F (0 to 1260 C).

Type K thermocouples employing either or both special thermoelements can be used with conventional extension wires at no sacrifice in guaranteed accuracy of the thermocouple-extension wire combination.

3.3.5 *Nickel-Molybdenum Types*

3.3.5.1 *19 and 20 Alloys*[8] *(nickel-nickel molybdenum alloys)*—1. The 19

---

[8] The 19 Alloy and the 20 Alloy thermocouple was developed by the General Electric Co. Since 1962, the Wilbur B. Driver Co. has been the sole manufacturer of the 19 Alloy and the 20 Alloy thermocouple.

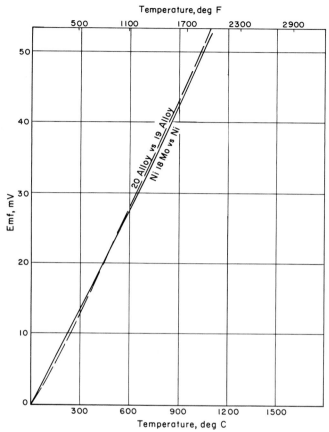

FIG. 15—*Thermal emf of nickel versus nickel-molybdenum alloys.*

Alloy-20 Alloy[7] thermocouple was developed for temperature sensing and control applications at elevated temperatures in hydrogen or other reducing atmospheres. The emf table of the 19 Alloy versus the 20 Alloy does not conform to the Type K or any existing base metal thermocouples designated by ASTM Specification E 230.

2. The 19 Alloy, which is the negative thermoelement, is essentially a Ni-1Co alloy. Its emf versus platinum values are somewhat more negative than those of the Type K negative thermoelement within the range of 32 to 2300 F (0 to 1260 C).

3. The 20 Alloy, which is the positive thermoelement, is essentially a Ni-18Mo alloy. Its emf versus platinum values are less positive within the range of 32 to 500 F (0 to 260 C) than the Type K positive thermoelement, but more positive within the range of 500 to 2300 F (260 to 1260 C). Figure 15 and Table 18 show the characteristics of these alloys.

TABLE 18—*Physical data and recommended applications of the 19 Alloy-20 Alloy thermocouple.*

	19 Alloy-20 Alloy
Nominal operating temperature range, in:	
Reducing atmosphere, (nonhydrogen)	1205 C (2200 F)
Wet hydrogen	1205 C (2200 F)
Dry hydrogen	1205 C (2200 F)
Inert atmosphere	1205 C (2200 F)
Oxidizing atmosphere	not recommended
Vacuum..	1205 C (2200 F)
Maximum short time temperature	1260 C (2300 F)
Approximate microvolts per degree:	
Mean, over nominal operating range	55 μV/C (31.0 μV/F) between the range of 1000 to 2300 F 59 μV/C
At top temperature of normal range................................	(32.9 μV/F)
Melting temperature, nominal:	
Positive leg	1430 C (2600 F)
Negative leg	1450 C (2640 F)
General stability with thermal cycling (good, fair, poor)	good
High temperature tensile properties (good, fair, poor)	good
Unaffected by mechanical working (good, fair, poor)	fair
Ductility (of most brittle leg) after use (good, fair, poor)	fair
Resistance to handling contamination (good, fair, poor)	good
Recommended extension wire:	
Positive leg	20 Alloy
Negative leg	19 Alloy

4. The 19 Alloy-20 Alloy thermocouple, when properly sealed in a protection tube, offers excellent emf stability at elevated temperatures in hydrogen or other reducing atmospheres.

5. The oxidation resistance of the 19 Alloy-20 Alloy thermocouple is not good. The 19 Alloy-20 Alloy thermocouples are not recommended for use in an oxidizing atmosphere above 1200 F (649 C).

6. 19 Alloy-20 Alloy extension wire should be used in connection with the 19 Alloy-20 Alloy thermocouple.

3.3.6 *Tungsten-Rhenium Types*

There are three tungsten-rhenium thermocouple systems available— tungsten versus tungsten-26 percent rhenium, doped tungsten-3 percent rhenium versus tungsten-25 percent rhenium and doped tungsten-5 percent rhenium versus tungsten-26 percent rhenium. The price of the first combination has the lowest cost. All have been employed to 2760 C (5000 F) but general use is below 2316 C (4200 F). Applications for these couples

have been found in space vehicles, nuclear reactors, and many high-temperature electronic, thermoelectric, industrial heating, and structural pro-
jects. However, when employed in a nuclear environment, the effect of
transmutation of the thermal emf of the couples should be considered.

The use of tungsten in certain applications as the positive element may
pose a problem, since heating tungsten to or above its recrystallization
temperature (approximately 1200 C) causes embrittlement resulting in a
loss of room-temperature ductility; an effect that is not experienced with
the alloy leg containing high rhenium. With proper handling and usage this
combination can be employed satisfactorily for long periods. One approach
to the brittleness problem is to add rhenium to the tungsten thermoelement.
Early research showed that the addition of 10 percent rhenium to the
tungsten element did much to retain ductility after recrystallization. This
much rhenium, however, greatly reduced the emf response for the thermocouple. Other techniques to retain room-temperature ductility are used by
manufacturers; these include special processing and doping with the addition of 5 percent or less rhenium to the tungsten thermoelement.

Doping usually consists of using additives during the process of pre-
paring the tungsten powder and results in a unique microstructure in the
finished wire. The additives essentially are eliminated during the subsequent
sintering of the tungsten-rhenium powder compact. In fact, presently
known analytical techniques do not disclose the presence of the additives
above the background level of such substances normally present as impurities in nondoped tungsten or tungsten-rhenium alloys.

The emf response of tungsten-3 percent rhenium and tungsten-5 percent
rhenium thermoelements used with thermoelements containing high per-
centages of rhenium is satisfactory. The thermoelectric power of the tungsten
versus tungsten-26 percent rhenium, tungsten-3 percent rhenium versus
tungsten-25 percent rhenium, and tungsten-5 percent rhenium versus
tungsten-26 percent rhenium is comparable at lower temperatures, but drops
off slightly for the latter two as the temperature is increased.

The tungsten thermoelement is not supplied to the user in a stabilized
(recrystallized) condition; therefore, a small change in emf is encountered
at the operating temperature. In the case of the doped W-3Re, doped
W-5Re, W-25Re and W-26Re thermoelements, these are supplied in a
stabilized (recrystallized) condition.

All three thermocouple combinations are supplied as matched pairs
guaranteed to meet the emf output of producer developed tables within
±1 percent. In addition, compensating extension wires are available for
each of the three combinations with maximum service temperatures as
high as 871 C (1600 F) for tungsten-5 percent rhenium versus tungsten-
26 percent rhenium.

Important factors controlling the performance at high temperatures are:
the diameter of the thermoelements (larger diameters are suitable for higher

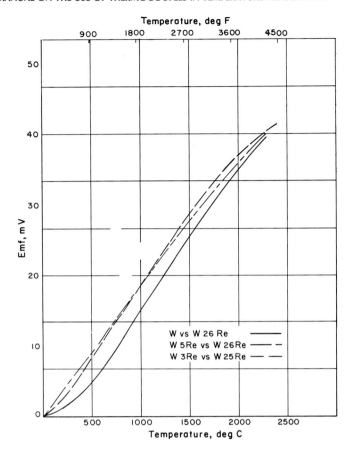

FIG. 16—*Thermal emf of tungsten-rhenium versus tungsten-rhenium thermocouples.*

temperatures), the atmosphere (vacuum, high-purity hydrogen, or high-purity inert atmospheres required), the insulation, and sheath material. Some evidence is at hand, however, which indicates the possibility of selective vaporization of rhenium at temperatures of the order of 1900 C and higher when bare (unsheathed) tungsten-rhenium thermocouples are used in vacuum. For this reason, the vapor pressure of rhenium should be considered when a bare couple is used in a high vacuum at high temperatures. This, of course, is not a problem when these couples are protected with a suitable refractory metal sheath.

Figure 16 and Table 19 show the characteristics of these alloys.

3.4 Compatibility Problems at High Temperatures

In order for thermocouples to have a long life at high temperatures, it is necessary to limit reactions between the metals, the atmosphere, and the

TABLE 19—*Tungsten-rhenium thermocouples.*

	W versus W-26Re	W-3Re versus W-25Re	W-5Re versus W-26Re
Nominal operating temperature range, in:			
Reducing atmosphere (nonhydrogen)	NR[a]	NR	NR
Wet hydrogen	NR	NR	NR
Dry hydrogen	2760 C (5000 F)	2760 C (5000 F)	2760 C (5000 F)
Inert atmosphere	2760 C (5000 F)	2760 C (5000 F)	2760 C (5000 F)
Oxidizing atmosphere	NR	NR	NR
Vacuum[b]	2760 C (5000 F)	2760 C (5000 F)	2760 C (5000 F)
Maximum short-time temperature	3000 C (5430 F)	3000 C (5430 F)	3000 C (5430 F)
Approximate microvolts per degree:			
Mean, over nominal operating range 0 C to 2316 C (32 F to 4200 F) ...	16.7/C (9.3/F)	17.1/C (9.5/F)	16.0/C (8.9/F)
At top temperature of normal range 2316 C (4200 F)	12.1/C (6.7/F)	9.9/C (5.5/F)	8.8/C (4.9/F)
Melting temperature, nominal:			
Positive thermoelement	3410 C (6170 F)	3360 C (6080 F)	3350 C (6062 F)
Negative thermoelement	3120 C (5648 F)	3120 C (5648 F)	3120 C (5648 F)
Stability with thermal cycling	good	good	good
High-temperature tensile properties ...	good	good	good
Stability under mechanical working ...	fair	fair	fair
Ductility (of most brittle thermoelement after use)	poor	poor to good depending on atmosphere or degree of vacuum	poor to good depending on atmosphere or degree of vacuum
Resistance to handling contamination	good	good	good
Extension wire	available	available	available

[a] NR = not recommended.
[b] Preferential vaporization of rhenium may occur when bare (unsheathed) couple is used at high temperatures and high vacuum. Check vapor pressure of rhenium at operating temperature and vacuum before using bare couple.

ceramic insulation. Such reactions may change the strength or corrosion-resistant properties of the alloys, the electrical output of the thermocouple, or the electrical insulation properties of the ceramic insulant.

At extremely high temperatures reactions can be expected between almost any two materials. Table 20 has been included to show the temperatures at which such reactions occur between pairs of metallic elements. At lower temperatures, certain reactions do not occur and such as do occur, proceed at a slower rate. Because of potential reactions, it is important to identify the impurities and trace elements as well as the major constituents of the thermocouple components. The "free energy of formation" (Gibbs free energy) for the oxides of each element at the temperatures of interest, can

TABLE 20[a] —Minimum melting temperatures of binary systems.

Element	Si	Ni	Co	Fe	Ti	Pt	Zr	Cr	V	Rh	Hf	Ru	Cb	Ir	Mo	Ta	Os	Re	W
Melting Point, deg F	2588	2651	2723	2802	3034	3224	3353	3407	3452	3560	4030	4082	4425	4428	4730	5423	5432	5724	6170
Element:																			
Si		1770	2183	2192	2426	1526	~2480	~2408	~2550		2102	2651	2372	2588	~2570	2525		~2057	2550
Ni			~2646	2588	1727	~2651	~1764	2444	2198		2214	~2550	2145	~2460	~2398	2480		2651	~2651
Co				~2696	1860	~2590	1789	2552	2264		2985	2802	2255		2444	2331	~2723	~2723	~2696
Fe					1985	~2730	1705	2743	2675				2480	2687	2624	~2462		~2802	2775
Ti						2390	~2876	2525	~2777				~3034		~3034	~3034		~3034	~3034
Pt							2165	2552					~3092		3224			~3224	~3224
Zr								~2372					3164	~3224	2740	3308		2912	~3020
Cr									3182				3020		3380	3092		3407	3038
V										~3224	3353	2453	3290		3452			3452	2974
Rh																		~3416	
Hf															3524				3488
Ru													3056		3506			4415	4001
Cb															3533	3578			~4425
Ir															~4250	~4425			
Mo																~4730	~4406	4424	~4730
Ta																	~4388	4874	~5423
Os																			4937
Re																			~5070
W																			

[a] Adapted from: *Constitution of Binary Alloys* by Rodney P. Elliot, McGraw-Hill, New York, 1965.

be determined to predict possible oxide reactions. Other reactions may occur and attention should be given to the possible formation of the carbides, nitrides, etc. of the various elements.

Helpful data may be obtained from published reports, but, because of the importance of trace elements and impurities, the sources should be treated with caution. In some cases, the amount and types of impurities in the materials used were unknown.

Certain reactions may be somewhat self-limiting in that the reaction product provides a protective film against further reaction. However, spalling or chipping off of the reaction product may occur because of thermal or physical stress. Thus, the reaction rate and the use of the corrosion product as protection can be ascertained only if tested under the desired operating conditions and times.

The use of oxygen-gettering material should be considered in instances where oxygen is present in limited amounts [2]. This method has been proven in sheath-type Chromel-Alumel thermocouples where it limits the preferential oxidation of Chromel. A thin tube (or sliver) of titanium at the hottest location of the thermocouple has been used.

3.5 References

[1] Caldwell, F. R., "Thermocouple Materials," NBS Monograph 40, National Bureau of Standards, 1 March 1962.
[2] Neswald, R. G., "Titanium for Realists," *Space/Aeronautics,* Vol. 48, No. 5, Oct. 1967, pp. 90–99.

4. THERMOCOUPLE HARDWARE AND FABRICATION

A complete thermocouple temperature sensing assembly, in accordance with the present state of the art, consists of the following;

A. Sensing element assembly including, in its most basic form, two dissimilar wires, supported by an electrical insulator and joined at one end to form a measuring junction. Such assemblies usually fall into one of three categories; those formed from wires have nonceramic insulation, those with hard-fired ceramic insulators, and those made from sheathed, compacted ceramic-insulated wires.

B. Protection tube—ceramic and metal protection tubes, sometimes referred to as thermowells, serve the purpose of protecting the sensing element assembly from the deleterious effects of corrosive or oxidizing or reducing atmospheres. In some cases, two concentrically arranged protection tubes may be used. The one closest to the sensing element assembly is designated the primary protection tube, while the outer tube is termed the secondary protection tube. Combinations such as an aluminum oxide primary tube and silicon carbide secondary tube often are used to obtain the beneficial characteristics of the combination, such as resistance to cutting flame action, and ability to resist thermal shock.

C. Connector—sensing element assembly wire terminations are made to either
- (a) Terminals
- (b) Connection head
  1. General purpose type
  2. Screw cover type
  3. Open type
- (c) Plug and jack quick-disconnect
- (d) Military standard (MS) type of connector

D. Miscellaneous hardware such as
- (a) Pipe nipple or adapter to join the protection tube to the head
- (b) Thermocouple gland—used primarily with sheathed, compacted ceramic-insulated thermocouple assemblies to serve the dual function of mounting and sealing-off pressure in the mounting hole (see Section 5).

4.1 Sensing Element Assemblies

Typical thermocouple element assemblies, shown in Fig. 17, A and B, illustrate common methods of forming the measuring junction, A—by twisting and welding and B by butt-welding. C shows an assembly using nonceramic insulation such as, asbestoc or fiber glass. D, E, and F show the use of various forms of hard fired ceramic insulators, double bore (D) fish-spine (E) and four-hole (F). Fish-spine provides flexibility, and four-hole provides for two independent sensing elements.

4.2 Nonceramic Insulation

The normal function of thermocouple insulation is to provide electrical insulation for the thermocouple assembly. If this function is not provided or is compromised in any way, the indicated temperature may be in error. Insulation is affected adversely by moisture, abrasion, flexing, temperature extremes, chemical attack, and nuclear radiation. Each type of insulation has its own limitations. A knowledge of these limitations is essential if accurate and reliable measurements are to be made.

Some insulations have a natural moisture resistance. Teflon, polyvinyl chloride (PVC), and polyimide are examples of this group. With the fiber type insulations, moisture protection results from impregnating with substances such as wax, resins, or silicone compounds. Once the impregnating materials have been vaporized off, there is no longer moisture protection. Typically, this occurs once the insulation has been exposed above 400 F (204 C). The moisture penetration problem is not confined to the sensing end of the thermocouple assembly. For example, if the thermocouple passes through hot and cold zones or through an area which is time-wise alternately hot and cold, condensation may produce errors in the indicated temperature, unless adequate moisture resistance is provided.

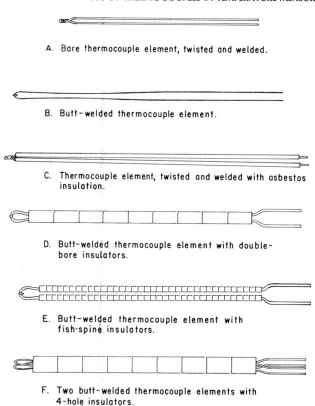

A. Bare thermocouple element, twisted and welded.

B. Butt-welded thermocouple element.

C. Thermocouple element, twisted and welded with asbestos insulation.

D. Butt-welded thermocouple element with double-bore insulators.

E. Butt-welded thermocouple element with fish-spine insulators.

F. Two butt-welded thermocouple elements with 4-hole insulators.

FIG. 17—*Typical thermocouple element assemblies.*

Protection from abrasion and flexing usually is provided by impregnating materials. However, one cycle over 400 F (204 C) usually results in a deterioration of this protection. After exposure to higher temperature (1100 to 1400 F) (593 to 760 C), whole sections of the insulation can fall off resulting in bare wire and a possible "short." Thermocouples in this condition should not be used if any flexing or abrasion is expected. It is recommended that they be discarded or that the exposed portion of the thermocouple assembly be cut off and another junction formed.

Insulations are rated for a maximum temperature both for continuous usage and for a single exposure. These distinctions should be observed when selecting an insulating material. At elevated temperatures even those insulations which remain physically intact become conductive. Under these conditions, the output of the thermocouple may be a function of the highest temperature to which the insulation is exposed, rather than the temperature of the measuring junction. The change in insulation resistance may be permanent if caused by deterioration of organic insulants or binders which leave a carbon residence. In considering the temperature to which the insulation is exposed, it should not be assumed that this is the temperature of the measuring junction. A thermocouple may be attached to a massive specimen which is exposed to a high temperature source to achieve a rapid heating rate. Parts of the thermocouple wires not in thermal contact with the specimen can be overheated severely while the junction remains within safe temperature limits. With this in mind, high quality insulation should be used when rapid heating rates are expected. Very little factual information is available on actual deterioration rates and magnitudes, but the problem exists so a conservative approach is recommended.

The basic types of elevated temperature insulations are fiber glass, fibrous silica, and asbestos. Of the three materials, fibrous silica has the best high-temperature electrical properties, although its handling and abrasion characteristics leave something to be desired. The next best high-temperature insulation is asbestos. Because this material has very poor mechanical properties a filler fiber or an impregnating material is added. In some instances, this filler is cotton or another organic compound which leaves a carbon residue after exposure to high temperature, and this results in a breakdown of electrical insulations. Asbestos loses its mechanical strength after exposure to elevated temperatures and tends to drop off even if the thermocouple assembly is not being handled. A more commonly used insulation is fiber glass. It can be impregnated to provide excellent moisture and mechanical characteristics. The main difficulty is that it becomes a conductor above 950 F (510 C). If one is willing to sacrifice the handling characteristics, a nonimpregnated fiber glass insulation is available which is good to 1200 F (649 C).

Chemical deterioration of insulation materials can produce a number of problems. If the environment reacts with the insulation, both the insulation and environment can be affected adversely. The insulation can be removed physically or made electrically conductive, and the process system can become contaminated. For example, some insulation materials are known to produce cracking in austenitic stainless steels.

In summary, an insulation should be selected only after considering possible exposure temperatures and heating rates, the number of temperature cycles, mechanical movement, moisture, routing of the thermocouple wire, and chemical deterioration (see Table 21).

Industry has established insulation color codes for various letter-designated thermocouple and extension wire types, as shown in Table 22.

4.3 Hard-Fired Ceramic Insulators

Hard-fired ceramic insulators most commonly used with bare thermocouple elements are mullite, aluminum-oxide, and steatite, the latter being the most common material where fish-spine insulators are concerned.

TABLE 21—*Insulation characteristics.*

Insulation	Continuous Use Temperature Limit, deg F	Single Exposure Temperature Limit, deg F	Moisture Resistance	Abrasion Resistance
Cotton	200	200	poor	fair
Polyvinyl	220	220	excellent	excellent
Enamel and cotton..................	200	200	fair	fair
Nylon[a]	260	260	good	good
Teflon[a]	400	600	excellent	excellent
Polyimide	600	750	excellent	good
Teflon and fiber glass[b]	600	700 to 1000	excellent to 600 F	good
Fiber glass-varnish or silicone impregnation............	900	1000	fair to 400 F, poor above 400 F	fair to 400 F, poor above 400 F
Fiber glass, nonimpregnated......	1000	1200	poor	fair
Asbestos and fiber glass with silicone[c]	900	1200[d]	good to 400 F	fair to 400 F, poor above 400 F
Felted asbestos	1000	1200	poor	poor
Asbestos over asbestos	1000	1200	poor	poor
Refrasil[e]	1600	2000	very poor	very poor

[a] Trademark–E. I. Du Pont de Nemours & Co.
[b] The Teflon vaporizes at 600 F with toxic effects.
[c] Individual wires are asbestos and overbraid is fiber glass.
[d] At 1400 F, the wire may be contaminated after a short exposure.
[e] Trademark–H. I. Thompson Co.

Single, double, and multibore insulators are available in a wide variety of sizes in both English and metric dimensions.

Lengths in the English dimensional units commonly stocked by many suppliers are 1, 2, 3, 6, 12, 18, 24, and 36 in., with longer lengths to as great as 72 in. being available on special order.

It is usually advisable, especially in the case of precious metal thermocouple element assemblies, to keep the insulator in one piece to minimize contamination from the environment.

Hard-fired ceramic insulators are made in oval as well as circular cross-section examples of which are shown in Fig. 18. Properties of refractory oxides are tabulated in Table 23.

4.4 Protection Tubes

4.4.1 Factors Affecting Choice of Protection Tubes

Thermocouples must be protected from atmospheres that are not compatible with the thermocouple alloys. Protection tubes serve the double purpose of guarding the thermocouple against mechanical damage and

Oval Double Bore Insulator

Round Double Bore Insulator

Round Four Bore Insulator

FIG. 18—*Cross-section examples of oval and circular hard-fired ceramic insulators.*

TABLE 22—*Color code of duplex thermocouple wire and extension wire insulators.*

Thermocouple Wire			Thermocouple Extension Wire		
Name and Symbol	Duplex Color	Polarity and Overall Covering	Name and Symbol	Single Conductor	Duplex Color
Copper	blue	+	Copper	blue	blue
constantan,	red	−	constantan,	red-blue trace	red
T	brown	overall	TX		blue
Chromel	yellow	+	Chromel	yellow	yellow
Alumel,	red	−	Alumel,	red-yellow trace	red
K	brown	overall	KX		yellow
Iron	white	+	Iron	white	white
constantan,	red	−	constantan,	red-white trace	red
J	brown	overall	JX		black
Chromel	purple	+	Chromel	purple	purple
constantan,	red	−	constantan,	red-purple trace	red
E	brown	overall	EX		purple
Pt-10Rh		+	Copper	black	black
platinum,		−	alloy ⅙11,	red-black trace	red
S		overall	SX		green
Pt-13Rh		+	Copper	black	black
platinum,		−	alloy ⅙11,	red-black trace	red
R		overall	SX		green

TABLE 23 — *Properties of refractory oxides[a].*

Composition	Porosity, volume %	Fusion Temperature, deg C	Maximum Normal Use Temperature, deg C	Density, Bulk (b), True (t), (g/cm³)	Specific Heat Capacity (cal/g/C) 20 to 1000 C	Linear Expansion (10⁻⁶ in./in. C) 20 to 1000 C	Thermal Conductivity (cal s⁻¹ C⁻¹ cm⁻² cm) At 100 C	At 1000 C	Modulus of Rupture, psi At 20 C	At 1000 C	Modulus of Elasticity, (10⁶ psi)	Thermal Stress Resistance
Sapphire crystal .... 99.9 Al₂O₃	0	2030	1950	3.97(t)	0.26	8.6	0.072	0.019	40 000- 150 000	30 000- 100 000	55	very good
Sintered alumina .... 99.8 Al₂O₃	3 to 7	2030	1900	3.97(t)	0.26	8.6	0.069	0.014	30 000	22 000	53	good
Sintered beryllia .... 99.8 BeO	3 to 7	2570	1900	3.03(t)	0.50	8.9	0.500	0.046	20 000	10 000	45	excellent
Sintered calcia .... 99.8 CaO	5 to 10	2600	2000	3.32(t)	0.23	13.0	0.033	0.017	...	...	...	fair-poor
Chrome-alumina cermet (Haynes Stellite LT-1) ...... 77 Cr, 23 Al₂O₃	2	1850	1300	5.9(b)	0.16	8.9	0.08	0.05	45 000	20 000	37.5	excellent
Sintered magnesia .... 99.8 MgO	3 to 7	2800	1900	3.58(t)	0.25	13.5	0.082	0.016	14 000	12 000	30.5	fair-poor
Sintered mullite .... 72 Al₂O₃, 28 SiO₂	3 to 10	1810	1750	3.03(t)	0.25	5.3	0.013	0.008	12 000	7 000	21	good
Sintered forsterite .... 99.5 Mg₂SiO₄	4 to 12	1885	1750	3.22(t)	0.23	10.6	0.010	0.005	10 000	...	...	fair-poor
Sintered spinel .... 99.8 MgAl₂O₄	3 to 10	2135	1850	3.58(t)	0.25	8.8	0.033	0.013	12 300	11 000	34.5	fair
Sintered titania .... 99.5 TiO₂	3 to 7	1840	1600	4.24(t)	0.20	8.7	0.015	0.008	8 000	6 000	...	fair-poor
Sintered thoria .... 99.8 ThO₂	3 to 7	3050	2500	10.50(t)	0.06	9.0	0.022	0.007	12 000	7 000	21	fair-poor
Sintered yttria .... 99.8 Y₂O₃	2 to 5	2410	2000	4.50(t)	0.13	9.3	(0.02)	...	...	...	...	fair-poor
Sintered urania .... 99.8 UO₂	3 to 10	2800	2200	10.96(t)	0.06	10.0	0.020	0.007	12 000	18 000	25	fair-poor
Sintered stabilized zirconia ... 92 ZrO₂, 4 HfO₂, 4 CaO	3 to 10	2550	2200	5.6(t)	0.14	10.0	0.005	0.005	20 000	15 000	22	fair-good
Sintered zircon .... 99.5 ZrSIO₄	5 to 15	2420	1800	4.7(t)	0.16	4.2	0.015	0.008	12 000	6 000	30	good
Silica glass .... 99.8 SiO₂	0	1710	1100	2.20(t)	0.18	0.5	0.004	0.012	15 500	...	10.5	excellent
Mullite porcelain.......... 70 Al₂O₃, 27 SiO₂, 3 Mo + M₂O	2 to 10	1750	1400	2.8(b)	0.25	5.5	0.007	0.006	10 000	6 000	10	good
High alumina porcelain .... 90-95 Al₂O₃, 4-7 SiO₂, 1-4 Mo + M₂O	2 to 5	1800	1500	3.75(b)	0.26	7.8	0.05	0.015	50 000	...	53	very good

[a] Kingery, W. D., "Oxides for High Temperature Applications," *Proceedings of the International Symposium on High Temperature Technology,* McGraw-Hill, New York, 1960.

interposing a shield between the thermocouple and its surroundings so as to maintain it as nearly as possible in its best atmosphere.

The agencies that must be excluded are: (*a*) metals (solid, liquid, or vapor) which, coming into contact with the thermocouple, would alter its chemical composition; (*b*) furnace gases and fumes which may attack the thermocouple materials (sulphur and its compounds are particularly deleterious); (*c*) materials such as silica and some of its metallic oxides, which, in contact with the thermocouple in a reducing atmosphere, are reduced, and combine with the thermocouple to attack it; and (*d*) electrolytes which would attack the thermocouple material.

The choice of the proper protection tube is governed by the conditions of use and by the tolerable life of the thermocouple. There may be occasions when the strength of the protection tube may be more important than the long term thermoelectric stability of the thermocouple. On the other hand, gas tightness and resistance to thermal shock may be of paramount importance. In other cases, chemical compatibility of the protection tube with the process may be the deciding factor.

4.4.2 *Common Forms of Protection Tubes*

4.4.2.1 *Metal Tubes*—Metal tubes offer adequate mechanical protection for base metal thermocouples at temperatures to 2100 F (1150 C). It must be remembered that all metallic tubes are somewhat porous at temperatures exceeding 1500 F (815 C) so that, in some cases, it may be necessary to provide an inner tube of ceramic material.

(*a*) Carbon steels can be used to 1300 F (700 C) usually in oxidizing atmospheres.

(*b*) Austenitic stainless steels (300 series) can be used to 1600 F (870 C), mostly oxidizing although Types 316, 317, and 318 can be used in some reducing atmospheres.

(*c*) Ferritic stainless steels (400 series) can be used from 1800 to 2100 F (975 to 1150 C) in both oxidizing and reducing atmospheres.

(*d*) High nickel alloys, Nichrome, Inconel, etc., can be used to 2100 F (1150 C) in oxidizing atmospheres.

Where the protection tube is subject to high pressure or flow-induced stresses or both a drilled thermowell often is recommended. Although less expensive metal tubes, fabricated by plugging the end of the protection tube, may satisfy application requirements, more stringent specifications usually dictate the choice of gun-drilled bar stock, polished and hydrostatically tested as a precaution against failure.

Examples of drilled thermowells are shown in Fig. 19.

4.4.2.2 *Ceramic Tubes*—Ceramic tubes are used usually at temperatures beyond the ranges of metal tubes although they are sometimes used at lower temperatures in atmospheres harmful to metal tubes.

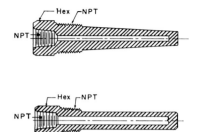

Straight and Tapered
Drilled Thermowells

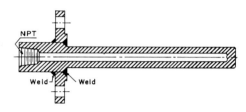

Flanged Thermowell

FIG. 19—*Examples of drilled thermowells.*

The ceramic tube most widely used has a Mullite base with certain additives to give the best combination of mechanical and thermal shock properties. Upper temperature limit 3000 F (1650 C).

Silicon carbide tubes are used as secondary protection tubes. This material resists the cutting action of flames. It is not impermeable to gases and, where a dense tube is required, a nitride-bonded type material can be obtained so that the permeability is greatly reduced.

Fused alumina tubes can be used as primary or secondary protection tubes or both where temperatures to 3600 F (1980 C) are expected and when a gas tight tube is essential. Fused alumina tubes and insulators should be used with platinum-rhodium, platinum thermocouples above 2200 F (1200 C) in order to ensure long life and attain maximum accuracy. (The Mullite types contain impurities which can contaminate platinum above 2200 F (1220 C). The alumina tubes are more expensive than the Mullite base tubes, but can be obtained impervious to most gases to 3300 F (1815 C).

4.4.2.3 *Metal-Ceramic Tubes*—"Cermets" are combinations of metals and metallic oxides which, after proper treatment, form dense, high-strength, corrosion-resistant tubes usable to about 2600 F (1425 C) in most atmospheres.

4.5 Circuit Connections

The electrical output of a thermocouple assembly is taken at the end of the extension wires which are most often a separate component of the temperature sensing system. Thermocouples, usually of the sheathed mineral insulated variety, can incorporate a transition fitting inside which the extension wires are welded or brazed to the ends of the thermocouple wires in what usually is referred to as an intermediate junction. Thus the sensing element and extension wires are an integral part of the temperature sensing system.

Where hard-fired ceramic insulators are used, the thermocouple wires usually terminate in a head which is attached firmly to the protection tube in which the insulator is housed—or it may be supported by the insulator itself.

The connection between thermocouple and extension wires is made by means of a screw clamp or binding post to permit easy replacement of the thermocouple. The user should be aware of the need to keep the temperature of the termination below 400 F in the interest of overall accuracy. In any case, the termination should never exceed the temperature limitation of the extension wires.

Highest system accuracy is achieved by running the thermocouple conductors to the reference junction and avoiding the use of extension wires. If circuit connectors are necessary, they should be of thermoelectric materials. Connectors of the quick-disconnect type usually are so designed. For example, a Chromel-Alumel thermocouple connector incorporates Chromel and Alumel pins or sockets or both.

Industrial types of thermocouple assemblies using heads of the general purpose or screw cover type contain a terminal block assembly consisting of a ceramic block and floating brass terminals. One assumes in using such a component that the temperature of the terminals is uniform and that no temperature gradient exists along their length. In accordance with the law of intermediate metals, no error should result providing this condition results.

Closed and open type heads are generally of cast iron body construction and cast or sheet metal formed covers incorporating an asbestos composition gasket seal. Cast aluminum or zinc alloy die castings are also available. The terminal block assembly may be a phenolic molding which would limit its use to about 170 C (338 F). Higher temperature capability would dictate the use of a ceramic terminal block assembly.

4.6 Complete Assemblies

Figure 20 shows complete assemblies of the components which have been described in the foregoing sections. Many other combinations are possible. Manufacturers' catalogs may be consulted for details.

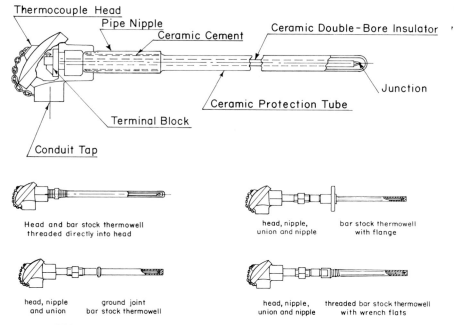

FIG. 20—*Typical examples of hard-fired ceramic-insulated thermocouples.*

4.7 Selection Guide for Protection Tubes

The following information has been extracted from various manufacturer's literature. It is offered as a guide to the selection of protection tubes. Caution should be exercised in applying this information to specific situations.

Application	Protection Tube Material
Heat treating:	
Annealing	
Up to 1300 F (704 C)	wrought iron
Over 1300 F (704 C)	28% chrome iron or Inconel[a]
Carburizing hardening	
Up to 1500 F (816 C)	Wrought iron or 28% chrome iron
1500 to 2000 F (1093 C)	28% chrome iron or Inconel
Over 2000 F (1093 C)	ceramic
Nitriding salt baths	28% chrome iron
Cyanide	nickel
Neutral	28% chrome iron
High speed	ceramic
Iron and steel:	
Basic oxygen furnace	quartz
Blast furnaces	
Downcomer	Inconel, 28% chrome iron
Stove Dome	silicon carbide

[a] Trademark–International Nickel Co.

Application	Protection Tube Material
Hot blast main	Inconel
Stove trunk	Inconel
Stove outlet flue	wrought iron
Open hearth	
Flues and stack	Inconel, 28% chrome iron
Checkers	Inconel, Cermet
Waste heat boiler	28% chrome iron, Inconel
Billet heating slab heating and butt welding	
Up to 2000 F (1093 C)	28% chrome iron, Inconel
Over 2000 F (1093 C)	ceramic, silicon carbide
Bright annealing batch	
Top work temperature	not required (use bare Type J thermocouple)
Bottom work temperature	28% chrome iron
Continuous furnace section	Inconel, ceramic
Forging	silicon carbide, ceramic
Soaking pits	
Up to 2000 F (1093 C)	Inconel
Over 2000 F (1093 C)	ceramic, silicon carbide
Nonferrous metals:	
Aluminum	
Melting	cast iron (white-washed)
Heat treating	wrought iron
Brass or bronze	not required (use dip-type thermocouple)
Lead	28% chrome iron, wrought iron
Magnesium	wrought iron, cast iron
Tin	extra heavy carbon steel
Zinc	extra heavy carbon steel
Pickling tanks	chemical lead
Cement:	
Exit flues	Inconel, 28% chrome iron
Kilns—heating zone	Inconel
Ceramic:	
Kilns	ceramic and silicon carbide
Dryers	wrought iron, silicon carbide
Vitreous enameling	Inconel, 28% chrome iron
Glass:	
Fore hearths and feeders	platinum thimble
Lehrs	wrought iron
Tanks	
Roof and wall	ceramic
Flues and checkers	28% chrome iron, Inconel
Paper:	
Digesters	Type 316 stainless steel., 28% chrome iron
Petroleum:	
Dewaxing	Type 304 stainless steel or carbon steel
Towers	Type 304 stainless steel or carbon steel
Transfer lines	Type 304 stainless steel or carbon steel
Factionating column	Type 304 stainless steel or carbon steel
Bridgewall	Type 304 stainless steel or carbon steel
Power:	
Coal-air mixtures	Type 304 stainless steel
Flue gases	wrought iron or 28% chrome iron
Preheaters	wrought iron or 28% chrome iron
Steel lines	Type 347 or 316 stainless steel
Water lines	carbon steel
Boiler tubes	Type 309 or 310 stainless steel
Gas Producers:	
Producer gas	28% chrome iron

Application	Protection Tube Material

Water gas
  Carburetor .... Inconel, 28% chrome iron
  Superheater .... Inconel, 28% chrome iron
  Tar stills .... carbon steel

Incinerators:
  Up to 2000 F (1093 C) .... 28% chrome iron, Inconel
  Over 2000 F (1093 C) .... ceramic (primary), silicon carbide (secondary)

Food:
  Baking ovens .... wrought iron
  Charretort, sugar........................... wrought iron
  Vegetables and fruit .... Type 304 stainless steel

Chemical:
  Acetic Acid
    10 to 50%, 70 F .... Type 304 stainless steel
    50%, 212 F .... Type 316 stainless steel
    99%, 70 to 212 F .... Type 430 stainless steel
  Alcohol, ethyl, methyl
    70 to 212 F .... Type 304 stainless steel
  Ammonia
    All concentration, 70 F .... Type 304 stainless steel
  Ammonium chloride
    All concentration, 212 F .... Type 316 stainless steel
  Ammonium Nitrate
    All concentration, 70 to 212 F .... Type 304 stainless steel
  Ammonium Sulphate
    10% to saturated, 212 F .... Type 316 stainless steel
  Barium Chloride
    All concentration, 70 F .... Monel[b]
  Barium Hydroxide
    All concentration, 70 F .... carbon steel
  Barium sulphite .... Nichrome[c]
  Brines .... Monel
  Bromine .... tantalum
  Butadiene .... Type 304 stainless steel
  Butane .... Type 304 stainless steel
  Butylacetate .... Monel
  Butyl Alcohol .... copper
  Calcium Chlorate
    Dilute, 70 to 150 F .... Type 304 stainless steel
  Calcium Hydroxide
    10 to 20%, 212 F .... Type 304 stainless steel
    50%, 212 F .... Type 316 stainless steel
  Carbolic Acid
    All, 212 F .... Type 316 stainless steel
  Carbon Dioxide
    wet or dry................................ 2017-T4 aluminum, Monel
  Chlorine Gas
    Dry, 70 F .... Type 316 stainless steel
    Moist, 20 to 212 F .... Hastelloy C[d]
  Chromic Acid
    10 to 50%, 212 F .... Type 316 stainless steel
  Citric Acid
    15%, 70 F .... Type 304 stainless steel
    15%, 212 F .... Type 316 stainless steel
    Concentrated, 212 F .... Type 316 stainless steel

[b] Trademark–Driver-Harris Co.
[c] Trademark–Union Carbide Corp.

[d] Trademark–International Nickle Co.

Application	Protection Tube Material
Copper Nitrate	Type 304 stainless steel
Copper Sulphate	Type 304 stainless steel
Cresols	Type 304 stainless steel
Cyanogen gas	Type 304 stainless steel
Dow therm[e]	carbon steel
Ether..	Type 304 stainless steel
Ethyl acetate	Monel
Ethyl chloride	
70 F	Type 304 stainless steel
Ethyl Sulphate	
70 F	Monel
Ferric chloride	
5%, 70 F to boiling	tantalum
Ferric Sulphate	
5%, 70 F	Type 304 stainless steel
Ferrous Sulphate	
Dilute, 70 F	Type 304 stainless steel
Formaldehyde	Type 304 stainless steel
Formic Acid	
5%, 70 to 150 F	Type 316 stainless steel
Freon	Monel
Gallic Acid	
5%, 70 to 150 F	Monel
Gasoline	
70 F	Type 304 stainless steel
Glucose	
70 F	Type 304 stainless steel
Glycerine	
70 F	Type 304 stainless steel
Glycerol	Type 304 stainless steel
Hydrobromic Acid	
98%, 212 F	Hastelloy B
Hydrochloric Acid	
1%, 5%; 70 F	Hastelloy C
1%, 5%; 212 F.............................	Hastelloy B
25%, 70 to 212 F	Hastelloy B
Hydrofluoric Acid	Hastelloy C
Hydrogen Peroxide	
70 to 212 F	Type 316 stainless steel
Hydrogen Sulphide	
Wet and dry................................	Type 316 stainless steel
Iodine	
70 F	tantalum
Lactic Acid	
5%, 70 F	Type 304 stainless steel
5%, 150 F.................................	Type 316 stainless steel
10%, 212 F	tantalum
Magnesium Chloride	
5%, 70 F	Monel
5%, 212 F.................................	nickel
Magnesium Sulphate	
Hot and cold	Monel
Muriatic Acid	
70 F	tantalum
Naphtha	
70 F	Type 304 stainless steel

[e] Trademark–Dow Chemical Corp.

Application	Protection Tube Material
Natural Gas	
70 F	Type 304 stainless steel
Nickel Chloride	
70 F	Type 304 stainless steel
Nickel Sulphate	
Hot and cold	Type 304 stainless steel
Nitric Acid	
5%, 70 F	Type 304 stainless steel
20%, 70 F	Type 304 stainless steel
50%, 70 F	Type 304 stainless steel
50%, 212 F	Type 304 stainless steel
65%, 212 F	Type 316 stainless steel
Concentrated, 70 F	Type 304 stainless steel
Concentrated, 212 F	Tantalum
Nitrobenzene	
70 F	Type 304 stainless steel
Oleic Acid	
70 F	Type 316 stainless steel
Oleum	
70 F	Type 316 stainless steel
Oxalic acid	
5%, hot and cold	Type 304 stainless steel
10%, 212 F	Monel
Oxygen	
70 F	steel
Liquid	stainless steel
Elevated temperatures	stainless steel
Palmitic acid	Type 316 stainless steel
Pentane..	Type 304 stainless steel
Phenol	Type 304 stainless steel
Phosphoric acid	
1%, 5%; 70 F	Type 304 stainless steel
10%, 70 F	Type 316 stainless steel
10%, 212 F	Hastelloy C
30%, 70 F, 212 F	Hastelloy B
85%, 70 F, 212 F	Hastelloy B
Picric Acid	
70 F	Type 304 stainless steel
Potassium bromide	
70 F	Type 316 stainless steel
Potassium carbonate	
1%, 70 F	Type 304 stainless steel
Potassium chlorate	
70 F	Type 304 stainless steel
Potassium hydroxide	
5%, 70 F	Type 304 stainless steel
25%, 212 F	Type 304 stainless steel
50%, 212 F	Type 316 stainless steel
Potassium nitrate	
5%, 70 F	Type 304 stainless steel
5%, 212 F	Type 304 stainless steel
Potassium permanganate	
5%, 70 F	Type 304 stainless steel
Potassium sulphate	
5%, 70 F	Type 304 stainless steel
Potassium sulphide	
70 F	Type 304 stainless steel

Application	Protection Tube Material
Propane	Type 304 stainless steel
Pyrogallic acid	Type 304 stainless steel
Quinine bisulphate	
Dry	Type 316 stainless steel
Quinine sulphate	
Dry	Type 304 stainless steel
Sea water	Monel
Salicylic acid	nickel
Sodium bicarbonate	
All concentration, 70 F	Type 304 stainless steel
5%, 150 F	Type 304 stainless steel
Sodium carbonate	
5%, 70 to 150 F	Type 304 stainless steel
Sodium chloride	
5%, 70 to 150 F	Type 316 stainless steel
Saturated, 70 to 212 F	Type 316 stainless steel
Sodium fluoride	
5%, 70 F	Monel
Sodium hydroxide	Type 304 stainless steel
Sodium hypochlorite	
5% still	Type 316 stainless steel
Sodium nitrate	
Fused	Type 316 stainless steel
Sodium peroxide	Type 304 stainless steel
Sodium sulphate	
70 F	Type 304 stainless steel
Sodium sulphide	
70 F	Type 316 stainless steel
Sodium sulphite	
150 F	Type 304 stainless steel
Sulphur dioxide	
Moist gas, 70 F	Type 316 stainless steel
Gas, 575 F	Type 304 stainless steel
Sulphur	
Dry-molten	Type 304 stainless steel
Wet	Type 316 stainless steel
Sulphuric acid	
5%, 70 to 212 F	Hastelloy B
10%, 70 to 212 F	Hastelloy B
50%, 70 to 212 F	Hastelloy B
90%, 70 F	Hastelloy B
90%, 212 F	Hastelloy D
Tannic acid	
70 F	Type 304 stainless steel
Tartaric acid	
70 F	Type 304 stainless steel
150 F	Type 316 stainless steel
Toluene	2017-T4 aluminum
Turpentine	Type 304 stainless steel
Whiskey and wine	Type 304 stainless steel
Xylene	copper
Zinc chloride	Monel
Zinc sulphate	
5%, 70 F	Type 304 stainless steel
Saturated, 70 F	Type 304 stainless steel
25%, 212 F	Type 304 stainless steel

4.8 Bibliography

White, F. J., "Accuracy of Thermocouples in Radiant Heat Testing," *Experimental Mechanics,* Vol. 2, July 1962, p. 204.
Baker, H. D., Ryder, E. A., and Baker, N. H., *Temperature Measurement in Engineering,* Vol. 1, Wiley, New York, 1953.

5. SHEATHED, CERAMIC-INSULATED THERMOCOUPLES

5.1 General Considerations

When data were being gathered in 1961 for National Bureau of Standards Monograph 40 [*3*] on Thermocouple Materials, a section was included to cover ceramic packed thermocouple stock "because of the wide use and increasing popularity." New uses continue to be found for this unique hetrogeneous materials combination. Compacted ceramic insulated thermocouple material consists of three parts as shown in Fig. 21.

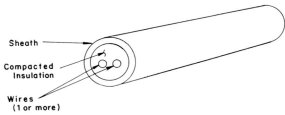

Sheath

Compacted
Insulation

Wires
(1 or more)

FIG. 21—*Compacted ceramic insulated thermocouple showing its three parts.*

The advantages of this configuration are:

1. It isolates the thermocouple wires from environments that may cause rapid deterioration.

2. It reduces long term calibration drift.

3. It lessens temperature versus wire size problems.

4. It provides an excellent high temperature insulation for thermocouple wires.

5. The sheath can be made of a metal compatible with the process in which it is being used and bears the brunt of the environmental effects.

6. It is easy to use:

   a. Forms easily and retains the bent configuration.

   b. The sheath can be welded without loss of insulation.

   c. Available in a wide variety of sizes and materials.

   d. Readily fabricated into finished thermocouple assemblies with minimum of technique and equipment.

   e. Useful at high pressures or high temperatures or both.

7. It is inexpensive as a finished thermocouple.

5.2 Construction

All compacted types of thermocouples are made by similar processes: they begin with matched thermocouple wires surrounded by noncompacted ceramic insulating material held within a metal tube. By swaging, roll-forming, or other mechanical reduction processes the tube is reduced in diameter, the insulation is compacted around the wires, and the assembly is elongated.

Several options are available to the designer of the sheathed material depending upon the material combinations selected for the temperature measurement application.

Briefly, these are the factors considered in making a design, based on a finished product specification.

1. Type of wire material.
2. Number of wires and their diameter.
3. Type of sheath material.
4. Type of insulation.
5. Length, outside diameter, and wall thickness.

A ductile sheath and refractory (brittle) wire combination requires a design wherein the starting tubing diameter is only slightly larger than the finished size and only large enough on the inside diameter to accommodate a crushable preformed ceramic insulation with the wire(s) strung through the insulator hole(s). This combination then would be reduced to the final diameter by one of the compaction methods usually in a single reduction pass. The design is such that the wire is neither elongated nor reduced in diameter because of its brittle nature at room temperature.

A brittle sheath/brittle wire combination does not lend itself to a compacted insulation design and, therefore, is assembled as a tube-insulator-wire combination without the subsequent sheath reduction.

Ductile wire/ductile sheath combinations cover the widest range of commonly used materials and offer the designer the widest choice of design approaches.

He may elect to make his design using the approach outlined for the ductile sheath/brittle wire combination. He also may choose to use powder loading of the insulating material using a centralizing device and vibration as a variation on the method for making the sheathed wire assembly. Isostatic pressing of the combination is another approach occasionally used; however, by far, the most common technique is to use crushable preformed insulators.

The multiple pass approach as generally described by Siede and Edison [4] lends itself to initial starting material dimensions appreciably larger than the finished diameter of the compacted thermocouple. Since there is a constant volume relationship between the initial compacted tube diameter (D_0) and length (L_0) and the final tube diameter (D_F) and length (L_F), the final lengths achieved are

$$L_F = L_0 \left(\frac{D_0}{D_F}\right)^2$$

and are restricted only by the capabilities of the reduction and annealing facilities available and by the ceramic and metallurgical capabilities of the designer, the design team, or the material suppliers.

The reduction processes, as investigated by Rautio [5], indicate that the sheath diameter, wall, insulation spacing, and wire diameter are all reduced proportionately once the compaction point of the insulation is reached.

For design insight into the reduction effects on the tube diameter and wall thickness, Mohrnheim's [6] paper and its associated references are recommended. LaVan [31] compares the various reduction methods in his paper.

Nominal physical dimensions of sheathed ceramic insulated thermocouples are shown in Fig. 22 [7]. The ratios of sheath outside diameter to wire size and to sheath wall thickness offer a balance between maximum wall thickness (for protection of the sheath compacted insulant) and suitable insulation spacing for effective insulation resistance at elevated temperature.

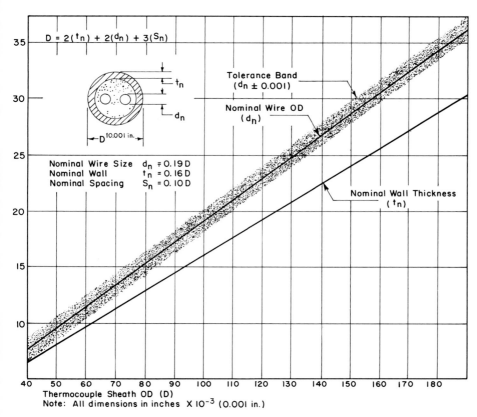

FIG. 22—*Graph showing idealized relationship between thermocouple sheath outside diameter and nominal internal dimensions for reactor grade thermocouples.*

A decrease in the diameter of the thermoelements could be advantageous in regard to insulation resistance, but it is detrimental structurally according to the work done by Ihnat [8]. Hansen [9] reported when wire diameters were one half of the nominal sizes (Fig. 22) multiple reductions of the sheathed assembly were restricted because of wire fracture.

Ihnat reported on the factors affecting the choice of insulation materials on the basis of resistivity, mechanical strength, resistance to thermal shock, vibration stress, thermal expansion, chemical compatibility, and cost.

Other factors such as relative hardness, compacted density, thermal conductivity, and neutron radiation effects have been reported by Zysk [10].

It is estimated that 90 percent of all sheathed thermocouples produced to date have used magnesium oxide (MgO) as the insulation material. Magnesium oxide is popular as a thermocouple insulator because of its overall compatibility with standard thermoelements and sheathing materials, its relative low cost, and availability.

Aluminum, beryllium, and thorium oxide insulations are also available from suppliers for use with certain wire and sheath combinations. The latter two materials are combined usually with refractory sheath and thermoelements.

Because many applications of ceramic insulated thermocouples are at temperatures above 400 F, much attention has been given to cleanliness and chemical and metallurgical purity of the components. Bliss [11] points up many of the fabrication and application difficulties that can be experienced if extreme care is not exercised by the supplier. A warning heeded by the sailors and frontiersmen of early America can be applied today to the sheathed thermocouple manufacturer: Keep your powder clean and dry!

5.3 Insulation

For most practical purposes the sheathed thermocouple material should have a minimum insulation resistance of 100 megohms at 500 V dc for diameters larger than $\frac{1}{16}$ in. outside diameter. This readily is obtained by dry, uncontaminated compacted ceramic. The addition of moisture by hygroscopic action is moved by capillary attraction through exposed ends; the unwitting capture of oil, humid air, perspiration, oil vapors, or lint will reduce the insulation resistance. Also the insulation resistance of all ceramics, compacted and uncompacted, reduces with an increase in temperature.

For insulation resistance greater than 1000 megohms, special techniques may be required to obtain and maintain these values. It is entirely feasible that 1000 megohms stock laying exposed to 70 F air and a relative humidity above 50 percent will experience a degradation of insulation resistance to less than 0.1 megohm in 15 min. Higher humidity will cause a more rapid degradation. Therefore, the following precautions should be exercised when handling compacted ceramic insulated thermocouples:

1. Do not leave an end exposed for periods longer than 2 or 3 min. Immediately seal ends.

2. Expose ends only in a region of low relative humidity.

3. Store assembly in a warm (above 100 F) and dry (relative humidity less than 25 percent) area.

Some of the characteristics of the more common compacted ceramics are shown in Tables 24 and 25.

TABLE 24—*Characteristics of insulating materials used in ceramic-packed thermocouple stock.*

Insulator	Minimum Purity, %	Melting Point, deg F	Approximate Usable Temperature, deg F
Magnesia (MgO)	99.4	5 050	3 000
Alumina (Al₂O₃)	99.5	3 650	2 800
Zirconia (ZrO₂)	99.4	4 500	1 200
Beryllia (BeO)	99.8	4 550	4 200
Thoria (ThO₂)	99.5	5 950	5 000

TABLE 25—*Thermal expansion coefficient of refractory insulating materials and three common metals.*

Material	Average Coefficient of Expansion $\times 10^{-6}$ (25 to 700 C), deg C
Copper	16.5
Stainless steels	13.9 to 16.4
Aluminum	9.6
Magnesium oxide	12.9
Beryllium oxide	8.1
Aluminum oxide	7.1 to 8.0
Zirconium oxide	4.2 to 5.2

5.4 Wire

The wire is the primary functioning part of the assembly. It is the exposure of the junction of the dissimilar materials to heat that generates a commensurate voltage. From established tables (see Section 10) the voltages can be converted into temperature. Deviation from these tables, as given in Section 10, are equally applicable to sheathed thermocouples.

Since the wires are contained in a protective sheath and firmly held, small diameter wires can be exposed to high temperatures for long periods of time without serious deterioration.

5.5 Sheath

The sheath material performs several functions in the ceramic-insulated thermocouple system:

TABLE 26—Sheath materials of ceramic-packed thermocouple stock and some of their properties [3].

Material	Melting Point, deg F	Recommended Maximum in Air, deg F	Recommended Operating Atmosphere	Recommended Continuous Maximum Temperature, deg F	Tensile Strength,[c] psi At 200 F	Tensile Strength,[c] psi At 1600 F
Stainless steel:						
304	2 560	1 920	ORNV	1 650	68 000	
309			ORNV	2 000		
310	2 560	2 000	ORNV	2 100	87 000	23 000
316	2 500	1 650	ORNV	1 700	75 000	23 000
321	2 550	1 650	ORNV	1 600	70 000	17 000
347	2 600	1 680	ORNV	1 600	75 000	
430	2 700	1 550	ORNV	1 200		
446	2 700	2 000	ORNV	2 000		
Inconel	2 550	2 000	ONV[b]	2 100	93 000	5 000
Inconel X	2 620	1 500	ONV[b]	2 200	150 000	11 000
Incoloy	2 500	1 640			77 000	3 000
Hastelloy X	2 350	2 300			106 000	7 000
Hastelloy C	2 310	1 820			136 000	64 000
Haynes 25	2 425	1 820			147 000	13 000
Hastelloy B	2 375	1 400			125 000	51 000
Monel	2 460	1 640	ONV			
Chromel	2 600	2 100			90 000	21 000
Copper	1 980	600	O[a]RNV	600		
Brass	1 850	700				
Aluminum	1 220	800	ORNV	700		
Nichrome	2 550	2 200		2 000		

Alumel	2 550	2 100	ONV		82 000	19 000
Nickel.............	2 647	1 100			
Iron...............	2 798	600			
Zircalloy..........	3 350	1 400			
Platinum...........	3 217	3 000	ON^b	3 000	
Pt-Rh 10%.........	3 362	3 100	ON	3 100	110 000
Columbium	4 474	1 600	VN	3 800	137 000	30 000
Molybdenum	4 730	400	VNR		
Molybdenum disilicized		3 100	ON	3 000	
Molybdenum chromalized		3 100	ON	3 000	
Tantalum	5 425	750	V	5 000	96 000	22 000
Titanium	3 035	600	VN	2 000	

NOTE—Symbols describing atmospheres are: O—Oxidizing; R—Reducing; N—Neutral; V—Vacuum.
^{a} Scales readily in oxidizing atmosphere.
^{b} Very sensitive to sulfur corrosion.
^{c} After exposure to temperature of 100 h except for stainless steels; Haynes 25, W, Mo, Ta, and Cb.

TABLE 27—Recommended sheath diameter for long term service in air [27].

Sheath diameter	0.040	0.062	0.125	0.188	0.250
Nominal wall	0.007	0.010	0.020	0.025	0.032
Chromel-Alumel............	1 400 F	1 600 F	1 600 F	1 600 F	1 800 F
Iron-constantan	1 000 F	1 200 F	1 400 F	1 400 F	1 600 F
Chromel-constantan	1 200 F	1 400 F	1 400 F	1 600 F	1 700 F

1. It holds the oxide in compaction.
2. It shields the oxide and thermoelements from the environment.
3. It furnishes mechanical strength to the assembly.

Since no one sheath material is suitable for all environments a wide variety of different materials are offered. Table 26 shows a group of materials and some of their characteristics that can be used as sheaths.

For maximum life a sheath diameter should be selected that offers the heaviest wall in relationship to the maximum use temperature.

Table 27 illustrates recommended temperatures for various sheath diameters and corresponding wall thicknesses. With the proper selection of sheath material for the environment, the thermocouple can be exposed to temperatures as high as 2000 F for any diameter but at some decrease in life.

5.6 Combinations of Sheath, Insulation, and Wire

Precautions to be observed in selecting combinations are:

1. Be sure the sheath material will survive the environment (see Tables 26 and 27). Do not fail to realize that a thin sheath will deteriorate more rapidly than a thick one.

2. Be sure the assembly of sheath and wire is fully annealed for maximum sheath life and stability of wire calibration.

TABLE 28—*Capability of wire and sheath material* [13].

Wire \ Sheath	304	310	316	321	347	440	Platinum Alloys	Hastelloy X	Copper	Aluminum	Inconel 600	Inconel 702	Tantalum	Columbium Alloys
Chromel-Alumel	1	1	1	1	1	1	1	1	4	4	1	1	3	3
Iron-constantan	1z	1	1	1	1	3	4	4	4	4	3	3	4	4
Copper-constantan	3	3	3	3	3	3	4	4	1	4	3	3	4	4
Chromel-constantan	1	1	1	1	1	1	1	1	4	4	1	1	3	3
Platinum rhodium-platinum	1	1	1	1	1	1	1	1	4	4	1	1	4	4
Tungsten rhenium alloys	3	3	3	3	3	3	3	1	4	4	3	3	3	3
Iridium rhodium alloys	3	3	3	3	3	3	3	1	4	4	3	3	3	3
Copper	3	4	3	3	3	4	4	4	1	2	3	4	2	2
Nickel	1	1	1	1	1	1	1	1	4	4	1	1	3	3
Aluminum	4	4	4	4	4	4	4	4	4	1	4	4	4	4
Nichrome[a]	1	1	1	1	1	1	1	1	4	4	1	1	3	3

NOTE—1. Easy manufacturing and good operational compatibility.
2. Easy manufacturing but poor operational compatibility.
3. Difficult manufacturing but good operational compatibility.
4. Difficult manufacturing and poor operational compatibility.

[a] Trademark—Driver Harris Co.

3. Select a sheath and wire material of similar coefficients of linear expansion. For example: Platinum-rhodium/platinum has about one half the expansion of stainless or Inconel; a grounded hot junction will pull apart unless sufficient expansion loops are allowed in the wire. Insulators often are used for this combination that are hard fired, and either noncompacted or isolated junctions also are used.

4. Select a sheath and wire material of similar annealing characteristics.

5. The combination of Chromel-Alumel, iron-constantan or Chromel-constantan with aluminum or copper sheath are not desirable. Table 28 indicates capability of wire and sheath materials.

6. Table 29 gives dimensions of available ceramic-packed stock.

TABLE 29—*Dimensions and wire sizes of available ceramic-packed material.*

(a)

Sheath Outside Diameter, in.	Outside Diameter Tolerance, ±in.	Nominal Wall Thickness, in,	Approximate Wire, B&S gage	Nominal Production Length, ft.
0.010	0.001	0.0015	48	250
0.020	0.001	0.003	38	250
0.032	0.001	0.005	36	250
0.040	0.001	0.007	34	250
0.062	0.002	0.010	29	150
0.090	0.002	0.014	26	125
0.125	0.002	0.018	24	100
0.188	0.003	0.025	18	60
0.250	0.003	0.032	17	40
0.313	0.003	0.040	16	40
0.375	0.003	0.049	14	30
0.430	0.003	0.065	13	30
0.500	0.003	0.065	12	30

(b)

Sheath Diameter, in.	Nominal Conductor Diameters, in.			
	1-Wire	2-Wire	3-Wire	4-Wire
0.313	0.064	0.051	0.040	0.040
0.250	0.051	0.040	0.032	0.032
0.188	0.040	0.032	0.022	0.022
0.125	0.032	0.022	0.011	0.011
0.062	0.022	0.011	0.006	0.006
0.040	0.011	0.006	...	...
0.025	0.006	0.004	...	...

5.7 Limitations of the Basic Material

1. The sheath can be bent around a mandrel twice the sheath diameter without damage.

2. The life of material having 0.032 in. diameter or less is limited due to grain growth in the sheath wall.

3. Four wires in 0.062 in. sheath diameter and smaller are not practical to handle in the field.

4. Two wires in 0.032 in. sheath diameter and smaller are difficult to handle in the field but are used in laboratory environments.

5. All crushed ceramics are like sponges and tend to capture moisture at the exposed surfaces. The tendency is reduced greatly in air at temperatures of 120 F and greater. If moisture does get into the assembly (as indicated by a reduction of insulation resistance), it can be substantially removed from the ends of the sheath by the following process:

A. Select a point about 6 in. from the exposed end and apply heat until the sheath is dull red.

B. Gradually work this slightly incandescent zone towards the exposed ends.

C. When end has cooled below 250 F, apply a sealant, that is,

(a) Dow Corning DC 803,

(b) General Electric Glyptal,

(c) wax, or

(d) proprietory sealants.

6. Compacted ceramic material usually is coiled for shipping as shown in Table 30.

TABLE 30—*Compacted ceramic material coiled for shipping.*

Diameter, in.	0.025	0.040	0.062	0.125	0.188	0.250	0.313	0.375
Size of coil, in.	18	18	24	30	48	48	60	60
Weight lb/100 ft.	0.2	0.5	0.8	3.0	6.0	11.0	16.0	24.0

7. Stock material and completed thermocouples usually are supplied to the end user in the fully annealed state with proper metallurgical grain size and no surface corrosion.

8. Therefore, the following precautions should be exercised when specifying or handling compacted ceramic insulated thermocouples:

A. Ensure that the sheath material is annealed fully.

B. If the material is coiled to a diameter which causes a permanent set, subsequent annealing may be required.

C. Avoid repeated bending at same location as this work hardens the sheath and may change the thermocouple calibration.

D. Do not expose the sheath to deleterious temperatures or atmospheres (see Tables 26 and 27).

5.8 Testing

Many tests have been devised to evaluate sheathed ceramic insulated thermocouples. These tests cover the physical and metallurgical properties of the sheath and thermoelements, the isothermal electrical properties of the insulation, and the thermoelectric properties of the thermoelements.

ASTM has issued a specification E 235-67 [14] on sheathed thermocouples for nuclear and other high reliability applications that covers Type K thermoelements in various sheath materials with options available for other sheath materials; alumina or magnesia insulations and alternative thermoelements. In Aerospace Information Report 46 [15] the Society of Automotive Engineers has covered the preparation and use of Type K thermoelements for turbojet engines, while the military has issued Mil T 22300 [16] (Ships) and Mil T 23234 [17] (Ships) for nuclear application of the sheathed thermocouples. In addition many government and commercial laboratories, thermocouple fabricators, and thermocouple users have issued specifications covering specific tests and test procedures for evaluating sheathed thermocouple materials.

Scadron [18] outlined some of the tests and tolerances that are applied for the verification of physical, electrical, and thermoelectric integrity of the sheathed thermocouple. Most manufacturer's catalogs cover some aspects of the tests, guarantees, and procedures used to verify product integrity.

Fenton et al [33] were concerned with inhomogeneity in bare and sheathed thermoelements and developed procedures and apparatus for testing for drift and inhomogeneity.

National Bureau of Standards Circular 590 [19] covers thermocouple testing. While this work was devised for the bare thermoelements it is applicable to the metal sheathed, ceramic insulated stock for evaluation of the electrical characteristics.

Sections 3 and 10 cover thermoelectric characteristics as well as recommendations for upper temperature limits and limits or error for most thermoelement materials. A notable exception for sheathed thermocouples is made in Table 2.

Increased life and accuracy was achieved using sheathed thermocouples by McElroy and Potts [20] in their study for the Oak Ridge National Laboratories of drift characteristics of Type K elements. A report from Stroud [21] indicates that a sealed sheathed thermocouple with $\frac{1}{8}$-in.-diameter Inconel sheath, MgO insulation, and Type J thermoelements was exposed for more than 3500 h at 2000 F (1093 C) in air without falling outside the acceptable calibration error band. This is an example of one of the benefits of the sheathed construction. Zysk [10] points up that small dimensions, flexibility, and complete resistance to thermal shock are other advantages of this material.

TABLE 31 — *Various characteristics tests, and the source of testing procedure applicable to sheath ceramic-insulated thermocouples.*

Characteristics	Tests	Procedure	Comments
	PHYSICAL AND METALLURGICAL		
1. OD of sheath	ring gage, micrometer, or met mounts for high precision	good mechanical practice or transverse section per ASTM Method E 3-62	tolerances per Table 29 sample preparation: requires care to prevent smear
2. Roundness of sheath	rotate for total indicated reading	Vee block and dial indicator	important for tube fitting use
3. Sheath surface finish	profilometer or roughness gage	ANSI Standard B46.1	also check clean finish for welding and brazing
4. Sheath wall thickness.........	thread gage micrometer or metallurgical mount	transverse section per ASTM Method E 3-62	concerns machining, bending, welding, life: can be checked only at ends
5. Insulation spacing	comparator or metallurgical mount	transverse section per ASTM Method E 3-62	see electrical tests
6. Insulation compactness	(A) tap test, (B) helium pressure, (C) dye absorption, (D) compaction density	(A) supplier catalog, (B) MIL-T-22300, (C) and (D) ASTM Method D 2771-69	relative values must be related to application
7. Wire diameter.........	micrometer or metallurgical mount	visual or transverse section per ASTM Method E 3-62	±1 wire gage size (Table 29)
8. Wire surface and roundness	micrometer and visual	3-point diameter check	10% out of round max showing slight embedment of insulating material into wire
9. Ductility and formability	bend and visual or metallurgical mount	ASTM E 235-67	bent 90 deg only for ease of making metallurgical mount

10. Sheath integrity	(A) water immersion (B) hydrostatic (C) helium leak, (D) dye penetrant	(A) water soak with cross protected[18], (B) ASTM Method E 165, (C) ASTM Method E 235-67, (D) ASTM Method E 235-67	(A) may use hydrostatic pressure or steam for some applications: seal weld to protect insulation
11. Junction integrity	radiographs	ASTM Method E 235-67	application for junction and 4 in. of sheath
12. Material compatability	thermal cycling	ASTM Method E 235-67	thermal expansion for grounded junction integrity check
13. Metallurgical integrity	metallurgical mount	ASTM Method E 235-67	checks grain size
14. Response time	thermal response test	gas: SAE 46, liquid [24]	assesses thermal mass and welding technique
15. Uniformity of properties	tests 1, 2, 3, 10, 11, and 13 full length	as noted	use sampling procedure for large runs
16. Stability under service conditions	application experience or life tests	as required	check with other users in same field

CONSTANT TEMPERATURE ELECTRICAL

17. Insulation resistance	megohm meter between wires and sheath and wires	ASTM Method E 235-67	50 V for OD less than 0.061, 500 V for greater than 0.062
18. Electrical resistance (loop)	Wheatstone bridge	see section 3	tolerances ±1 gage size based on wire type, length and dia: also checks continuity
19. High temperature insulation resistance	megohm meter or Wheatstone bridge	immerse full length except for end in temperature chamber: test with ungrounded junction or before junction is made	relative to operational requirement, effective means of testing insulation quality at expected operating temperature of thermocouple

TABLE 31—*Various characteristics tests, and the source of testing procedure applicable to sheath ceramic-insulated thermocouples (continued).*

Characteristics	Tests	Procedure	Comments
		Constant Temperature Electrical (continued)	
20. Dielectric strength	high potential generator	ac voltage (current limited) applied at room temperature for 1 min 80 V per mil of insulation [32]	seldom used check for contamination and voids in insulation: voltage usually 500 V for 062 OD and larger
		Isothermal Electrical	
21. Capacitance	capacitance bridge		test is usually not applied to thermocouples
22. Uniformity along length	Test 17, 18, 19, 20, 21 for full length	check samples or full lengths coiled if necessary	all tests are seldom necessary for single application
23. Stability under service conditions	application experience or life tests	as required	check for experience by other users
		Thermoelectric Properties	
24. Emf versus temperature	potentiometer and heat or cold source	Section 6	test at one or more points up to maximum use temperature
25. Homogeneity	heat source and galvanometer	NBC Circular 590 and Ref 33	max of ± 100 μV for J, K, T; ±25 μV for B, R, S thermoelements
26. Uniformity full length	Test 24 and 25 for full length	apply sampling procedure for large quantities	random sampling is most economical approach
27. Stability under service conditions	experience or life tests	actual tests	no substitute for actual

Table 31 shows various characteristics, tests, and the source of testing procedure which are applicable to sheathed ceramic-insulated thermocouples.

5.9 Measuring Junction

Measuring or "Hot" Junction

Numerous variations in measuring junction construction are possible with this type of material. The application dictates the most desirable method.

1. *Exposed or Bare Wire Junction*—In this type of a junction the sheath and insulating material are removed to expose the thermocouple wires. These wires are joined to form a measuring junction. The junction may be of the twist-and-weld or butt-weld type.

(*a*) Fast response (see Table 32).
(*b*) Exposed magnesia will pick up moisture.
(*c*) Not pressure tight.
(*d*) Wires subject to mechanical damage.
(*e*) Wires subject to environment and usually will have a very short life.
(*f*) Useful life shortened as a result of rapid calibration drift.

FIG. 23—*Exposed or bare wire junction.*

2. *Grounded Junction*—A closure is made by welding in an inert atmosphere so that the two thermocouple wires become an integral part of the sheath weld closure. Thus, the wires are "grounded" to the sheath.

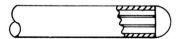

(*a*) Slower response than exposed wire (see Table 32).
(*b*) Pressure tight to above 100,000 psi.
(*c*) Wires protected from mechanical damage.
(*d*) Wires not exposed to environment and will have a longer life.
(*e*) Coefficient of expansion wire must be close to that of sheath to prevent pulling apart of hot junction. There are proprietary type of hot junctions that permit the use of materials with differing coefficients of expansion.

FIG. 24—*Grounded junction.*

3. *Ungrounded or Isolated Junction*—This type is similar to the grounded junction except that the thermocouple wires are first made into a junction which is then insulated from the sheath and the sheath closure. The closure is formed by welding without incorporating the thermocouple wires. Thus, the thermocouple is "ungrounded" in relation to the sheath material.

(*a*) Slower response than grounded hot junction (see Table 32).
(*b*) Pressure tight to above 100,000 psi.
(*c*) Wires protected from mechanical damage.
(*d*) Wires not exposed to environment and will have a longer life.
(*e*) Differential expansion between wires and sheath is minimized.
(*f*) Most expensive.

FIG. 25—*Ungrounded or isolated junction.*

4. *Reduced Diameter Junction—*

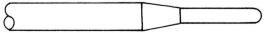

(*a*) May be either grounded or insulated (2 and 3 mentioned previously).
(*b*) Applied where fast response is required at junction and heavier sheath or wires are desired for strength, life, or lower circuit resistance over the balance of the unit.

FIG. 26—*Reduced diameter junction.*

TABLE 32—*Time constant characteristics, time in seconds* [27].

	Sheath Dimension, in.					
	0.040	0.062	0.125	0.188	0.250	0.313
Exposed Junction	0.005	0.02	0.03	0.07	0.1	0.15
Water	0.005	0.02	0.03	0.07	0.1	0.15
Gas: 50 lb/sft²	0.03	0.15	0.3	0.6	0.8	1.6
Grounded Junction: water	0.1	0.2	0.7	1.1	2.0	2.5
Gas: 50 lb/sft²	0.8	1.6	4.5	8.3	12.5	17.5
Ungrounded Junction: water	0.3	0.5	1.3	2.2	4.5	7.2
Gas: 50 lb/sft²	1.6	3.2	9.0	24.0	38.5	48.0

5.10 Terminations

There are numerous ways in which thermocouples of this nature can be treated at the "cold" end. The most common treatments are as follows:

(*a*) *Square Cut and Sealed*—Wire is procured in this manner when the user intends to fabricate his own cold end assembly. The seal inhibits moisture absorption.

(*b*) *Wires, Exposed and Sealed*—Here, the sheath and the insulation is removed leaving the bare thermocouple wires exposed for a specified length. The insulating material is then sealed to inhibit moisture absorption.

(*c*) *Transition Fitting*—The terminal end is provided with a fitting wherein the thermocouple wires are joined to more suitable wires. In this fitting the necessary sealant for the mineral insulant also is provided.

(*d*) *Terminals or Connectors*—Various types of fittings are available to facilitate external electrical connections. These include screw terminal heads, open or enclosed, and plug or jack connections.

Termination consists of attaching flexible connecting wires (see Figs. 27 and 28).

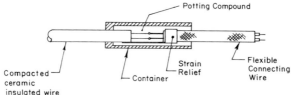

1. Useful where permanent connections are desired.
2. Least expensive.

FIG. 27—*Termination consisting of attaching flexible connecting wires.*

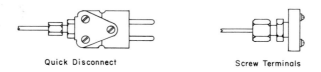

Quick Disconnect   Screw Terminals

FIG. 28—*Connectors consisting of a proprietory quick disconnect at the end of the compacted ceramic insulated cable.*

5.11 Installation of the Finished Thermocouple

Many types of installation are possible with sheathed thermocouples. Typical installations are shown in Figs. 29, 30, and 31.

Other special flanges and adaptors are available from many thermocouple manufacturers.

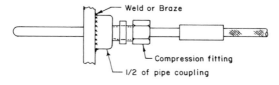

FIG. 29—*Fittings to adapt into process line. Up to 5000 psi.*

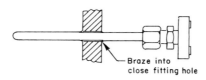

FIG. 30—*Braze for higher pressure operation. Up to 100,000 psi.*

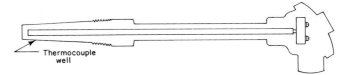

FIG. 31—*Element in thermocouple wells.*

5.12 Sheathed Thermocouple Applications

Application information for sheathed thermocouples has been well covered in the literature. Many suggested applications are made in the various suppliers catalogs, and Sannes article [*23*] on the application of the smaller sheath diameters is useful. Section 9.2 of this manual is an excellent reference on surface temperature measurement. A boiler tube application of sheathed thermocouples is discussed in Ref *25*. Gas stream performance is well documented [*26–29*] because of the applications to jet engines and rockets.

Nuclear reactor applications have used sheathed thermocouples extensively for monitoring and control. Johannessen discussed the reliability assurance [*30*] of such applications, and other aspects also are covered in the same reference.

In the event of failure of sheathed thermocouples, Table 33 [*22*] has been included to aid in analyzing the cause.

TABLE 33—*Failure mode and effects analysis.*

Component	Mode of Failure	Possible Causes of Failure
Sheath	longitudinal splits	excessive cold work, improper heat treat, improper drawing speed, or insufficient wall thickness
	rupture at high temperature	excessive vapor pressure due to presence of helium after leak check, moisture in insulation, or insufficient wall thickness
	galling, inclusions and pits	incomplete lubrication, improper reduction method, improper die configuration, contaminated lubrication
	discoloration	improper heat treat atmosphere or temperature, improper cleaning of sheath prior to heat treatment
	brittle material, carbide precipitates, large grain size	improper heat treat time or temperature or quenching or all three
Insulation	low insulation resistance	moisture, contamination or excessive migration of conductors
	corona, arcing or breakdown at dielectric potential	voids, moisture, contamination, conductor decentralization, excessive dielectric potential, loose pack
	loose pack	inadequate sheath reduction, poor initial design, low tensile sheath material, or insufficient sheath wall thickness

TABLE 33—*Failure mode and effects analysis (continued)*.

Component	Mode of Failure	Possible Causes of Failure
	discoloration	inherent contamination, contamination due to reaction with sheath or conductors or both, improper or unclean manufacturing facilities
Conductors	open circuit	excessive elongation, improper initial design, improper heat treatment: defective starting wires
	short circuit (conductor to conductor or conductor to sheath)	loose pack, voids, bend radius too small, improper assembly, conductor decentralization and contaminated insulation
	embrittlement	excessive cold working, improper heat treatment, improper initial condition of conductor
	open circuit-high temperature	nonuniform reduction of area, temperature above wire melting point: defective initial wires (microcracks)
	out of calibration	improper heat treatment, initial conductor out of calibration, cold work effects not removed, nonhomogeneous section
	poor conductor finish	improper heat treatment on initial conductor, excessive insulation grain size, insulation crushability, insulation grain configuration and hardness, excessive swaging or rolling
	temperature/emf variation with length	nonhomogeneous conductor, nonuniform heat treat or unrelieved cold working
	oblated conductors	worn swaging or rolling dies, or improper die setup
	microscopic conductor fractures	temperature above melting point, excessive cold work and improper heat treat, defective initial wires

Thermocouple Failure Mode	Possible Causes of Failure
Open circuit at room temperature ...	brittle wires, excessive elongation in manufacture, improper heat treat or excessive cold work: initially defective wires
Open circuit at high temperature or during cycling	brittle wires, nonuniform elongation causing necking of conductors, improper heat treat, excessive temperature for conductors, differential thermal expansion between conductors and sheath: defective initial wires
Short circuit................................	loose insulation, voids, insulation contamination, conductors twisted, conductor decentralization, moisture
Low insulation resistance	insulation contamination, moisture absorption, improper sealing at ends
High potential breakdown...............	contamination of insulation, conductor decentralization, loose insulation, insufficient sheath wall thickness, voids, excessive potential

TABLE 33—*Failure mode and effects analysis (continued).*

Thermocouple Failure Mode	Possible Causes of Failure
Sheath fracture during forming.........	bend radius too small, brittle sheath material, insufficient sheath wall thickness, inadequate or improper heat treat
Sheath fracture under vibration	support points too far apart, insufficient sheath wall thickness, brittle sheath due to inadequate or improper heat treat, excessive cold work
Sheath burn-through	temperature reaction with atmosphere lowering effective melting point, insufficient sheath wall thickness, improper sheath material for application
Loose insulation material	improper design and insufficient sheath wall thickness

5.13 References

[3] Caldwell, F. R., "Thermocouple Materials," NBS Monograph 40, National Bureau of Standards, 1 March 1962.

[4] Siede, O. A. and Edison, L. R., "Mechanics of Drawing and Swaging Ceramic Filled Tubes," paper presented at ASTM 69th Annual Meeting, Atlantic City, N.J., 30 June 1966.

[5] Rautio, W. S., "Metal-Sheathed Ceramic-Insulated Thermocouples-Swage or Draw?" *Instruments and Control Systems,* May 1962.

[6] Mohrnheim, F., "Definitions and Mathematical Relations of Tube Drawing Processes," *Wire,* Dec. 1959.

[7] MacKenzie, D. J., "Graphs Showing Idealized Relationship between Thermocouple Sheath O. D. and Internal Dimensions for Reactor Grade Thermocouples," American Standard, Inc., Internal Report, March 1963.

[8] Ihnat, M. E., "A Jet Engine Thermocouple System for Measuring Temperatures up to 2300 F," WADC Technical Report 57-744, Jan. 1959.

[9] Hansen, P. F., Internal Report Aero Research Instrument Department, American Standard, Inc., Project 91042, March 1968.

[10] Zysk, E. D., Technical Bulletin, Engelhard Industries, Vol. V, No. 3, Dec. 1964.

[11] Bliss, P., "Post Fabrication Testing of Sheathed Thermocouples," Pratt and Whitney Internal Report.

[12] AEC Report 1067, Atomic Energy Commission, Washington, D.C.

[13] Scadron, M. D., "Compacted Ceramic Insulated Thermocouples," 14th Annual Instrument Maintenance Clinic, ISA Meeting, Houston, Tex., Feb. 1966.

[14] ASTM Method E 235-67, "Thermocouples Sheathed, Type K for Nuclear or for Other High Reliability Applications," *1968 Book of ASTM Standards,* Part 30, p. 731.

[15] Committee AE-2, Society of Automotive Engineers, "The Preparation and Use of Chromel-Alumel Thermocouples for Turbojet Engines," Aerospace Information Report No. 46, 15 March 1956.

[16] Military Specification, MIL-T-22300(Ships) "Thermocouples, Corrosion Resistant Metal Sheathed, for Nuclear Reactors," 11 Jan. 1960

[17] Military Specification, MIL-T-23234(Ships), confidential.

[18] Scadron, L., "Ceramic Insulated Thermocouples," *Instruments and Control Systems,* May 1961.

[19] Roeser W. F. and Lonberg, S. T., "Methods of Testing Thermocouples and Thermocouples Materials," NBS Circular 590, National Bureau of Standards, 6 Feb. 1958.

[20] McElroy, D. L. and Potts, J. F., "Thermocouple Research to 1000 F," Final Report OR-NL2773, Oak Ridge National Laboratories, 30 June 1959.

[21] Stroud, R. C., Leeds & Northrup, Philadelphia, Pa., private communication, June 1960.
[22] MacKenzie, D. J., "Failure Data and Analysis," American Standard, Inc., Internal Report, Jan. 1963.
[23] Sannes, T., "Tiny Thermocouples," *Instrumentation Technology,* March 1967.
[24] Rall D. C. and Hombaker, D. R., "A Rational Approach to the Definition of a Meaningful Response Time for Surface Temperature Transducers," 21st Annual ISA Conference and Exhibit, Oct. 1966.
[25] "Nickel Base Alloy Protection Tubes Aid Monitoring of Boiler Temperatures," *Inco Nickel Topics,* Vol. 18, No. 7, 1965.
[26] Faul, J. C., "Thermocouple Performance in Gas Streams," *Instruments and Control Systems,* Vol. 35, Dec. 1962.
[27] MacKenzie, D. J. and Scadron, M. D., "Selection of Thermocouples for High Gas Temperature Measurement," 524E National Aeronautic Meeting, Society of Automotive Engineers, April 1962.
[28] Moffat, R. J., "Designing Thermocouples for Response Rate," *Transactions,* American Society of Mechanical Engineers, Feb. 1958.
[29] Ihnat, M. E. and Hagel, W. C., "A Thermocouple System for Measuring Turbine Inlet Temperatures," *Transactions,* American Society of Mechanical Engineers, *Journal of Basic Engineering,* March 1960.
[30] Johannessen, H. G., "Reliability Assurance in Sheathed Thermocouples for Nuclear Reactors," High Temperature Thermometry Seminar, TID7586, Oak Ridge Naval Laboratory, Oak Ridge, Aug. 1960.
[31] LaVan, J. F., "ARI Evaluation Report on Ceramic Insulated, Metallic Sheathed Thermocouple Reducing Methods and Associated Equipment," PIR 4.1, Aero Research Instrument Co., Inc., 1961.
[32] Faul, J. C., personal communication.
[33] Fenton, A. W., Dacey, R., and Evans, E. J., "Thermocouples: Instabilities of Seebeck Coefficient," United Kingdom Atomic Energy Authority, TRG Report 1447 (R) 1967.

6. EMF MEASUREMENTS

6.1 General Considerations

The basic principle of thermoelectric thermometry is that a thermocouple develops an emf which is a function of the difference in temperature of its measuring junction and reference junction. If the temperature of the reference junction is known, the temperature of the measuring junction can be determined by measuring the emf generated in the circuit. The use of a thermocouple in temperature measurements, therefore, requires the use of an instrument capable of measuring emf.

There are two types of emf measuring instruments in use in industry-deflection meters (millivoltmeters) and potentiometers. The digital voltmeter, which is now being used extensively, is a type of potentiometer. Because of its limitations, the deflection meter is not used for precise measurements.

The deflection meter consists of a galvanometer with a rigid pointer which moves over a scale graduated in millivolts or degrees. The galvanometer indicates by its deflection the magnitude of the current passing through it, and if the circuit in which it is placed includes a thermocouple, it measures the current I generated by the thermocouple in the circuit. If the circuit has a resistance R and the emf is E, by Ohm's law, $E = RI$. If R is kept constant, I is proportional to E, then the scale can be calibrated in terms of millivolts rather than in milli- or microamperes. This calibration holds as long as R remains constant. Any change in R introduces an error in the indicated value of E. Changes in resistance may result from changes in temperature of the thermocouple or its extension wires or of the copper galvanometer coil, from corrosion of the thermocouple wires, from changes in the depth of immersion of the thermocouple, or from changes in contact resistance at switches or binding posts.

In general, the reference junctions of the thermocouple measuring circuit are located near the meter movement. The effect of temperature changes on these junctions is compensated for by a bimetallic spiral attached to one of the control springs of the pointer. This system maintains accurate meter calibration during changes in ambient temperature.

In spite of its limitations, the deflection meter serves a very useful purpose in a great variety of industrial measurements of temperature where the precision required does not demand the use of the potentiometric method of measurement.

Where precision is required in the measurement of thermal emfs, the potentiometer invariably is used. Because of its reliability and freedom from the uncertainties arising from changes in circuit resistance, meter calibration, etc., as well as its much greater openness of scale, it is used in many cases where the precision needed might not seem to justify the higher cost.

Recording potentiometers are used widely for industrial process temperature measurement and control. When reliability and reproducibility are the prime concern, such instruments are satisfactory. However, because of potential inaccuracies in charts caused by printing limitations and humidity effects and practical limits on chart widths and scale lengths resulting in inadequate readability, they are not used where precision is the criteria.

6.2 Potentiometer Theory

Accurate measurement is usually a matter of comparing an unknown quantity against a known quantity or standard—the more direct the comparison, the better. Accurate weighing, for example, often is accomplished by direct comparison against standard weights using a mechanical balance. If the measured weights are too heavy for direct comparison, lever arms may be used to multiply the forces.

The potentiometer, as the term is used here, serves a similar function in the measurement of voltage, and, in fact, may be called a "voltage balance," the standard voltage being furnished by a standard cell, the "lever" being resistance ratios, and the galvanometer serving as the balance indicator. Since no current is drawn from the standard cell or the measured source at balance, the measurement is independent of external circuit resistance, except to the extent that this affects galvanometer or balancing mechanism sensitivity.

6.2.1 *Potentiometer Circuits*

Figure 32 shows a simple potentiometer circuit which includes a resistor *R*, a standard cell *S*, a battery *B*, a galvanometer *G*, and a rheostat *r*. *R* may be a calibrated slidewire, with a known resistance, and *R′* a fixed resistor such that *R* and *R′* is some simple multiple of the emf of *S*. If *e* is taken as 1.019 V, the sum of *R* and *R′* may be chosen as 101.9. If the switch *K* is turned to the standard cell position, the galvanometer, in general, will deflect. The setting of rheostat *r*, is adjusted until the galvanometer remains at rest when *K* is closed. Then the drop of potential of the battery current through *R* and *R′* is exactly equal to the emf of *S*, so that if the current in *R* and *R′*, *T* = 1.019/101.9 A, or 10 mA, through each ohm of *R* there is then a drop of potential 0.010 V. If *R* is 20 ohms the total drop through the slidewire is 0.2 V.

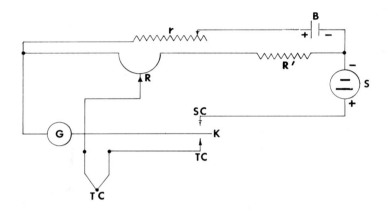

FIG. 32—*A simple potentiometer circuit.*

Now if *K* is turned to the "thermocouple" position, a setting may be found for the sliding contact on *R* at which there will be no deflection of the galvanometer. At this position, the drop of potential through *R* of the contact is equal to the emf of the thermocouple. If balance occurs at the 5 ohm point, the emf of the thermocouple is 0.01 × 5 = 0.05 V–50 mV.

In this measurement the galvanometer has been used only as a means of detecting the presence of a current, and readings are made only when no perceptible current is passing through it. Therefore, it is not calibrated, and it is only called upon to indicate on balance with sufficient sensitivity to give the desired precision of setting of the sliding contact. An increase in the resistance of the thermocouple circuit can increase only the limits of positions of the contact between which there is no perceptible deflection, but does not affect the position of balance, nor the measured value of the emf. Since the galvanometer is used only to indicate the existence and direction of current, it is necessary to design it to give a linear relationship between current and deflection. Zero stability is not extremely important since it is only necessary to look for departure from any equilibrium position when the key is closed to detect a need for adjustment and the direction in which it should be made. The galvanometer is usually of the suspension type, even in portable potentiometers because of sensitivity required, although electronic null balance detectors are available with similar sensitivities.

The calibration of an instrument of this type is stable since resistors and slidewires can be made with a high degree of stability. The emf of an unsaturated standard cell, as used for potentiometric work, does not change more than about 0.01 percent per year and has a negligible temperature coefficient.

The usefulness of an elementary potentiometer of this sort, having its entire range across the slidewire, is limited by the resolution possible in setting and reading the position of the contact on R.

6.2.2 *Standard Cell*

While the Zener diode is becoming used more widely as a reference, especially in digital voltmeters, the standard of reference for all voltage measurements is the Weston standard cell. Two types are available—the saturated and the unsaturated.

The saturated cell containing undissolved cadmium sulfate crystals is used in standardizing laboratories such as the National Bureau of Standards to represent the value of the absolute volt. Its emf is 1.01864 absolute volts at 20 C, reproducible within a few microvolts and reasonably constant over long periods of time. From 10 to 40 C its temperature coefficient is negative, the emf decreasing by about 40 μV per deg C rise in temperature.

In potentiometric work the unsaturated cell is used. It is more portable than the saturated type, and its temperature coefficient is very near zero. When new, its emf is about 1.019 V. This voltage tends to decrease about 100 mV (0.01 percent) per year, giving a useful life of about ten years.

A standard cell must never be short circuited, nor should its emf be measured with a voltmeter. In precise measurements, the balances should be made with a resistance of at least 10,000 ohms in series with the cell, until the balance is well within range of the detector scale.

Ambient temperatures near the cell should remain between +4 C and +40 C if the cell emf is to remain within ±0.01 percent of its certified value.

6.3 **Precision Potentiometry**

There are three types of precision-indicating potentiometers presently available; they are (1) laboratory high precision, (2) plant precision, and

(3) portable precision. There are three subtypes of the laboratory high precision instruments; (1a) six-dial, (1b) three-dial automatic balancing, and (1c) three-dial manual balancing.

6.3.1 Laboratory High Precision

The six-dial potentiometer meets the most exacting requirements for standardizing, research, and testing laboratories. It is particularly useful for precise temperature measurements in studies of specific heats, melting and boiling points, and for calibrating less precise potentiometers. It also is used for precise measurements such as required in calorimetry cryogenics. It measures emfs in the range of 0 to 1.6 V with 1 μV, or less, limit of error. It is essentially free of thermal emfs, and transients amount to less than 0.1 μV.

The three-dial automatic null balancing potentiometer indicator has a range of 0 to 70.1 mV, is readable to 1 μV, and has a limit of error of 0.02 percent, or 3 μV, whichever is greater. The balancing operation is automatic once the two-dial switches are set manually to place the unknown emf within the scale range. This automatic feature facilitates the rapid measurement of a large number of emfs with a minimum of effort and skill on the part of the operator, and with a high degree of accuracy.

The three-dial, manually operated potentiometer, also is used for precise measurements of voltage. As applied to the measurement of thermal emfs, its characteristics are such as to make it the most generally used of all the potentiometers in this field when accuracy and convenience of measurement are demanded. It is not portable in the sense in which the term is applied to self-contained potentiometers, since to attain the results of which it is capable, some parts of the circuit must be mounted separately. In particular, when highest precision is required, a very sensitive reflecting galvanometer must be mounted in a vibration-free support.

6.3.2 Plant Precision Potentiometer

This potentiometer is a self-contained, two-dial automatic null balancing indicator. It has a range of 0 to 40 mV, is readable to 2.5 μV, and has a limit of error of 0.5 percent below six mV and 0.33 percent above 6 mV. The balancing operation is automatic once the dial switch is set manually to place the unknown emf within the scale range.

6.3.3 Portable Precision Potentiometer

This potentiometer is a self-contained, three-dial, manually balancing instrument. It is used for precision checking of thermocouple pyrometers in laboratory and plant and for general temperature measurements. It has two ranges, -10.1 to 0 mV and 0 to 100.1 mV. It is readable to 1 μV. Limits of error are: (a) $\pm(0.05$ percent of reading plus 3 μV) without internal reference junction compensation and (b) $\pm(0.05$ percent of reading plus 6 μV) with reference junction compensation.

An example of (a) when using an external reference junction (see Section 7) is as follows:

> A Chromel-Alumel (Type K) thermocouple develops 41.314 mV at 1000 C (1832 F). The thermocouple error ($\pm\frac{3}{4}$ percent) can be 7.5 C; the instrument error will be $0.0005 \times 41.314 + 0.003 = 0.024$ mV or 0.56 C.

The reference junction compensation (b) is provided by an additional slidewire. This slidewire is used to set the reference junction voltage to the value which corresponds to the temperature of the reference junction (obtain millivolt value from reference tables, Section 10).

The portable precision potentiometer is designed to measure thermocouple emfs with a precision adequate for all but the more refined laboratory applications. It includes a built-in lamp and scale galvanometer.

6.4 *Semiprecision Potentiometry*

Potentiometers are available similar to those used for precision potentiometry except for limits of error. They are essentially of the two-dial type with self-contained galvanometer, standard cell, and battery. They are used primarily for general temperature measurements by means of thermocouples, checking thermocouple pyrometers in laboratory and plant, and as a calibrated source of voltage.

One model has two ranges; -10.1 to 0 mV and 0 to 100.1 mV, it is readable to 1 μV. The limit of error is $\pm(0.05$ percent of reading $+20$ μV) without reference junction compensator and $\pm(0.05$ percent of reading $+40$ μV) with reference junction compensator.

7. REFERENCE JUNCTIONS

7.1 **General Considerations**

A thermocouple circuit is by its nature a differential measuring device, producing an emf which is a function of the temperatures of its two junctions. One of these junctions is at the temperature which is to be measured and is referred to as the measuring junction. The other junction is maintained at a known temperature and is referred to as the reference junction. (In a practical thermocouple circuit (see Section 2.3) copper wires are often connected to the thermocouple alloy conductors at the reference junction. The term reference junctions will be used to refer to this situation.) If these junctions are both at the same temperature, the presence of the copper "intermediate metal" introduces no change in the thermocouple's emf. If they are not at the same temperature the analysis of the circuit is complicated. Moffat [34] gives a helpful analytical method.

The Seebeck coefficient (thermoelectric power) of many common thermocouples is approximately constant from the ice point to the upper temperature limit of the materials (see Section 3.1.1). For such thermocouples, an uncertainty in the temperature of the reference junction will reflect a similar uncertainty in the deduced temperature of the measuring junction. However this situation does not exist for all thermocouple pairs. Notable exceptions occur in the case of the high rhodium-in-platinum alloy thermocouples [35, 36]. In particular, if the reference junction of a platinum-30 percent rhodium versus platinum-6 percent rhodium (Type B) thermocouple lies within the range 0 to 50 C (32 to 122 F), a 0 C (32 F) reference junction may be assumed, and the error will not exceed 3 mV. This represents about 0.3 C (0.5 F) error in high-temperature measurements. [37]

7.2 Reference Junction Techniques

Three basic methods are used to take account of the reference junctions of the thermocouple circuit: (1) the junction is maintained at a fixed temperature, (2) the temperature of the reference junction is allowed to vary, and a compensating emf is introduced into the circuit or accounted for by calculation, (3) the temperature of the reference junction is allowed to vary, and compensation is made by a mechanical adjustment of the readout instrument.

Some variations of these techniques will be described in the following sections. Sources of error which are common to several techniques will be discussed in Section 7.3.

7.2.1 Fixed Reference Temperature

7.2.1.1 *Triple Point of Water*—A cell can be constructed in which there is an equilibrium between ice, water, and water vapor [38]. The temperature of this triple point is +0.01 C on the International Practical Temperature Scale of 1948, and it is reproducible to about 0.0001 C. Williams [39] describes a commercially available cell which is not affected by factors such as air saturation and pressure which can cause several millidegrees fluctuation in the temperature of an ice bath. To utilize such precision, extreme attention to immersion error and galvanic error is required, and the measurement system must be of the highest quality.

7.2.1.2 *Ice Point*—An ice bath consisting of a mixture of melting shaved ice and water forms an easy way of bringing the reference junctions of a thermocouple to 0 C (32 F). If a proper technique is used, the uncertainty in the reference junction temperature can be made negligibly small. With extreme care the ice point can be reproduced to 0.0001 C [40].

A recommended form of ice bath is described in Ref 41, and is illustrated

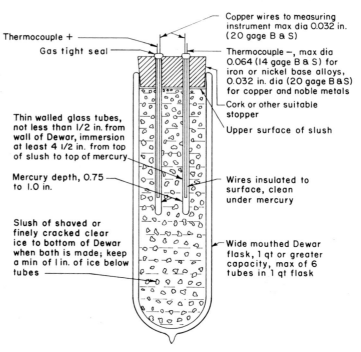

Thermocouple +

Gas tight seal

Copper wires to measuring
instrument max dia 0.032 in.
(20 gage B & S)

Thermocouple −, max dia
0.064 (14 gage B & S) for
iron or nickel base alloys,
0.032 in. dia (20 gage B & S)
for copper and noble metals

Cork or other suitable
stopper

Upper surface of slush

Thin walled glass tubes,
not less than 1/2 in. from
wall of Dewar, immersion
at least 4 1/2 in. from top
of slush to top of mercury

Mercury depth, 0.75
to 1.0 in.

Wires insulated to
surface, clean
under mercury

Slush of shaved or
finely cracked clear
ice to bottom of Dewar
when bath is made; keep
a min of 1 in. of ice below
tubes

Wide mouthed Dewar
flask, 1 qt or greater
capacity, max of 6
tubes in 1 qt flask

FIG. 33—*Recommended ice bath for reference junction.*

in Fig. 33. Using the recommended construction with maximum diameters of 14 gage B&S for iron or nickel base alloys and 20 gage B&S for copper and noble metals, immersed at least 4½ in. in the water-ice mixture, Caldwell [*42*] found the immersion error to be less than 0.05 C (see Section 7.3.1). Finch [*43*] describes an alternative construction which reduces the immersion error when large conductors must be used.

If improper technique is used, serious errors can be encountered. The largest error which is likely to occur arises due to melting of the ice at the bottom of a bath until the reference junctions are below the ice level and surrounded by water alone. This water may be as much as 4 C above the ice point. While an ice bath is being used excess water should be removed periodically and more ice added, so that the ice level is maintained safely below the reference junctions [*41*].

If the ice used to prepare the bath has been stored in a freezer at a temperature below 0 C, it may freeze the surrounding water and remain at a temperature below 0 C for some time. To avoid this condition the bath should be examined to confirm that a slush of ice shavings in water actually exists.

If appreciable concentrations of salts are present in the water used to make the ice bath, the melting point can be affected. McElroy [*44*] investigated various combinations of tap and distilled water to determine the error introduced. He observed bath temperatures of $+0.013$ C using distilled water with tap ice and -0.006 C using tap water and tap ice. It is probable that the use of tap water will not introduce significant errors unless accuracies better than 0.01 C are required [42].

Another possible source of error is galvanic action which is discussed in Section 7.3.2.

Although the ice bath is an easy way of achieving a convenient fixed point for the reference junctions in the laboratory, the need for constant attention makes it unsuitable for industrial application where some form of automatic reference junction is desirable.

7.2.1.3 *Automatic Ice Point*—The development of the thermoelectric refrigerator (Peltier cooler) has enabled the production of practical devices in which an equilibrium between ice and water is constantly maintained [45, 46]. Since water expands 9 percent on freezing, the change of volume may be used to control the heat transfer. Substances which undergo a phase change do so without change of temperature. This is an important advantage since the system behaves as if it had an infinite thermal capacity.

This device can provide a reference medium which is maintained at a precise temperature, but careful design is necessary if this precision is to be utilized fully. The system is subject to immersion error (Section 7.3.1) and galvanic error (7.3.2) due to condensation.

Some commercially available devices provide wells into which the user may insert reference junctions formed from his own calibrated wire. Others are provided with many reference junction pairs brought out to terminals which the user may connect into his system. The latter type are subject to the wire matching error (Section 7.3.3).

If the potential errors have been successfully overcome, the error introduced into a system by these devices may be less than 0.1 C.

7.2.1.4 *Constant Temperature Ovens*—A thermostatically controlled oven provides a way of holding a reference junction at an approximately constant temperature. The major advantage of this method is that one oven can be used with a large number of circuits while maintaining isolation between the circuits without the need for providing separate power supplies for each circuit.

The oven must be maintained at a temperature above the highest anticipated ambient temperature. The need for temperature uniformity within the oven and a precise temperature control system are inherent complications, in addition to the immersion error (Section 7.3.1) and the wire matching error (Section 7.3.2).

Ovens are available with rated accuracies ranging from 0.1 to 0.05 C. A common commercially used junction temperature is 150 F, although other values may be specified.

To use reference junctions held at elevated temperatures with tables or an instrument based on a 0 C reference junction temperature, a constant amount must be added to the thermal emf.

A device using two ovens is available which makes this correction automatically. To generate the necessary emf, (for example, 2.662 mV for a Type K system with a 150 F (66 C) reference junction temperature) an additional internal thermocouple is connected in series with the external thermocouple circuit. This internal thermocouple has its reference junction in the first oven and its measuring junction in the second oven. The second oven is maintained at a suitable temperature above that of the first so that the required emf is generated. The external thermocouple circuit then behaves as if its reference junction were at 0 C. The manufacturers' stated accuracies of these devices are similar to those of single oven units.

7.2.2 Electrical Compensation

If the temperature of the reference junction of a thermocouple is allowed to change, the output emf will vary in accordance with the Seebeck coefficient of the couple at the reference temperature. A compensating circuit containing a source of current and a combination of fixed resistors and a temperature sensitive resistor (TSR) can be designed which will have a similar variation of emf as the temperature of the TSR is varied. If the reference junction of the thermocouple is coupled thermally to the TSR and the compensating circuit is connected in series with the thermocouple so that its temperature-variable emf opposes that caused by the reference junction, the thermocouple behaves as if the reference junction temperature were held constant. In addition, by suitable choice of the fixed resistors, any fixed reference junction temperature may be simulated. Since the TSR is at the temperature of the reference junction, no warm-up or stabilization time is involved. Muth [47] has given a more extended description of these circuits.

The disadvantages of this arrangement include the need for a stable power source for each thermocouple circuit, the difficulty of exactly matching the Seebeck coefficient over an extended temperature range, and the addition of series resistance in the thermocouple circuit.

This principle is used in almost all self-balancing recording thermocouple potentiometers. Here the power source already is present as part of the potentiometer circuit. The Seebeck coefficient is matched adequately to allow the accuracy of the entire instrument to be typically 0.25 percent of full scale over a reasonable range of ambient temperatures.

Electric reference junction compensators are also available as small circuit modules with self-contained battery power sources or for connection to a-c power. A typical specification for a battery powered unit is compensation to ±0.5 F over an ambient of +55 to +90 F. Improved specifications are quoted for more elaborate devices.

Some portable manually balanced potentiometers are provided with a thermometer to read the reference junction temperature and an adjustable circuit calibrated in millivolts or temperature. The control must be set manually to add the required emf to simulate an ice-point reference junction temperature.

7.2.2.1 *Zone Box*—In systems employing many thermocouples an alternative method of dealing with the thermocouple reference junctions is sometimes used. All of the thermocouples are routed to a device called a zone box where each of the thermocouple conductors is joined to a copper wire which is routed to the emf measuring instrument [*48, 49*].

Within the zone box all of the reference junctions between the thermocouple conductors and the copper wires are insulated electrically but kept in good mutual thermal contact with each other and with a single transducer which compensates for the temperature within the zone box. This transducer may be a thermocouple of the same type as the measuring thermocouples having any of the types of reference junction described above. In this case its emf may be added to each of the measuring thermocouples in turn as they are switched to the emf measuring instrument. Alternatively the transducer may be a thermistor or resistance temperature detector which needs no reference temperature. In this case, the analyzing instrument determines the necessary correction to the thermal emfs, based on the measured reference junction temperature. Several variations of this technique are described in Ref *49*.

The advantage of this arrangement is the simplicity of the zone box which generally requires no heaters, controls, or power supplies. Since the zone box is approximately at the ambient temperature, the immersion error (Section 7.3.1) can be made easily negligible with a moderate amount of thermal insulation and care to avoid locations having extreme thermal gradients. The wire matching error (Section 7.3.3) should receive attention because the reference junctions are at ambient temperature and the calibration error at this temperature must be accounted for.

7.2.3 *Mechanical Reference Compensation*

To complete this account of methods used to compensate for the temperature of thermocouple reference junctions, a device used on millivolt pyrometers must be included. The millivolt pyrometer measures the thermal emf of a thermocouple circuit by measuring the current produced in a circuit of fixed and known resistance. The current operates a galvanometer with a rigid pointer which moves over a scale graduated in degrees [*43*]. The reference junction is at the temperature of the instrument, and hence the available thermal emf is a function of the temperature of the instrument. Compensation is often accomplished by attaching one of the hairsprings of the D'Arsonval galvanometer movement to a bimetallic thermometer element so that the electrical zero of the instrument is adjusted to correspond to the temperature of the instrument. This system is subject to the wire matching error (Section 7.3.3), but the precision of the pyrometer seldom justifies making corrections.

7.3 Sources of Error

Several sources of error which may disturb the control or measurement of the reference junction temperature are discussed in this section.

7.3.1 *Immersion Error*

Whenever reference junctions are being maintained at a temperature which differs from the ambient, heat is transferred between the reference temperature medium (oven, ice bath, etc.) and the ambient via the electrical insulation which separates the junctions from the medium and via the wires which emerge from the reference junctions. Thus the temperature of the junctions always differs from that of the reference medium to a greater or lesser degree. Caldwell [42] gives data which allow the error to be estimated for the standard type of ice bath. For other situations the error may be calculated by methods outlined in Refs *50* and *51,* if the coefficients governing heat flow between the medium, the wires, and the ambient can be evaluated.

With careful design the immersion error usually can be made negligible.

7.3.2 *Galvanic Error*

If water is allowed to contact the thermocouple alloy and copper wires of the reference junction, a galvanic cell may be set up, causing voltage drops which disturb the thermal emfs. If the reference junction is at a temperature below the dew point, the water may appear due to condensation. Insulation on both wires and precautions to avoid the accumulation of water in contact with the wires normally will prevent this error [*42, 44*].

7.3.3 *Wire Matching Error*

Thermocouple wire normally is calibrated with its reference junction at 0 C, and corrections are determined to enable accurate measurements at elevated temperatures. The calibration deviation at ambient temperature is seldom of interest and usually is not determined.

Many reference junction devices are equipped with thermocouple alloy wires, and provision is made for interconnection with the user's thermocouples at ambient temperatures. If the calibration of the wire supplied with the device differs from that of the thermocouples at the ambient temperature, a significant error can result due to the interconnection of the wires. This source of error often is overlooked, since it is assumed that if the interconnection of both wires of a thermocouple pair occurs at the same temperature no error is introduced. The existence of this error is visualized easily if the circuit is analyzed using Moffat's method [*34*].

A simple correction for the wire matching error can be made if the ambient temperature deviation of the wire supplied with the reference junction device is known and the user's thermocouples are calibrated at ambient temperature.

The wire matching error is avoided in those reference junction devices in which the user's wire is used to form the reference junctions.

7.4 References

[*34*] Moffat, R. J., "The Gradient Approach to Thermocouple Circuitry," *Temperature, Its Measurement and Control in Science and Industry,* Vol. 3, Part 2, Reinhold, New York, 1962, pp. 33–38.

[*35*] Caldwell, F. R., "Thermocouple Materials," *Temperature, Its Measurement and Control in Science and Industry,* Vol. 3, Part 2, Reinhold, New York, 1962, pp. 81–134.

[*36*] Zysk, E. D., "Noble Metals in Thermometry, Recent Developments," *Technical Bulletin, Englehard Industries, Inc.,* Vol. 5, No. 3, 1964.

[*37*] Burns, G. W. and Gallagher, J. S. "Reference Tables for the Pt-30 Percent Rh Versus Pt-6 Percent Rh Thermocouple," *Journal of Research, NBS-C Engineering and Instrumentation,* Vol. 70C, No. 2, 1966.

[*38*] Stimson, H. F., "Precision Resistance Thermometry and Fixed Points," *Temperature, Its Measurement and Control in Science and Industry,* Vol. 2, Reinhold, N. Y., 1955, pp. 141–168.

[*39*] Williams, S. B., "Triple Point of Water Temperature Reference," *Instruments and Control Systems,* Vol. 33, 1960.

[*40*] Thomas, J. L., "Reproducibility of the Ice Point," *Temperature, Its Measurement and Control in Science and Industry,* Vol. 1, Reinhold, New York, 1941, pp. 159–161.

[*41*] "Recommended Ice Bath for Reference Junctions," SAE Aerospace Recommended Practice No. 691, Committee AE-2, Temperature Measurement Sensing, Society of Automotive Engineers, Inc., 1964.

[*42*] Caldwell, F. R., "Temperatures of Thermocouple Reference Junctions in an Ice Bath," *Journal of Research, NBS Engineering and Instrumentation,* Vol. 69C, No. 2, 1965.

[*43*] Finch, D. I., "General Principles of Thermoelectric Thermometry," *Temperature, Its Measurement and Control in Science and Industry,* Vol. 3, Part 2, Reinhold, New York, 1962, pp. 3–32.

[*44*] McElroy, D. I., "Progress Report I. Thermocouple Research Report for the Period, November 1, 1956 to October 31, 1957," ORNL-2467.

[*45*] Morgan, W. A., "Close Temperature Control of Small Volumes, A New Approach," Preprint Number 11.7-2-64, Instrument Society of America, 1964.

[*46*] Feldman, C. L., "Automatic Ice-Point Thermocouple Reference Junction," *Instruments and Control Systems,* Jan. 1965, pp. 101–103.

[*47*] Muth, S., Jr., "Reference Junctions," *Instruments and Control Systems,* May 1967, pp. 133–134.

[*48*] Roeser, W. F., "Thermoelectric Thermometry," *Temperature, Its Measurement and Control in Science and Industry,* Vol. 1, Reinhold, New York, 1941, pp. 202.

[*49*] Claggett, T. J., "External Thermocouple Compensation," *Instrumentation,* Vol. 18, No. 2, Second Quarter, 1965.

[*50*] Baker, H. D., Ryder, M. E., and Baker, M. A., *Temperature Measurement in Engineering,* Vol. 1, Section 7, Wiley, 1953.

[*51*] Bauerle, J. E., "Analysis of Immersed Thermocouple Error," *Review of Scientific Instruments,* Vol. 32, No. 3, Mar. 1961, pp. 313.

8. CALIBRATION OF THERMOCOUPLES

The calibration of a thermocouple consists of the determination of its electromotive force (emf) at a sufficient number of known temperatures so that, with some accepted means of interpolation, its emf will be known over the entire temperature range in which it is to be used. The process requires a standard thermometer to indicate temperatures on a standard scale, a means for measuring the emf of the thermocouple, and a controlled environment in which the thermocouple and the standard can be brought to the same temperature. Some of the more commonly used techniques for accomplishing such calibrations will be discussed in this chapter.

Much of this material is based upon National Bureau of Standards Circular 590, *Methods of Testing Thermocouples and Thermocouple Materials,* and the calibration methods appearing in Chapter 19, Part 3, *Temperature Measurement,* of the American Society of Mechanical Engineers Power Test Codes.

8.1 General Considerations

8.1.1 *Temperature Scale*

The international Practical Temperature Scale of 1948 (IPTS) is realized by the National Bureau of Standards to provide a standard scale of temperature for use by science and industry in the United States. This scale [52, 53] was adopted by the General Conference on Weights and Measures in 1960 [52] and has general international acceptance.[9] The unit of temperature on the scale is the degree Celsius written deg C (Int. 1948) formerly called Centigrade. The Fahrenheit Scale is reproduced using the conversion formula

$$\text{deg F} = 9/5 \text{ deg C} + 32$$

The IPTS is a practical, standard, empirical scale designed to conform closely to the Thermodynamic Kelvin Temperature Scale (absolute scale). It is a practical scale because values of temperature on it are more reproducible and easily determined than values on the Thermodynamic Scale. The scale is based on six reproducible temperatures (defining fixed points), to which numerical values are assigned, and on formulas establishing the relation between temperature and the indications of instruments calibrated by means of values assigned to the six defining fixed points. These fixed points are defined by specified equilibrium states, each of which, except for the triple point of water (see Section 7 on Reference Junctions), is under a pressure of 101,325 N/m^2 (1 standard atm).

---

[9] As of the time of writing, a new temperature scale has been defined known as IPTS 68 (see Refs 53 and 54).

The fixed points of the scale and the exact numerical values assigned to them are given in Table 34.

TABLE 34—*Defining fixed points.*

Temperature of equilibrium between	temperature,[a] deg C (Int. 1948)
Liquid oxygen and its vapor (oxygen point)	−182.97
Ice, liquid water, and water vapor (triple point of water)	+0.01
Liquid water and its vapor (steam point)	100
Liquid sulfur and its vapor (sulfur point)	444.6[b]
Solid silver and liquid silver (silver point)	960.8
Solid gold and liquid gold (gold point)	1063

[a] Exact values assigned. The pressure is 1 standard atm, except for the triple point of water.
[b] In place of the sulfur point, it is recommended to use the temperature of equilibrium between solid zinc and liquid zinc (zinc point) with the value 419.505 C (Int. 1948). The zinc point is more reproducible than the sulfur point, and the value which is assigned to it has been so chosen that its use leads to the same values of temperature on the International Practical Temperature Scale as does the use of the sulfur point.

The procedures for interpolation lead to a division of the scale into four parts.

(a) From 0 C to 630.5 C (antimony point) the temperature t is defined by the formula

$$R_t = R_0(1 + At + Bt^2),$$

where R_t is the resistance at temperature t of the platinum wire resistor of a standard resistance thermometer, and R_0 is the resistance at 0 C. The constants R_0, A, and B are to be determined from the values of R_t at the triple point of water, at the steam point, and at the sulfur point (or the zinc point). The platinum wire of a standard resistance thermometer shall be annealed and its purity shall be such that R_{100}/R_0 is not less than 1.3920.

(b) From the oxygen point to 0 C, the temperature t is defined by the formula

$$R_t = R_0[1 + At + Bt^2 + C(t - t_{100})t^3],$$

where R_0, A, and B are determined in the same manner as in (a) above, the constant C is to be determined from the value of R_t at the oxygen point, and $t_{100} = 100$ C.

(c) From 630.5 C to the gold point the temperature t is defined by the formula

$$E = a + bt + ct^2,$$

where E is the electromotive force of a standard Type S thermocouple, when one of the junctions is at 0 C and the other at the temperature t. The constants a, b, and c are to be determined from the values of E at 630.5 C, at the silver point, and at the gold point. The value of the electromotive force at 630.5 C is to be determined by measuring this temperature with a standard resistance thermometer.

(d) Above the gold point the temperature t is defined by the formula

$$\frac{J_t}{J_{Au}} = \frac{\exp\left[\dfrac{C_2}{\lambda(t_{Au} + T_0)}\right] - 1}{\exp\left[\dfrac{C_2}{\lambda(t + T_0)}\right] - 1},$$

where J_t and J_{Au} are the radiant energies per unit wavelength interval at wavelength λ emitted per unit time per unit solid angle per unit area of a blackbody at the temperature t and the gold point, respectively, C_2 is the second radiation constant with the value $C_2 = 0.01438$ m-deg C, λ is in meters, $T_0 = 273.15$ deg, and $\exp[x]$ is defined as e^x.

In addition to the defining points of the Scale, given in Table 34, certain other points may be useful for calibration purposes. Some of these and their reported temperatures are given in Table 35. Except for the triple point of benzoic acid, each temperature is for a system in equilibrium under a pressure of 1 standard atm.

TABLE 35—*Secondary reference points.*
The pressure is 1 standard atm, except for the triple point of benzoic acid.

	deg C	deg F
Boiling point of helium	−268.935	−452.083
Boiling point of equilibrium hydrogen	−252.883	−423.189
Sublimation point of carbon dioxide	−78.5	−109.3
Freezing point of mercury[a]	−38.86	−37.95
Freezing point of water	0.00	32.00
Triple point of benzoic acid	122.36	252.25
Freezing point of indium	156.61	313.90
Freezing point of tin	231.91	449.44
Freezing point of bismuth	271.37	520.47
Freezing point of cadmium	321.03	609.86
Freezing point of lead	327.43	621.37
Freezing point of antimony	630.5	1166.9
Freezing point of aluminum	660.1	1220.2
Freezing point of copper	1083	1981
Freezing point of palladium	1552	2826
Freezing point of platinum	1769	3216

[a] Ref 55.

8.1.2 *Working Standards*

Any one of several types of thermometers, calibrated in terms of the IPTS, may be used as a working standard for the calibration of thermocouples. The choice will depend upon the temperature range covered, whether a laboratory furnace or a stirred liquid bath is used, the accuracy expected of the calibration, or in cases where more than one type will suffice, the convenience or preference of the calibrating laboratory.

8.1.2.1 *Resistance Thermometers*—The standard platinum resistance thermometer is the most accurate standard for use from approximately −253 C (−423 F) to 630.5 C (1167 F). In cases where an uncertainty approaching 0.1 C is necessary at temperatures below −56 C (−69 F) or above about 200 C (392 F) there are few alternatives to the use of resistance thermometers as standards.

8.1.2.2 *Liquid-in-Glass Thermometers*—This type of thermometer may be used from approximately −183 C (−297 F) to 400 C (752 F), or even higher with special types. Generally, the accuracy of these thermometers is less below −56 C (−69 F) where organic thermometric fluids are used, and above 300 C (572 F) where instability of the bulb glass may require frequent calibration. Specifications for ASTM liquid-in-glass thermometers are given in Ref 56.

8.1.2.3 *Types E and T Thermocouples*—Either of these types of thermocouples may be used down to a temperature of −183 C (−297 F) or lower, but the attainable accuracy may be limited by the accuracy of the emf measurements and the inhomogeneity of the wire at low temperatures. The stability of the larger sizes of wire is greater than that of smaller wires under the same conditions. Twenty-four gage wire is a useful compromise between the lesser stability of smaller wire and the greater thermal conduction (greater required depth of immersion) of larger wire. Recommended upper limits are 425 C (797 F) for the Type E and 200 C (392 F) for Type T.

8.1.2.4 *Types R and S Thermocouples*—Type S or Type R thermocouple is the most satisfactory working standard for use in the range from 630.5 C (1167 F) up to about 1200 C (2192 F). Its use may be extended down to room temperature if it is desired to use the same standard over a wide range, but its sensitivity falls off appreciably as temperatures below 200 C (392 F) are reached. Twenty-four gage wire most commonly is used for these standards.

8.1.2.5 *High Temperature Standards*—The IPTS above 1063 C (1945 F) is defined in terms of ratios of radiant energy, the ratios usually being measured by means of an optical pyrometer. The optical pyrometer [57], sighted on a blackbody cavity built into the calibration furnace, therefore, can serve as a working standard for all temperatures above 1063 C (1946 F). On the other hand, thermocouples, calibrated on the optical pyrometer scale, can be used themselves as standards. The Type B thermocouple [58] is useful up to about 1600 C (2912 F). Tungsten rhenium alloys can be used to higher temperatures, but the optical pyrometer more commonly is used.

8.1.3 *Annealing*

Practically all base-metal thermocouple wire is annealed or given a "stabilizing heat treatment" by the manufacturer. Such treatment generally is considered sufficient, and seldom is it found advisable to further anneal the wire before testing.

Although new platinum and platinum-rhodium thermocouple wire as sold by some manufacturers already is annealed, it has become regular practice in many laboratories to anneal all Types R and S thermocouples, whether new or previously used, before attempting an accurate calibration. This is accomplished usually by heating the thermocouple electrically in air. The entire thermocouple is supported between two binding posts, which should be close together, so that the tension in the wires and stretching while hot are kept at a minimum. The temperature of the wire is determined most conveniently with an optical pyrometer.[10]

There are some questions as to the optimum temperature and length of time at which such thermocouples should be annealed to produce the most constant characteristics in later use [59] and as to whether annealing for more than a few minutes is harmful or beneficial. Most of the mechanical strains are relieved during the first few minutes of heatings at 1400 to 1500 C (2552 to 2732 F), but it has been claimed that the changes in the thermal emf of a couple in later use will be smaller if the wires are heated for several hours before calibration and use. The principal objection to annealing thermocouples for a long time at high temperatures, aside from the changes in emf taking place, is that the wires are weakened mechanically as a result of grain growth. It has been found that annealing at temperatures much above 1500 C (2732 F) produces rapid changes in the emf and leaves the wire very weak mechanically. The National Bureau of Standards has adopted the procedure of annealing Types R and S thermocouples for 1 h at 1450 C (2642 F).

It has not been demonstrated conclusively that Types R and S thermocouples after contamination can be improved materially in homogeneity by prolonged heating in air, although it is logical to suppose that certain impurities can be driven off or, through oxidation, rendered less detrimental.

8.1.4 *Measurement of Emf*

One of the factors in the accuracy of the calibration of a thermocouple

---

[10] The ordinary portable type of optical pyrometer is very satisfactory for this purpose. As commonly used, the magnification is too low for sighting on an object as small as the wires of noble-metal thermocouples, but this is remedied easily by lengthening the telescope tube or using an objective lens of shorter focal length.

is the accuracy of the instrument used to measure the emf. Fortunately, in most instances, an instrument is available whose performance is such that the accuracy of the calibration need not be limited by the accuracy of the emf measurements. For work of the highest accuracy it is advisable to use a potentiometer of the type designed by Diesselhorst [60], White [61], or Wenner [62], in which there are no slidewires and in which all the settings are made by means of dial switches. However, for most work, in which an accuracy of 5 μV will suffice, slidewire potentiometers of the laboratory type are sufficiently accurate. Portable potentiometers accurate within 40 to 100 μV also are available. For a more detailed consideration of emf measurements see Section 6.

8.1.5 *Homogeneity*

The emf developed by a thermocouple made from homogeneous wires will be a function of the temperature difference between the measuring and the reference junction. If, however, the wires are not homogeneous, and the inhomogeneity is present in a region where a temperature gradient exists, extraneous emf's will be developed, and the output of the thermocouple will depend upon factors in addition to the temperature difference between the two junctions. The inhomogeneity of the thermocouple wire, therefore, is an important factor in accurate measurements.

Thermocouple wire now being produced is usually sufficiently homogeneous in chemical composition for most purposes. Occasionally inhomogeneity in a thermocouple may be traced to the manufacturer, but such cases are rare. More often it is introduced in the wires during tests or use. It usually is not necessary, therefore, to examine new thermocouples for inhomogeneity, but thermocouples that have been used for some time should be so examined before an accurate calibration is attempted.

While rather simple methods are available for detecting thermoelectric inhomogeneity, no satisfactory method has been devised for quantitatively determining it or the resulting errors in the measurement of temperatures. Abrupt changes in the thermoelectric power may be detected by connecting the two ends of the wire to a sensitive galvanometer and slowly moving a source of heat, such as a bunsen burner or small electric furnace, along the wire. This method is not satisfactory for detecting gradual changes in the thermoelectric power along the length of the wire. Inhomogeneity of this nature may be detected by doubling the wire and inserting it to various depths in a uniformly heated furnace, the two ends of the wire being connected to a galvanometer as before. If, for example, the doubled end of the wire is immersed 10 in. in a furnace with a sharp temperature gradient so that two points on the wire 20 in. apart are in the temperature gradient, the emf determined with the galvanometer is a measure of the difference in the thermoelectric properties of the wire at these two points.

After reasonable homogeneity of one sample of wire has been established, it may be used in testing the homogeneity of similar wires by welding the two together and inserting the junction into a heated furnace. The resulting emf at various depths of immersion may be measured by any convenient method. Other similar methods have been described for detecting inhomogeneity [63].

Tests such as those described above will indicate the uncertainty in temperature measurements due to inhomogeneity in the wires. For example, if a difference in emf of 10 μV is detected along either element of a platinum-rhodium couple by heating various parts of the wire to 600 C (1112 F), measurements made with it are subject to an uncertainty of the order of 1 deg to 500 C, or 2 deg at 1200 C. Similarly, if an emf of 10 μV is detected along an element of a base-metal couple with a source of heat at 100 C, measurements made with it are subject to an uncertainty of the order of 0.2 deg at this temperature. The effects of inhomogeneity in both wires may be either additive or subtractive, and, as the emf developed along an inhomogeneous wire depends upon the temperature distribution, it is evident that corrections for inhomogeneity are impracticable if not impossible.

8.1.6. *General Calibration Methods*

The temperature-emf relation of a homogeneous [64] thermocouple is a definite physical property and therefore does not depend upon the details of the apparatus or method employed in determining this relation. Consequently, there are numerous methods of calibrating thermocouples, the choice of which depends upon the type of thermocouple, temperature range, accuracy required, size of wires, apparatus available, and personal perference. However, the emf of a thermocouple with its measuring junction at a specified temperature depends upon the temperature difference between its measuring and reference junctions. Therefore, whatever method of calibration is used, the reference junction must be maintained constant at some known temperature (see Section 7), and this temperature must be stated as a necessary part of the calibration results.

Thermocouple calibrations are required with various degrees of accuracy ranging from 0.1 to 5 or 10 C. For an accuracy of 0.1 deg, agreement with the IPTS and methods of interpolating between the calibration points become problems of prime importance, but for an accuracy of about 10 deg comparatively simple methods of calibration usually will suffice. The most accurate calibrations in the range −183 C (−297 F) to 300 C (572 F) are made by comparing the couples directly with a standard platinum-resistance thermometer in a stirred liquid bath. In the range 300 to 630.5 C (572 to 1167 F) (and below if a platinum-resistance thermometer and stirred liquid bath is not available) thermocouples are calibrated most accurately at the freezing or boiling points of pure substances. Between 630.5 and 1063

C (1167 and 1945 F), Type S thermocouple calibrated at 630.5 C and the freezing points of gold and silver, serves to define the IPTS, and other types of thermocouples are calibrated most accurately in this range by direct comparison with the standard thermocouple calibrated as specified. Other thermocouples may be calibrated just as accurately at the fixed points as the Type S thermocouple, but interpolated values at intermediate points may depart slightly from the IPTS. Above 1063 C (1945 F), the most basic calibrations are made by observing the emf when one junction of the thermocouple is in a blackbody furnace, the temperature of which is measured with an optical pyrometer. However, the difficulties encountered in bringing a blackbody furnace to a uniform temperature make the direct comparison of these two types of instruments by no means a simple matter.

Although the Type S thermocouple serves to define the IPTS only in the range 630.5 to 1063 C, this type of thermocouple calibrated at fixed points is used extensively both above and below this range as a working standard in the calibration of other thermocouples. For most industrial purposes a calibration accurate to 2 or 3 C in the range room temperature to 1200 C (2192 F) is sufficient. Other thermocouples can be calibrated by comparison with such working standards almost as accurately as the calibration of the standard is known. However, it might be pointed out that outside the range 630.5 to 1063 C any type of thermocouple suitable for the purpose, and calibrated to agree with the resistance thermometer or optical pyrometer in their respective ranges, has as much claim to yielding temperatures on the IPTS as the Type S thermocouple. In fact, at the lower temperatures certain types of base-metal couples are definitely better adapted for precise measurements.

The calibration of couples then may be divided into two general classes, depending upon the method of determining the temperature of the measuring junction: (1) calibration at fixed points and (2) calibration by comparison with standard instruments such as thermocouples, resistance thermometers, etc.

In order to obtain the high accuracies referred to above and usually associated with calibrations at fixed points, it is necessary to follow certain prescribed methods and to take the special precautions described in detail in the following paragraphs, but for an accuracy of about 5 C the more elaborate apparatus to be described need not be employed.

8.1.7. *Calibration Uncertainties*

The several factors which contribute to the uncertainties in the emf versus temperature relationship for a particular thermocouple as determined by calibration may be grouped into two kinds; those influencing the observations at calibration points, and those arising from any added uncertainty as a result of interpolation between the calibration points. Errors from either of these sources of uncertainty can be reduced materially, within limits, through use of well designed equipment and careful techniques; hence, the required accuracy should be clearly understood when choosing calibration facilities.

Estimates of the accuracies attainable in the calibration of homogeneous thermocouples by different techniques are given in Tables 36, 37, 38, 39, and 40. The estimates assume that reasonable care is exercised in the work. More or less accurate results are possible using the same methods, depending upon soundness of the techniques used. While excessive care is a waste when relatively crude measurements are sufficient, it should be emphasized that inadequate attention to possible sources of error is more often found to be the practice than the converse. In the following some of the important considerations associated with the various calibration methods are emphasized briefly.

8.1.7.1 *Uncertainties Using Fixed Points*—The equilibrium temperatures listed in Table 35 (with the possible exception of the sublimation point of carbon dioxide) are sufficiently exact, and the materials are readily available in high enough purity, that accurate work can be done using these fixed points with no significant error being introduced by accepting the temperatures listed. Using freezing points, however, good designs of freezing point cells and furnaces are important for controlling the freezes and for providing sufficient immersion for the thermocouple, if the full potential of the method is to be realized.

Although uncertainties of the order of ± 1 C in the temperatures are assigned to the freezing points (and hence by implication to the melting points) of palladium and platinum, these contribute in only a minor way overall uncertainties of calibrations using freezing point techniques.

8.1.7.2. *Uncertainties Using Comparison Methods*—The accuracy attained at each calibration point using the comparison method will depend upon the degree to which the standard and the test thermocouple are maintained at the same temperature and the accuracy of the standard used. Comparison measurements made in stirred liquid baths usually present no special problems provided that sufficient immersion is used. Because of the high-thermal conductivity of copper, special attention should be given to the problem of immersion when calibrating Type T thermocouples.

As higher and higher temperatures are used the difficulties of maintaining the test thermocouple and the standard at the same measured temperature are magnified whether a tube furnace, an oven with moderating block, or whatever means is used for maintaining the desired temperature. In addition, at temperatures of about 1500 C (2732 F), and higher, the choice of insulating materials becomes very important (see Section 4). Special attention must be paid to possible errors arising from contamination from the insulators or protection tube and from electrical leakage.

TABLE 36—*Accuracies attainable using fixed point techniques.*

| | | | Calibration Uncertainty | |
| | Temperature Range, deg C | Calibration Points[a] | At Observed. Points, deg C | Of Interpolated Values, deg C |
Type				
S	0 to 1100	Zn, Sb[b], Ag, Au	0.2	0.3
R	0 to 1100	Sn, Zn, Al, Cu-Ag, Cu	0.2	0.5
E	0 to 870	Sn, Zn, Al, Cu-Ag	0.2	0.5
J....................	0 to 760	Sn, Zn, Al	0.2	1.0
K	0 to 1100	Sn, Zn, Al, Cu-Ag, Cu	0.2	1.0

[a] Metal freezing points.
[b] Temperature measured by standard platinum resistance thermometer.

TABLE 37—*Accuracies attainable using comparison techniques in laboratory furnaces (Type R or S standard).*

| | | | Calibration Uncertainty | |
| | Temperature Range, deg C | Calibration Points | At Observed Points, deg C | Of Interpolated Values, deg C |
Type				
R or S	0 to 1100	about every 100 C	0.3	0.5
E	0 to 870	about every 100 C	0.5	0.5
J....................	0 to 760	about every 100 C	0.5	1.0
K	0 to 1100	about every 100 C	0.5	1.0

TABLE 38—*Accuracies attainable using comparison techniques in stirred liquid baths.*

| | | | | Calibration Uncertainty | |
| | Temperature Range, deg C | Calibration Points | Type of Standard[a] | At Observed Points | Of Interpolated Values |
Type					
E	−196 to 425	about every 100 C	PRT	0.1	0.2
	−196 to 435	about every 50 C	PRT	0.1	0.1
	−196 to 435	about every 50 C	E or T	0.2	0.2
	−56 to 200	about every 50 C	LIG	0.1	0.1
T	−196 to 250	about every 100 C	PRT	0.1	0.2
	−196 to 250	about every 50 C	PRT	0.1	0.1
	−196 to 250	about every 50 C	E or T	0.2	0.2
	−56 to 200	about every 50 C	LIG	0.1	0.1

[a] PRT = standard platinum resistance thermometer; E or T = Type E or T thermocouple; and LIG = liquid-in-glass thermometer.

TABLE 39—*Tungsten-rhenium type thermocouples.*
(Maximum calibration uncertainties for range 1000 to 2000 C using melting points by wire or disk method).

Calibration Uncertainty	
At Observed Points	Of Interpolated Values[a]
Gold (1063 C) .... ±0.5 C	1000 to 1453 C, ±2.7 C
Nickel (1453 C) .... ±3.5 C	1453 to 1552 C, ±4.0 C
Palladium (1552 C) .... ±3.0 C	1552 to 1769 C, ±4.0 C
Platinum (1769 C) .... ±3.0 C	1769 to 2000 C, ±7.0 C
Rhodium (1960 C) .... ±5.0 C	

[a] These values apply only when all five observed points are taken.

TABLE 40—*Accuracies attainable using comparison techniques in special furnaces (optical pyrometer standard).*

Type	Temperature Range, deg C	Calibration Uncertainty	
		At Observed Points, deg C	Of Interpolated Values,[a] deg C
IrRh versus Ir[b]	1000 to 1300	2	3
IrRh versus Ir[b]	1300 to 1600	3	4
IrRh versus Ir[b]	1600 to 2000	5	8
W versus WRe[c]	1000 to 1300	2	3
W versus WRe[c]	1300 to 1600	3	4
W versus WRe[c]	1600 to 2000	5	8
30 versus 6[d]	1000 to 1550	2	3
30 versus 6[d]	1500 to 1750	3	5

[a] Using difference curve from reference table with calibration points spaced every 200 C.
[b] 40Ir60Rh versus Ir, 50Ir50Rh versus Ir, or 60Ir40Rh versus Ir.
[c] W versus 74W26Re, 97W3Re versus 75W25Re, or 95W5Re versus 74W26Re.
[d] 70Pt30Rh versus 94Pt6Rh.

When an optical pyrometer is used as the temperature measuring standard, a good blackbody must be used, and the design must be such that the test thermocouple is at the same temperature as the blackbody.

8.2 Calibration Using Fixed Points

The indications of the Type S thermocouple calibrated at 630.5 C and the silver and gold points, as mentioned in Section 8.1.1, define the IPTS between 630.5 C (1167 F) and 1063 C (1945 F). If such a thermocouple is calibrated also at the zinc point, a working standard will result which is accurate to about 0.3 C in the range 0 to 1100 C (32 to 2012 F). While the fixed-point calibration is prescribed for defining the IPTS, similar methods are also useful in the calibration of other types of thermocouples. Fixed points can be used with various degrees of accuracy, ranging from 0.1 to 5 C, for the calibration of various types of thermocouples in the range -253 C (-423 F) to the melting point of platinum at 1769 C (3216 F). Some of the fixed points for which values have been determined accurately are listed in Section 8.1.1, Table 35. Because of experimental difficulties, fixed points at temperatures higher than the freezing point of copper usually are realized as melting points rather than freezing points, as described later.

8.2.1. Freezing Points

The emf developed by a homogeneous thermocouple at the freezing point of a metal is constant and reproducible if all of the following conditions are fulfilled: (1) the thermocouple is protected from contamination; (2) the thermocouple is immersed in the freezing-point sample sufficiently far to eliminate heating or cooling of the junction by heat flow along the wires and protection tube; (3) the reference junctions are maintained at a constant and reproducible temperature; (4) the freezing-point sample is pure; and (5) the metal is maintained at essentially a uniform temperature during freezing.

Techniques for achieving these conditions are well developed [63, 65, 66]. Many of the metals listed in Table 35 of Section 8.1.1 are available commercially in high purity (99.999 percent or better) and can be used assuming the freezing point temperatures given in the table. It is essential, however, that protection tubes and crucibles be chosen of such material (see Section 4) that the pure metals will not be contaminated. Copper and silver must be protected from oxygen contamination [63], and it is also advisable to protect aluminum and antimony; this is done by using covered crucibles and covering the freezing point metals with powdered graphite. The choice of a suitable furnace is also important. The furnace must provide uniform heating in the region of the freezing point sample, and have adequate controls to bring the sample slowly into its freeze. Complete units consisting of freezing point sample, crucible, and furnace are available commercially. Freezing point standard samples of tin, lead, zinc, aluminum, and copper may be purchased from the National Bureau of Standards.

8.2.2. Melting Points

The emf of a thermocouple at the melting point of a metal may be determined with the same apparatus as that described above for freezing points, but the use of the freeze is usually more satisfactory. Melting points are used to advantage, however, when only a limited amount of material is available or at high temperatures where experimental techniques with freezing points are difficult. To apply this method [67, 68, 69, 70], a short length of metal whose melting point is known is joined between the end of the two wires of the thermocouple and placed in an electrically heated furnace the temperature of which is raised slowly. When the melting point of the metal is reached, the emf of the thermocouple remains steady for a few minutes and then drops to zero as the fused metal drops away from the junction. With good technique[11] the method can give results with no greater uncertainty than that with which the IPTS is realized above 1063 C (1946 F) by optical pyrometry.

8.3 Calibration Using Comparison Methods

The calibration of a thermocouple by comparison with a working standard [71] is sufficiently accurate for most purposes and can be done conveniently in most industrial and technical laboratories. The success of this method usually depends upon the ability of the observer to bring the measuring junction of the thermocouple to the same temperature as the actuating element of the standard, such as the measuring junction of a standard thermocouple or the bulb of a resistance or liquid-in-glass thermometer. The accuracy obtained is further limited by the accuracy of the standard. Of course, the reference junction temperature must be known, but this can be controlled, as described in Section 7. The method of bringing the measuring junction of the thermocouple to the same temperature as that of the actuating element of the standard depends upon the type of thermocouple, type of standard, and the method of heating.

8.3.1 *Laboratory Furnaces*

The calibration procedure consists of measuring the emf of the thermocouple being calibrated at selected calibration points, the temperature of each point being measured with a working standard. The number and choice of calibration points will depend on the type of thermocouple, the temperature range covered, and the accuracy required (see Sections 8.1.6 and 8.4).

8.3.1.1. *Platinum-Rhodium Versus Platinum Thermocouples*—Thermocouples employing platinum and platinum rhodium alloys seldom are used for accurate measurements below 300 C (572 F) because the sensitivity of these thermocouples decreases rapidly at low temperatures.

These thermocouples usually are calibrated at temperatures up to 1200 C by comparison with either a Type S or Type R working standard in electrically heated furnaces. Above 1200 C (2192 F) the Type B thermocouple is a preferred working standard because of its greater stability at high temperatures. This thermocouple may be used to 1600 C (2912 F) or higher.

One method for the comparison of two such thermocouples is based upon the simultaneous reading of the emf of the standard and the test

---

[11] This method is not well adapted to metals that oxidize rapidly, and, if used with materials whose melting temperature is altered by the oxide, the metal should be melted in a neutral atmosphere.

thermocouple without waiting for the furnace to stabilize at any given temperature. The measuring junctions are maintained always at close to the same temperature by welding them into a common bead or by wrapping them together with platinum wire or ribbon. A separate potentiometer is used to measure each emf, one connected to each thermocouple, and each potentiometer is provided with a reflecting galvanometer. The two spots of light are reflected into a single scale, the galvanometers being set in such a position that the spots coincide at the zero point on the scale when the circuits are open, and therefore also when the potentiometers are set to balance the emf of each thermocouple. Simultaneous readings are obtained by setting one potentiometer to a desired value and adjusting the other so that both spots of light pass across the zero of the scale together as the temperature of the furnace is raised or lowered.

By making observations first with a rising and then with a falling temperature, the rates of rise and fall being approximately equal, and taking the mean of the results found, several minor errors such as those due to differences in the periods of the galvanometers, etc., are eliminated or greatly reduced.

This method is particularly adapted to the calibration of thermocouples at any number of selected points. For example, if it is desired to determine the temperature of a thermocouple corresponding to 10.0 mV, this emf is set up on the potentiometer connected to the thermocouple, the emf of the standard thermocouple observed as desired above, and the temperature obtained from the emf of the standard. If it is desired to determine the emf of a thermocouple corresponding to 1000 C (1832 F), the emf of the standard corresponding to this temperature is set up on the potentiometer connected to the standard, and the emf of the thermocouple being calibrated is observed directly with the second potentiometer. To reduce the time required to calibrate by this method the furnace should be so constructed that it will heat or cool rapidly. Fast response is obtained in one furnace design which employs a nickel-chromium tube as the heating element [63].

A similar furnace using a silicon carbide tube as the heating element can be used to extend the calibration range upward [58]. At temperatures above 1063 C (1945 F) the IPTS is defined in terms of ratios of radiation (Section 8.1.1) usually measured with an optical or a photoelectric pyrometer. If the test thermocouple is inserted into the back of a blackbody cavity built into the furnace, a pyrometer may be used directly as the working standard. Alternatively, the Type B thermocouple can be used as the working standard after it has been calibrated against a pyrometer.

The thermocouples are insulated and protected by suitable ceramic tubes (Section 4). It is essential that good insulation be maintained between the two potentiometers and thermocouple circuits except at the point where the junctions are welded together. The reference junctions are maintained at a known temperature (Section 7).

Variations of the two potentiometer method may be used to automate the calibration process when the thermocouple being calibrated and the standard thermocouple are of the same type [71]. If the emf of the standard is read with one potentiometer and the emf difference between the standard and the unknown are read with the second potentiometer the calibration data may be recorded automatically [72, 73].

If two potentiometers are not available for taking simultaneous readings, the furnace may be brought to essentially a constant temperature, and the emf of each thermocouple read alternately on one instrument [71].

When the thermocouples are calibrated by welding or wrapping the junctions together, both would be expected to be close to the same temperature even when the temperature of the furnace is changing. If it is necessary or advisable to calibrate the thermocouples without removing them from the protection tubes, then the junctions of the thermocouple being tested and that of the standard should be brought as close together as possible in a uniformly heated portion of the furnace. In this case it is necessary that the furnace be brought to approximately a constant temperature before taking observations.

There are a number of other methods of heating and of bringing the junctions to approximately the same temperature, for example, inserting the thermocouples properly protected into a bath of molten metal or into holes drilled in a large metal block. The block of metal may be heated in a muffle furnace or, if made of a good thermal conductor such as copper, may be heated electrically. Tin, which has a low melting point, 232 C (450 F), and low volatility, makes a satisfactory bath material. The thermocouples should be immersed to the same depth with the junctions close together. Ceramic tubes are sufficient protection, but to avoid breakage by thermal shock when immersed in molten metal it is preferable to place them inside of secondary tubes of iron, nickel-chromium, graphite, or similar material. In all of these methods, particularly in those cases in which the junctions of the thermocouples are not brought into direct contact, it is important that the depth of immersion be sufficient to eliminate cooling or heating of the junctions by heat flow along the thermocouple and the insulating and protecting tubes. This can be determined by observing the change in the emf of the thermocouple as the depth of immersion is changed slightly. If proper precautions are taken, the accuracy yielded by any method of heating or bringing the junctions to the same temperature may be as great as that obtained by any other method.

8.3.1.2 *Base-Metal Thermocouples*—The methods of testing base-metal thermocouples above room temperature are generally the same as those just described for testing noble-metal thermocouples with the exception, in some cases, of the methods of bringing the junctions of the standard and the thermocouple being tested to the same temperature and the methods of protecting platinum-rhodium standards from contamination. One arrangement of bringing the junction of a platinum-rhodium standard to the same temperature as that of a large base-metal thermocouple for accurate calibration is to insert the junction of the standard into a small hole (about 0.06 in. in diameter) drilled in the hot junction of the base-metal thermocouple. The platinum-rhodium standard is protected by ceramic tubes to within a few hundredths of an inch of the hot junction, and the end of the ceramic tube is sealed to the thermocouple by pyrex glass or by a small amount of kaolin and water-glass cement. This prevents contamination of the standard thermocouple, with the exception of the small length of about 0.1 in., which is necessarily in contact with the base-metal thermocouple. If the furnace is uniformly heated in this region (and it is of little value to make such a test unless it is) contamination at this point will not cause any error. If the wire of the standard becomes brittle at the junction, this part of the wire may be cut off and enough wire drawn through the softened seal to form a new junction. The seal should be examined after each test and remade if it does not appear to be good. More than one base-metal thermocouple may be welded together and the hole drilled in the composite junction. The thermocouples should be clamped in place so that the junctions remain in contact. If two potentiometers are used for taking simultaneous readings, the temperature of the furnace may be changing as much as a few degrees per minute during an observation, but if a single instrument is used for measuring the emf, the furnace temperature should be maintained practically constant during observations. When wires, insulators, and protection tubes are large, tests should be made to ensure that the depth of immersion is sufficient.

8.3.2 *Stirred Liquid Baths*

At temperatures below 620 C (1148 F) stirred liquid baths provide an efficient medium for bringing a thermocouple and a temperature standard to the same temperature.

Water, petroleum oils, or other organic liquids, depending upon temperature range, are commonly used bath media. Molten salts or liquid tin are used at temperatures higher than are suitable for oil. Baths suitable for this work are described in Ref 74.

Base-metal thermocouples, either bare wire or insulated, may be calibrated accurately in such baths. Usually no special preparation of the thermocouple will be required other than to insert it to the bottom of a protection tube for immersion in the liquid bath. Borosilicate glass tubing, such as pyrex glass, is convenient for use up to 538 C (1000 F). Vitreous silica or ceramic tubing may be used to 620 C (1148 F). The tube should be closed at the immersed end and of an internal diameter such as to permit easy insertion of the thermocouple or thermocouples to be calibrated but no larger than necessary. Unfavorable heat transfer conditions in an unnecessarily large diameter tube will require a greater depth of immersion in the bath than would a close fitting tube. If a bare wire thermocouple is being calibrated, the wires must be provided with electrical insulation over the length inserted in the protection tube. Sheathed thermocouples may be immersed directly in the bath liquid in cases where the sheath material will not be attacked by the liquid. Salt baths for use at high temperature must be provided with suitable wells into which the thermocouple protection tubes and standard thermometers may be inserted for protection from the molten salt.

The standard thermometer may be a thermocouple standard inserted in the protection tube with the thermocouple being calibrated, or it may be a liquid-in-glass thermometer or resistance thermometer immersed in the bath close to the thermocouple protection tube. The choice of a standard thermometer will be governed principally by the degree of uncertainty which can be tolerated.

8.3.3 *Fixed Installations*

After thermocouples have been used for some time at high temperatures, it is difficult, if not impossible, to determine how much the calibrations are in error by removing them from an installation and testing in a laboratory furnace. The thermocouples are usually heterogeneous after such use and in such a condition that the emf developed by the thermocouples depends upon the temperature distribution along the wires [75]. If possible, such a thermocouple should be tested under the same conditions and in the same installation in which it is used. Although it is not usually possible to obtain as high a precision by testing the thermocouple in place as is obtained in laboratory tests, the result is far more useful in the sense of being representative of the behavior of the thermocouple [70]. The calibration is accomplished by comparing the thermocouple with a thermocouple standard.

In this case, as in the calibration of any thermocouple by comparison methods, the main objective is to bring the hot junction to the same temperature as that of the thermocouple being tested. One method is to drill a hole in the furnace, flue, etc., at the side of each thermocouple permanently installed, large enough to permit insertion of the checking thermocouples. The hole is kept plugged, except when tests are being made. The standard thermocouple is inserted through this hole to the same depth as the thermocouple being tested with the hot junction ends of the protection tubes as close together as possible. Preferably a potentiometer should be used with the standard thermocouple.

In many installations the base-metal thermocouple and protecting tube are mounted inside another protecting tube of iron, fire clay, carborundum, or some other refractory which is permanently cemented or fastened into the furnace wall. Frequently there is room to insert a small test thermocouple in this outer tube alongside of the fixed thermocouple. A third method, much less satisfactory, is to wait until the furnace, flue, etc., have reached a constant temperature and make observations with the thermocouple being tested, then remove this thermocouple and insert the standard thermocouple to the same depth.

If desired, comparisons can be made, preferably by either of the first or second methods at several temperatures, and a curve obtained for each permanently installed thermocouple showing the necessary corrections to be applied to its readings. Although testing a thermocouple at one temperature yields some information, it is not safe to assume that the changes in the emf of the thermocouple are proportional to the temperature or to the emf. For example, it has been observed that a thermocouple which had changed in use by the equivalent of 9 C at 315 C (16 F at 599 F) had changed only the equivalent of 6 C at 1100 C (11 F at 2012 F).

It may be thought that the method of checking thermocouples under working conditions is unsatisfactory because, in most furnaces used in industrial processes, large temperature gradients exist, and there is no certainty that the standard thermocouple is at the same temperature as the thermocouple being tested. This objection, however, is not serious, because if temperature gradients do exist of such a magnitude as to cause much difference in temperature between two similarly mounted thermocouples located close together, the reading of the standard thermocouple represents the temperature of the fixed thermocouple as closely as the temperature of the latter represents that of the furnace.

Another advantage of checking thermocouples in the same installation in which they are used is that the thermocouple, extension wires, and indicator are tested as a unit and under the conditions of use.

8.4 Interpolation Methods

An experimental thermocouple calibration consists of a series of voltage measurements determined at a finite number of known temperatures. If a test thermocouple were compared with a standard temperature instrument at 100 temperatures within a 10 F range, there would be little need for interpolation between the calibration points. However, if from 4 to 10 calibration points are all that can be afforded in a given range of interest, then what is needed to characterize an individual thermocouple is a continuous relation, by means of which temperatures can be approximated with a minimum uncertainty from voltage measurements at intermediate levels. Efforts to obtain such a continuous relation appear thwarted from the start because of the small number of discrete calibration points available. However, interpolation between the calibration points is possible since the emf changes only slowly and smoothly with temperature.

One can present raw calibration data directly in terms of temperature (T) and voltage (E_{couple}), on a scale so chosen that the information appears well represented by a single curve (see Fig. 34) or by a simple mathematical equation. For example, for the highest accuracy in the range 630.5 to

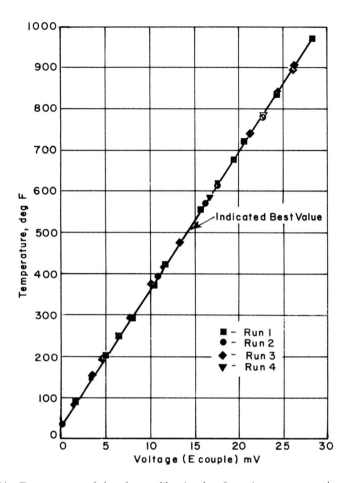

FIG. 34—*Temperature emf plot of raw calibration data for an iron-constantan thermocouple.*

1063.0 C with the Type S thermocouple, the method is that prescribed in Ref *63*, Page 13. An equation of the form $e = a + bt + ct^2$, is used where a, b, and c are constants determined by calibration at the freezing points of gold, silver, and antimony. By calibrating the thermocouple also at the freezing point of zinc and using an equation of the form $e = a + bt + ct^2 + dt^3$, the temperature range can be extended down to 400 C without introducing an uncertainty of more than 0.1 C in the range 630.5 to 1063.0 C. By calibrating the thermocouple at freezing points of gold, antimony, and zinc and using an equation of the form $e = a + bt + ct^2$, a calibration is obtained for the range 400 to 1100 C, which agrees with IPTS to 0.5 C. However, in general, this practice of directly representing thermocouple characteristics does not yield results within the required limits of uncertainty.

A better method [12] is based on the use of differences between observed values and values obtained from standard reference tables. Such reference tables and the mathematical means for generating them are presented in Section 10 of this Manual. The data of Fig. 34 are replotted in Fig. 35 in terms of differences from the proper reference table. The maximum spread between points taken at the same level (replication), but obtained in random order with respect to time and level (randomization) is taken as the uncertainty envelope. This information, taken from Fig. 35 is plotted in Fig. 36, and constitutes a vital bit of information about the particular thermocouple and the calibration system. In lieu of an experimental determination of the uncertainty, one must rely on judgment or on the current literature for this information.

Usually, only a single set of calibration points is available. Typical points would be those taken from one run shown in Figs. 34 or 35, and these are shown in Fig. 37 together with four of the many possible methods for representing the thermocouple difference characteristic. Although at first it appears that the most probable relation characterizing a given thermocouple is sensibly indeterminate from a single set of calibration points, it is an important fact that all experimental points must be continued within the uncertainty interval when the uncertainty interval is centered on the most probable interpolation equation.

Making use of this principle, together with the fact that overall experimental uncertainties are minimized by use of the least squares technique, one starts the search for the most probable interpolation equation by passing a least squares equation of the first degree through the experimental data. A check is then made to ascertain whether all experimental points are contained within the uncertainty envelope which is centered on the linear interpolation equation (see Fig. 38). One proceeds, according to the results of the foregoing check, to the next highest degree equation, stopping at the lowest degree least squares equation which satisfies the uncertainty requirements. For the example given here, a third degree interpolation equation is required (see Fig. 39). By obtaining voltage differences from the least squares fit of any set of calibration points, the uncertainty in the thermocouple difference characteristic will be within one half the uncertainty interval. Generally, the form of the uncertainty envelope and the degree of the most probable least squares interpolation equation are strongly dependent on the amount of calibration data available and on the temperature range under consideration. It is recommended that the number of distinct calibration points available should be at least 2 (degree $+1$). The factor two is arrived at from numerical analysis reasoning. A distinct calibration point is defined arbitrarily as one which is separated, temperaturewise, from all other points in the set by as much as one tenth the difference in temperature between the maximum and minimum temperatures

[12] Much of the material in this Section is based on Ref 76.

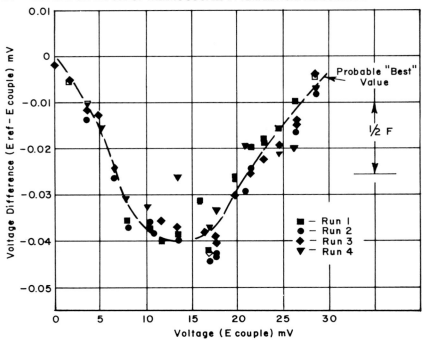

FIG. 35—*Difference plot of raw calibration data for an iron-constantan thermocouple.*

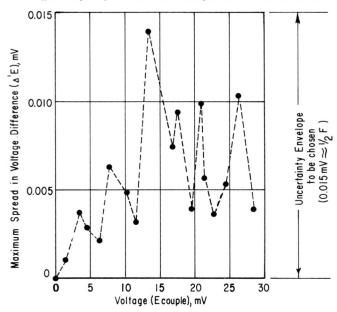

FIG. 36—*Typical determination of uncertainty envelope (from data of Fig. 35).*

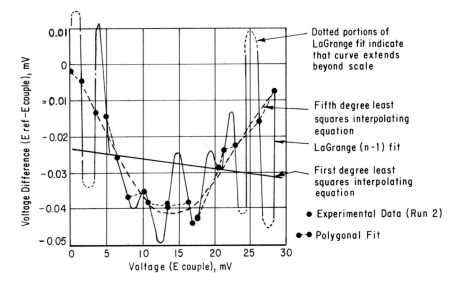

FIG. 37—*Various possible empirical representations of the thermocouple characteristic (based on a single calibration run).*

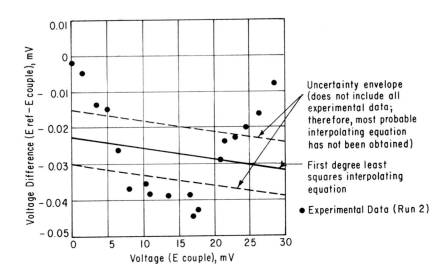

FIG. 38—*Uncertainty envelope method for determining degree of least squares interpolating equation for a single calibration run (linear).*

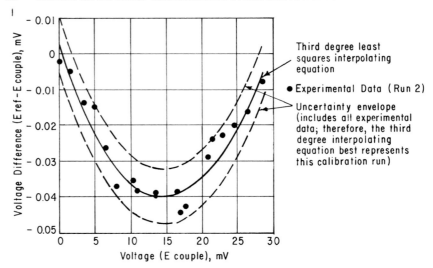

FIG. 39—*Uncertainty envelope method for determining degree of least squares interpolating equation for a single calibration run (cubic).*

of the particular run. The choice of one tenth presupposes a maximum practical degree of four for the least squares interpolation equation, in keeping with the low degree requirement of numerical analysis. Indeed, if the data cannot be represented by a fourth degree interpolation equation, one should increase the uncertainty interval and start the fitting procedure again.

Thus, in general, by using the proper reference table in conjunction with a difference curve, greater precision in temperature determination by means of thermocouples can be obtained from a given number of calibration points than from the use of the calibration data alone.

8.5 Single Thermoelement Materials

The standard method provided by ASTM for evaluating the emf characteristics of single thermoelement materials used in thermocouples for temperature measurement is designated as E 207 and is entitled "Standard Method of Thermal Emf Test of Single Thermoelement Materials by Comparison with a Secondary Standard of Similar Emf-Temperature Properties." The method covers the determination of the thermoelectromotive force of single thermoelement materials (thermoelements), against standard platinum, the cold junction being at the ice point, by comparison to the thermoelectromotive force of a working standard thermoelement of similar emf-temperature properties independently standardized with respect to the same standard platinum.

Summary of Method—The test thermoelements are welded to the working standard to form the test thermocouple. The method involves measuring the small electromotive force developed between the test thermoelement and the secondary working standard having emf-temperature properties similar to the thermoelement being evaluated. The thermoelectromotive force of the test thermoelement then is determined by algebraically adding this measured small emf to the known emf of the secondary working standard referenced to the standard platinum. The testing circuit is shown schematically in Fig. 40. Since the thermal emfs (against any reference material) of the test and standard thermoelements are similar, it is unnecessary to accurately control or measure the junction temperatures because the difference in emf changes insignificantly even for large changes in temperature at the junctions. Actually, the need for the accurate control of the measuring and reference junction temperatures is not eliminated, but merely is shifted to an accurate laboratory method which calibrates the secondary standard against the standard platinum [77].

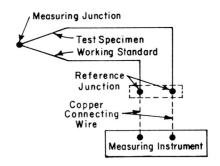

FIG. 40—*Circuit diagram for thermal emf test.*

8.5.1 *Test Specimen*

The test specimen is a length of wire, rod, ribbon, or strip of the coil or spool of the thermoelement material to be evaluated. The length is adequate to prevent the transfer of heat from the measuring junction to the reference junction during the period of test. A length of 2 to 4 ft, depending on the length of the testing medium and the transverse size of the thermoelement, is generally satisfactory. The transverse size of the specimen is limited only by the size of the test medium, the relative convenience of handling the specimens and working standard, and the maintenance of an isothermal test temperature junction.

8.5.2 *Working Standard*

The working standard is a thermoelement of emf-temperature properties similar to that of the test specimen, and which previously has been standardized thermoelectrically with respect to National Bureau of Standards Platinum No. 27. If a large amount of testing is anticipated, a coil or spool of working standard material is reserved. This working standard lot is selected on the basis of uniformity, and a minimum of three samples taken from the center and the ends of the lot is calibrated against the standard platinum. The working standard and the test specimen may differ in diameter, but it is convenient for their lengths to be about equal.

8.5.3 Reference Junction

The reference junction temperature of the test specimen and of the working standard is controlled during the period of test by maintaining it at the ice point (0 C). The ice-water junction is used because it is recognized as a convenient means for maintaining a constant temperature reference. The description and maintenance of the ice point are given in Section 7. A reference temperature other than the ice point may be used. However, the working standard should be calibrated against standard platinum using the alternate reference temperature, or the calibration data adjusted to correspond to the alternate reference temperature.

8.5.4 Measuring Junction

The measuring junction consists of a welded union of the test specimen and working standard. The weldment may be prepared by any method providing a good electrical connection which can be immersed fully in the uniform temperature region of the testing medium. Any number of test specimens and working standards may be welded together provided the resulting assembly does not introduce heat losses which prevent maintaining a uniform temperature region. If separation of the working standard and test specimen cannot be maintained during the test (except where they must make contact at the measuring junction), it is necessary that they be insulated from each other.

8.5.5 Test Temperature Medium

For temperatures up to 550 F (288 C), appropriate liquid baths may be used. For temperatures above 300 F (149 C), electrically heated tube or muffle type furnaces are recommended for comparison testing of base metal or noble metal thermoelement materials of similar emf-temperature properties. For convenience, a separate furnace may be controlled and available for each test temperature. This eliminates lost time in changing furnace temperatures when a large volume of testing is to be done. The length of each furnace is at least 15 in. so as to provide a minimum depth of immersion of 7 in. for the test thermocouple assembly. A constant immersion depth is maintained, whether single or multiple furnaces are used. The inside diameter of the furnace tube is approximately 1 to 3 in., the specific diameter depending upon the size and the number of the specimens to be included in the test thermocouple assembly. The furnace provides a uniform temperature zone extending at least 3 in. back from the measuring junction, or further if required to contain any inhomogeneity in the test thermocouple assembly. The temperature of each furnace is controlled manually or automatically to within ±20 F (±11 C) of the desired value which is ample for comparison testing of thermoelements having compositions similar to the secondary working standard.

8.5.6 Emf Indicator

The emf generated by the thermocouple, consisting of the test specimen and the working standard, is measured with instrumentation sensitive to ±0.001 mV with an accuracy over a 2 mV range of ±1 percent of the reading plus 0.003 mV. Any millivoltmeter, with circuitry errors taken into account, or potentiometer with a galvanometer or null indicator, providing measurements within these tolerances is acceptable. An indicator with a bidirectional scale (zero center) is convenient, but a unidirectional instrument may be used if a polarity switch is provided in the copper connecting circuit, or if the copper extensions to the instrument are exchanged whenever the polarity between the working standard and test specimen is reversed.

8.5.7 Procedure

For a furnace medium the test thermocouple assembly is inserted into the furnace so that the measuring junction extends at least 3 in. into the uniform temperature zone taking care there is no contact between the wires and the furnace wall. The free ends of the working standard and the test specimens are bent as required so they may be inserted into the mercury-containing glass tubes of the reference junction. If a horizontal furnace is used, the joined thermocouple assembly may be bent midway or towards the free ends in the form of an inverted "L." Care is exercised to minimize distorting the wires prior to testing because of the effect of cold work on emf output. After bringing the test temperature to the specified value, sufficient time is provided for the test assembly to reach steady state conditions before recording the emf generated between the test specimen and the working standard.

In a similar manner the emf generated between all other test specimens in the assembly is measured with respect to the working standard. Then the test temperature is raised to the next higher specified value, or the test assembly is advanced to the next furnace or bath having the next higher specified temperature if multiple furnaces or baths are used. A second set of readings is taken at the new temperature, and the procedure is repeated with readings taken at all specified temperatures. In all cases the readings are taken in sequence from the lowest to the highest temperature to minimize test variations between producer and consumer if any of the alloys are affected by differences in short time heating cycles. A working standard is used for one series of temperature changes only. However, if a portion considerably exceeding the region previously exposed to the uniform heating zone is discarded, the remainder of a base metal standard may be used for another test assembly. For noble metals and their alloys, the amount of reuse depends on the known stability of the material involved.

The polarity of the test thermoelement with respect to the working standard is determined as follows:

(1) If the test specimen is connected to the positive (+) terminal of an unidirectional potentiometer and balance can be achieved, the specimen is positive to the working standard.

(2) If the connections must be reversed to achieve balance, that is, the working standard must be connected to the positive terminal, the specimen is negative to the working standard.

(3) If an indicating potentiometer with a bidirectional scale is used, the specimen to the positive (+) terminal and the working standard to the negative (−) terminal are connected. The polarity of the specimen with respect to the working standard then will be indicated by the direction of balance of the instrument scale.

The emf of the test specimen with respect to Pt-27 is then reported for each test temperature after algebraically adding the measured emf between the test specimen and the standard, and the known emf between the standard and Pt-27.

8.6 References and Bibliography

References

[52] Stimson, H. F., "International Practical Temperature Scale of 1948, Text Revision of 1960," NBS Monograph 37, National Bureau of Standards, 1961.
[53] "The International Practical Temperature Scale of 1968," *Metrologia,* Vol. 5, No. 2, April 1969, pp. 34–49.
[54] Benedict, R. P., *Technical Journal,* Leeds & Northrup Co., Vol. 6, Summer Issue, 1969.
[55] Dengler, C. O., "A New Mercury Freezing Point Cell," *Proceedings,* International Standards Association, Vol. 17, Part 1, Paper No. 19.4.62.
[56] ASTM Method E 1-65, "Standard Specifications for ASTM Thermometers," *1966 Book of ASTM Standards,* Part 30, p. 73.
[57] Kostkowski, H. J. and Lee, R. D., "Theory and Methods of Optical Pyrometry," NBS Monograph 41, National Bureau of Standards, 1962.
[58] Burns, G. W. and Gallagher, J. S., "Reference Tables for the Pt-30% Rh Versus Pt-6% Rh Thermocouple," *Journal of Research,* National Bureau of Standards, Vol. 70C, April–June 1966.
[59] Corruccini, R. J., "Annealing of Platinum for Thermometry," *Journal of Research,* National Bureau of Standards, Vol. 47, No. 94, 1951, RP2232.
[60] Diesselhorst, H., "Thermokraftfreier Kompensationapparat mit funf Dekadeu und kustanter kleinem Widerstand," *Zeitschrift füs Instrumentenkunde,* Vol. 28, No. 1, 1908.
[61] White, W. P., "Thermokraftfreie Kompensationapparat e mit kleinem Widerstand und kustanter Galvanometereempfindlicheit," *Zeitschrift füs Instrumentenkunde,* Vol. 27, No. 210, 1907.
[62] Behr, L., "The Wenner Potentiometer," *Review of Scientific Instruments,* Vol. 3, No. 108, 1932.
[63] Roeser, W. F. and Lomberger, S. T., "Methods of Testing Thermocouple Materials," NBS Circular 590 National Bureau of Standards, 1958.
[64] Finch, D. I., "General Principles of Thermoelectric Thermometry," *Temperature, Its Measurement and Control in Science and Industry,* Vol. 3, Part 2, Reinhold, New York, 1962, p. 3.

[65] McLaren, E. H., "The Freezing Points of High Purity Metals as Precision Temperature Standards," *Temperature, Its Measurement and Control in Science and Industry,* Vol. 3, Part 1, Reinhold, New York, 1962, p. 185.

[66] Trabold, W. G., *Temperature, Its Measurement and Control in Science and Industry,* Vol. 3, Part 2, Reinhold, New York, p. 45.

[67] Fairchild, C. O., Hoover, W. H., and Peters, M. F., "A New Determination of the Melting Point of Palladium," *Journal of Research,* National Bureau of Standards, Vol. 2, No. 931, 1929, RP 65.

[68] Barber, C. R., "The Calibration of the Platinum/13% Rhodium-Platinum Thermocouple Over the Liquid Steel Temperature Range," *Journal of the Iron and Steel Institute,* Vol. 147, No. 205, 1943.

[69] Bedford, R. E., "Reference Tables for Platinum-20% Rhodium/Platinum-5% Rhodium Thermocouples," *Review of Scientific Instruments,* Vol. 35, No. 1177, 1964.

[70] Quigley, H. C., "Resume of Thermocouple Checking Procedures," *Instruments,* Vol. 25, No. 616, 1952.

[71] ASTM Method E 220, Standard Method for Comparison Calibration of Thermocouples," *1965 Book of ASTM Standards,* Part 30, p. 634.

[72] "Automatic Thermocouple Comparator," *Technical News Bulletin,* National Bureau of Standards, Vol. 47, No. 5, May 1963, p. 82.

[73] Toenshoff, D. A., "Automatic Calibration of Thermocouples," *Technical Bulletin,* Englehard Industries, Vol. 2, No. 3, Dec. 1961, p. 88.

[74] ASTM Method E 77-64, "Standard Method for Inspection, Test, and Standardization of Etched-Stem Liquid-in-Glass Thermometers," *1965 Book of ASTM Standards,* Part 30, p. 290.

[75] Dahl, A. I., "Stability of Base-Metal Thermocouples in Air from 800 to 2200 F," *Journal of Research,* National Bureau of Standards, Vol. 24, No. 205, 1940, RP 1278.

[76] Benedict, R. P. and Ashby, H. F., "Empirical Determination of Thermocouple Characteristics," *Transactions,* American Society of Mechanical Engineers for Power, Jan. 1963, p. 9.

[77] ASTM Method E 220-64, "Standard Method for Comparison Calibration of Thermocouples," *1965 Book of ASTM Standards,* Part 30, p. 634.

Bibliography

Temperature, Its Measurement and Control in Science and Industry, Vol. 1, Reinhold, New York, 1941, Vol. 3, Parts 1 and 2, 1962.

ASME Power Test Codes, Instruments and Apparatus Supplement, Part 3, Temperature Measurement, American Society of Mechanical Engineers, New York, 1961.

Dike, P. H., "Thermoelectric Thermometry," Leeds and Northrup Co., Philadelphia, 1954.

Benedict, R. P., *Fundamentals of Temperature, Pressure, and Flow Measurement,* Wiley, New York, 1969.

9. INSTALLATION EFFECTS

9.1 Temperature Measurement in Fluids

Fluids are divided readily into two types, compressible and incompressible, or more simply into gases and liquids. However, many concepts involved in the measurement of temperatures in fluids are common to both types, and these are discussed first.

9.1.1 *Response*

No instrument responds instantly to a change in its environment. Thus in a region where temperature is changing, a thermocouple will not be at the temperature of its environment and hence cannot indicate the true temperature. The simplified temperature changes considered here are the step change and the ramp change. In the step. change, the temperature of the environment shifts instantaneously from T_1 to T_2. In the ramp change, the environment temperature shifts linearly with time from T_1 to T_2.

It is common practice to characterize the response of a temperature sensor by a first order thermal time constant τ which is defined as:

$$\tau = \frac{w\mathbf{V}c}{hA} \qquad \dots \dots \dots \dots \dots \dots \dots \quad (41)$$

where w is specific weight, $\mathbf{V}$ is volume, and c is specific heat, all of the sensor; while h is the heat transfer coefficient, and A is the area of the fluid film surrounding the sensor.

A solution of the first order, first degree, linear, differential equation [78, 79] resulting from a heat balance between the fluid film surrounding the sensor and the sensor itself is

$$T = Ce^{-t/\tau} + \frac{1}{\tau} e^{-t/\tau} \int_0^t T_e e^{t'/\tau} dt \qquad \dots \dots \dots \dots \quad (42)$$

where T is the sensor temperature, and T_e is the environment temperature, both at time t, and C is a constant of integration.

For a ramp change in temperature (as is found in a furnace being heated at a uniform rate) Eq 42 reduces to

$$(T_e - T) = R\tau \qquad \dots \dots \dots \dots \dots \dots \quad (43)$$

Equation 43 states that if an element is immersed for a long time in an environment whose temperature is rising at a constant rate $R = dT_e/dt$, then τ is the interval between the time when the environment reaches a given temperature and the time when the element indicates this temperature.

For a step change in temperature (as when a thermocouple is plunged into a constant temperature bath), Eq 42 reduces to

$$(T_e - T) = (T_e - T_1)e^{-t/\tau} \qquad \dots \dots \dots \dots \quad (44)$$

Equation 44 states that if an element is plunged into a constant-temperature environment, τ is the time required for the temperature difference between the environment and the element to be reduced to $1/e$ of the initial difference. Note that for practical purposes the sensor will reach the new temperature after approximately 5 time constants. See Fig. 41 for a graphical presentation of these equations.

Below a Mach number of 0.4, the time constant hardly is affected by the fluid velocity. The size of the temperature change affects τ because physical properties are not necessarily linear functions of temperature. Wormser [80] considers these effects in greater detail. Scadron and Warshawsky

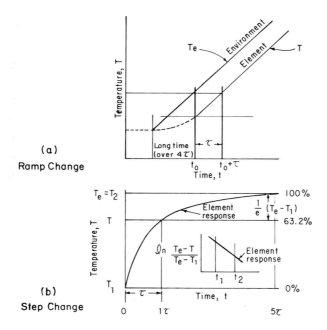

FIG. 41—*Graphical presentation of ramp and step changes.*

[81] present convenient nomographs for determining the time constant in the presence of heat transfer. Fluid turbulence tends to reduce the time constant by increasing the film coefficient.

If the sensor is contained in a thermowell the response time is increased because of the extra mass and the additional heat transfer coefficient involved. It may be necessary to consider the response as a second order function in which case a dead time is observed before the sensor responds to a step temperature change. However, Coon and Looney [82, 83] state that the time constant usually is represented adequately by the first order time constant τ.

To achieve the best time constant, the measuring junction should be in intimate contact with the well tip. Spring loading is sometimes used to accomplish this.

The response time usually is measured in liquids by plunging the thermocouple assembly into a well-stirred bath held at constant temperature. If the response time is of the same order as the immersion time, the velocity of immersion or depth of immersion or both become important parameters in the measurement. Reference *84* indicates a method for standardizing the liquid baths used in this determination.

For the response time in gases, the literature should be consulted [85, 86]. Manufacturers sometimes can supply this information for simple conditions.

Multijunctions [87] and electrical networks [88] have been used success-fully to improve the response of a thermocouple.

9.1.2 Recovery

Whenever a gas moves with an appreciable velocity, in addition to the thermal energy in the form of random translational kinetic energy of the molecules, some of its thermal energy is in the form of directed kinetic energy of fluid flow. The static temperature is a measure of the random kinetic energy, while the dynamic temperature is a measure of the directed kinetic energy. The total temperature is a concept (not a measurement) which sums the static and the dynamic temperatures. Such a total tempera-ture would be sensed by an adiabatic probe which completely stagnates an ideal gas. Thus

$$T_{adi,ideal} = T_t = T + \frac{V^2}{2Jgc_p} = T + T_V \qquad \ldots \ldots \ldots \ldots (45)$$

where:
- T_t = total temperature,
- T = static temperature
- T_V = dynamic temperature,
- V = directed fluid velocity,
- J = mechanical equivalent of heat,
- g = acceleration of gravity, and
- c_p = specific heat at constant pressure.

In real fluids, an adiabatic recovery factor (α) generally is defined such that

$$T_{adi} = T + \alpha T_V \qquad \ldots \ldots \ldots \ldots \ldots \ldots \ldots \ldots (46)$$

where α may be more or less than one depending on the relative importance of thermal conductance and thermal capacitance in the boundary layers surrounding the sensor. Since the Prandtl number is the ratio of these two effects, it is common to express recoveries in terms of the Prandtl number. See Refs 89, 90, 91, and 92 for more information on the recovery factor.

A real sensor immersed in a real fluid tends to radiate to its surroundings. Also there is a tendency for a conductive heat transfer along the probe stem. These two effects are balanced by a convective heat transfer between the probe and the fluid. In addition, real probes do not always stagnate a moving fluid effectively. To account for these realities in temperature measurement in moving fluids, a dynamic correction factor (K) is defined as

$$T_{probe} = T + K T_V \qquad \ldots \ldots \ldots \ldots \ldots \ldots \ldots (47)$$

where K corrects for impact, viscosity, and conductivity effects in the fluid, and radiation and conduction effects in the probe. K may take on any value depending on the relative importance of these effects. Variations of K between ± 35 cannot be ruled out. Therefore, the factor KT_V can far outweigh all other factors such as calibration deviations.

9.1.3 Thermowells

Protection tubes or thermowells or both often are used to separate the measuring junction of a thermocouple from the fluid whose temperature is of interest. Such devices are used to avoid contamination of the thermoelements, to provide safety in case of high pressure installations, to provide strength in the case of significant fluid bending forces in the thermocouple, etc. Thermocouples installed in wells can be withdrawn for inspection, calibration, and replacement. A thermocouple in a well responds more slowly to changes in fluid temperature. Typical wells and their strength requirements are defined in Ref 93.

The depth of immersion in the fluid is an important consideration. One method of checking for adequacy of immersion is to increase the depth of immersion of the thermocouple well assembly in a constant temperature bath until the thermocouple output becomes constant. A minimum immersion depth of ten times the well outside diameter is a rule of thumb often used.

9.1.4 Thermal Analysis of an Installation

A thermocouple installation may give an indication which differs from the fluid temperature which is to be measured because:

1. The boundary walls are at a temperature different from that of the fluid.

2. There may be a temperature gradient along the well.

3. The fluid may be flowing with an appreciable velocity.

4. The thermocouple may be improperly calibrated.

Item 4 has been covered in Section 8 and will not be considered further here.

Basically, the thermocouple temperature is the result of a heat balance between the various modes of heat transfer.

$$q_c = q_r + q_k \quad \dots\dots\dots\dots\dots\dots\dots \text{(48)}$$

where q indicates rate of heat transfer and the subscripts c, r, and k signify respectively convection, radiation, and conduction. An equation has been developed to describe this heat balance mathematically [94] as

$$\frac{d^2 T}{dx^2} + a_1(x)\frac{dT_x}{dx} - a_2(x, y)T_x = -a_2 a_3(x, y) \quad \dots\dots\dots \text{(49)}$$

where $a_1(x) = dA_k / A_k dx$ which indicates the effect of a change in cross-sectional area of the well; $a_2(x, y) = dA_c(h_r + h_c)/kA_k dx$ which indicates the effect of radiation coefficient (h_r), convection coefficient (h_c), conductivity (k), surface area for convection (A_c) and cross-sectional area for conduction (A_k); $a_3(x, y) = (h_c T_{adi} + h_r T_w)/(h_c + h_r)$ which relates the heat transfer coefficients to the adiabatic fluid temperature (T_{adi}) and the surrounding wall temperature (T_w).

Various solutions are possible for Eq 49 depending on the assumptions one is willing to make. Three simplified solutions are:

1. *Overall Linearization*—When the radiation coefficient is based on an average well temperature, the result is

$$\frac{T_x - a_3}{T_w - a_3} = \frac{e^{mx}}{1 + e^{2mL}} + \frac{e^{-mx}}{1 + e^{-2mL}} \quad \dots\dots\dots\dots (50)$$

where

$$m = \left(\frac{4D(h_r + h_c)}{k(D^2 - d^2)}\right)^{1/2}.$$

Typical values for h_r, h_c, and k are given in Ref 94. This approach leads to quick, approximate answers whenever the fluid can be considered transparent to radiation.

2. *Tip Solution*—When conduction effects are neglected along the well or protection tube, Eq 49 reduces to

$$T_{tip} = a_3 \quad \dots\dots\dots\dots\dots\dots\dots\dots (51)$$

which can be solved at once since h_r and h_c are available in the literature (see Ref 94). This approximation normally would give tip temperatures which are too high since conduction tends to reduce T_{tip}.

3. *Stepwise Linearization*—This is the usual solution to Eq 49. Detailed equations are beyond the scope of this manual, but briefly one divides the well, lengthwise, into a number of elements. The temperature at the center of each element is taken to represent the temperature of that entire element. The heat balance equation is applied successively to one element after another until a match between tip and base temperatures is achieved. Each installation is different. Each must be evaluated carefully to determine if the installation is capable of yielding temperatures within the allowable uncertainties.

9.2 Surface Temperature Measurement

9.2.1 General Remarks

There is no easy method of attaching a thermocouple to a surface so that it can be guaranteed to indicate the true surface temperature. To do this, it would be necessary to mount the measuring junction so that it could attain but not affect the surface temperature. In most cases, the presence of the thermocouple (or any alternative transducer) will cause a perturbation of the temperature distribution at the point of attachment, and thus it only will indicate the perturbed temperature.

9.2.1.1 *Measurement Error*—In many cases, a significant difference will exist between the indicated temperature and the "true" surface temperature, that is, the temperature that the surface would reach if no thermocouple were present. This difference is normally termed a "measurement error," but it should not be confused with calibration or extension wire errors which are common to all thermocouple measurements. The relationship between the indicated and true surface temperature is often defined by the equation:

$$Z = \frac{T_s - T_i}{T_s - T_a}$$

where:
Z = installation factor,
T_s = true surface temperature,
T_i = indicated surface temperature, and
T_a = temperature of the surroundings or coolant.
This equation expresses the measurement error $T_s - T_i$ as a fraction of the difference between the surface and ambient temperatures.

The value of Z for a particular installation may be calculated or found by experiment; however, as several simplifying assumptions normally are made in any theoretical derivation, experimental verification is necessary if an accurate value of Z is required.

9.2.1.2 *Installation Types*—There are two basic types of surface thermocouple installation: the permanent, which is used to give a continuous history of the surface temperature, and the temporary, normally made with a sensing probe in mechanical contact with the surface to obtain spot readings. The basic principles for accurate measurement are similar for both types, but the probe type of sensors are more susceptible to measurement errors and generally have a lower accuracy.

9.2.2 *Installation Methods*

The method of attaching the thermocouple to the surface is governed by considerations of the metallurgical and thermal properties of the materials, their relative sizes, and the modes of heat transfer at the surface. Common methods are shown in Fig. 42.

9.2.2.1 *Permanent Installations*—For thin materials, the thermocouple junction is attached either directly to the surface (Fig. 42*a*) or is mounted

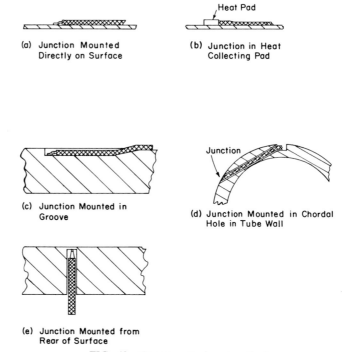

(a) Junction Mounted
Directly on Surface

(b) Junction in Heat
Collecting Pad

(c) Junction Mounted in
Groove

(d) Junction Mounted in Chordal
Hole in Tube Wall

(e) Junction Mounted from
Rear of Surface

FIG. 42—*Common attachment methods.*

in a heat collecting pad (Fig. 42*b*). It may be welded, brazed, cemented, or clamped to the surface. Good mechanical support of the leads is necessary so that no stresses are applied to the junction.

For thicker materials, the thermocouple junction may be peened into the surface or installed in a groove (Fig. 42*c*). The groove may be filled so that the surface is restored to its original profile. A thermocouple in a groove normally will have its junction below the surface and will indicate the subsurface temperature. A similar technique used with tubes is shown in Fig. 42*d*.

The configuration shown in Fig. 42*e* may be used where rapid response is required, as the junction can be made very thin by electroplating or mechanical polishing techniques [95-99].

Several installation methods are illustrated in the literature cited, particularly in Refs *100* and *101*.

Metal sheathed thermocouples are suited particularly to surface measurements, especially for severe environments. They combine good strength and small size, and the measuring junction may be reduced in diameter or flattened to achieve good response with small errors.

9.2.2.2 *Measuring Junctions*—The measuring junction may be formed in several ways, each having its own advantages and disadvantages.

The bead junction commonly is used. The temperature indicated is a function of the temperatures where the wires leave the bead [*102, 103*] so that the bead should be small, and the wires should leave the bead as close to the surface as possible. This may be accomplished by using a flattened bead. Good thermal contact between the bead and the surface is essential, especially if there are temperature gradients. If the surface is a material of poor thermal conductivity, it may be advantageous to mount the measuring junction in a heat collecting pad, or button, which has a good conductivity [*104*].

The simplest junction is shown in Fig. 43, in which a single wire is brought to the surface which acts as the second thermoelement. The circuit is completed with a wire of the same material as the surface. This technique usually involves calibration of the wire/surface-material thermocouple. This calibration may not be very reproducible since the surface material is probably not an alloy with controlled thermoelectric properties. It, however, does provide a junction exactly at the surface, and the perturbation errors (see Section 9.2.3) can be reduced to a very low value [*95, 105, 106*].

A common variation is the separated junction in which each wire is joined separately to the surface (which must be an electrical conductor). This type, which is really two series junctions, has the advantage that the two junctions form a part of the surface. The output of such a thermocouple is a weighted mean of the two individual junction temperatures, of the form

$$e_0 = e_m + (b_1 - b_2)\frac{(T_1 - T_2)}{2}$$

where:

T_1 and T_2 = junction temperatures,

e_0 = measured output,

e_m = output which would be measured if both junctions were at the mean temperature $\dfrac{T_1 + T_2}{2}$, and

b_1 and b_2 = Seebeck coefficients of the two thermocouple wires versus the surface material.

Moffat [*34*] gives a graphical method of analysis.

Thus the output will be greater or less than the mean depending upon which wire is at the higher temperature. The output generally cannot be calculated, as neither T_1, T_2, nor the relationship between each wire and the surface will be known. There will be, therefore, an uncertainty in the measured temperature if temperature gradients exist. This error is minimized

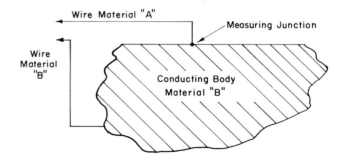

FIG. 43—*"Single wire" thermocouple.*

if the wires are bonded to the surface as close together as possible, to reduce $(T_1 - T_2)$. This type of junction has been shown to be more accurate than a bead junction [*107, 108*].

9.2.2.3 *Probes*—It is often desirable to know the temperature distribution over a surface or to make a spot check at one particular point. These measurements are made with a probe containing a thermocouple junction which is held in mechanical contact with the surface. The configuration of the junction is based on the intended application, and several types are commercially available. The probes are in most cases held normal to the surface, and for ease in use should be spring-loaded, which also reduces the error.

Probes are subject to the same errors as permanent installations, but the designer has no control over the conditions of use, and so the errors associated with this type of measurement may be significant. Since the probe provides a heat conducting path from the surface, thermal resistance due to oxide or dirt causes an additional error.

Correction factors [*109*] for several types of junctions range from 0.013 to 0.168, but in general must be determined for specific conditions.

The size of the junction should be as small as possible. Several types of junction are illustrated in Fig. 44. The junction types in order of decreasing measurement error are: grounded, exposed, button, and separated.

In order to reduce measurement errors, probes with an auxiliary heater have been used [*110, 111*]. The probe thermocouple is heated to the surface temperature so that no temperature gradients are set up when the probe

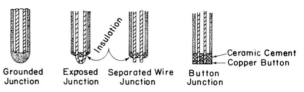

FIG. 44—*Types of junction using metal sheathed thermocouples.*

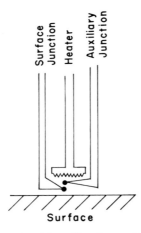

FIG. 45—*Thermocouple probe with auxiliary heater, diagramatic arrangement.*

is applied to the surface. One form of probe uses two thermocouples (Fig. 45). Equality of the auxiliary and surface junction temperatures indicates that no heat is being transferred along the probe and that the surface junction is at the surface temperature. With this type of probe, the two junctions must be very close, and the response to a change in heater power must be fast or an error can occur in transient measurements. A probe which is controlled automatically has been described recently [*112, 113*]. It is claimed to have an accuracy of ⅜ percent and can be used to 1000 F (538 C) on a variety of materials, with a measurement time of less than one second. Sasaki and Kamanda [*114*] used a different approach and eliminated the auxiliary junction. The surface junction was arranged to contact the surface at two second intervals only. The heater input was modulated over a twenty second period and adjusted so that the maxima and minima were above and below the surface temperature. At contact the surface junction temperature changed due to heat exchange unless the two temperatures were equal. With this method the surface temperature of glass bulbs was determined with an accuracy of 0.5 F.

9.2.2.4 *Moving Surfaces*—Surface temperatures of moving bodies are measured by several methods. A junction mounted in a probe may be held against the body [*115*], but this method results in errors caused by friction. In metal cutting investigations, the metal body and the cutting tool are used as the thermocouple materials. The output of this type of junction has been investigated extensively and has been analyzed by Shu et al [*116*]. Slip-rings to rotating members, intermittent and sliding contacts have been used also. A general review of such installations is given in Ref *101*.

9.2.2.5 *Current Carrying Surfaces*—A technique of eliminating errors caused by voltage drop in surfaces heated by the passage of d-c current has been described by Dutton and Lee [*117*]. A three-wire thermocouple

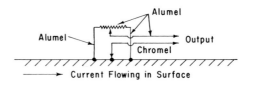

FIG. 46—*Three wire thermocouple to compensate for voltage drop induced by surface current.*

forming two junctions is used (Fig. 46), and the emf's due to the voltage drop in the surface are balanced out during successive reversals of current. When the balance is correct, the thermocouple output is constant regardless of the direction of the heating current.

This technique is also useful for surfaces carrying large alternating currents. In other cases, filters will suffice to attenuate the a-c component. Self-balancing potentiometers usually are affected adversely if the a-c pickup level is high. D-c indicating instruments and portable potentiometers are normally insensitive to a-c and present no problem unless the current is high enough to damage the coils. If the thermocouple junction is not isolated from the surface, voltages appearing between the surface and the instrument ground (common-mode voltages) cause an error with some instruments.

9.2.3 *Sources of Error*

When a thermocouple is attached to a surface, its presence alters the heat transfer characteristics of the surface and normally will change the temperature distribution. This causes an error which will be referred to as perturbation error.

9.2.3.1—The causes of perturbation error can be broken down into the following:

The heat transfer characteristics of the surface are changed by the installation, that is, the surface emissivity or effective thermal conductivity will be altered or the wires may act as fins providing additional heat transfer paths.

Thermal contact resistance between the junction and the surface will cause a temperature gradient which will prevent the measuring junction from attaining the correct temperature if heat is being exchanged between the surface and its surroundings.

If temperature gradients exist, there will be an error due to uncertainty in the exact position of the junction or junctions relative to the surface [*102, 103*].

The response time of the thermocouple installation introduces errors during transient conditions [*118, 119*] (see also Section 9.1.1). The response time will be that of the surface installation and not of the thermocouple alone.

Although not discussed in this chapter, the errors associated with any thermocouple measurement, such as deviations from standard emf and lead wire errors, must be taken also into consideration [120, 121].

9.2.4 Error Determination

The perturbation error must be determined if the surface temperature must be known with a high degree of accuracy. The installation should be designed to minimize this error, but in many cases materials and size make compromises necessary, so that an ideal design cannot be achieved. The error, and hence the correction factor, can be determined by the following methods.

9.2.4.1 *Steady State Conditions*—The direct calculation of the error involves solving the heat flow equation for the measuring junction and surface geometry. Normally simplifying assumptions are made, and the results must be interpreted in relation to them. The calculations show clearly the major sources of error and indicate means of reducing errors to a minimum [122, 123, 124, 125, 126].

Analog methods of solving the heat transfer equations using resistance or resistance-capacitance networks indicate the overall temperature distribution and show the perturbation effect of the thermocouple clearly. They are however difficult to make flexible, and so a number of analog models are required if a range of heat transfer conditions is to be studied [125, 126, 127, 128].

Relaxation methods of solving the heat transfer equations have been used to calculate the temperature distribution; this method is attractive if a computer is available to perform the considerable amount of arithmetic required.

Direct experimental measurement on the installation is often the only satisfactory way to accurately determine the error. Care must be taken to simulate the service conditions exactly as a change in a variable can significantly affect the error. The major problem is to determine the true surface temperature. This is discussed extensively in the literature [96, 128, 129, 130, 131, 132, 133].

9.2.4.2 *Transient Conditions*—If surface temperatures are changing, the response of the thermocouple attachment may cause a significant error. The response time of the thermocouple alone will have little significance for surface measurements if the heat transfer path between the surface and the measuring junction is poor or adds thermal mass.

The time required to change the measuring junction temperature causes the thermocouple output to lag the surface temperature in time and decreases its amplitude [108, 118, 119]. The time constant may be determined experimentally from the response to step or ramp functions of surface temperature change.

A. *Insulated Thermocouple Normal to an Electrically Heated Surface*— For surfaces with changing temperatures a bead thermocouple attached normal to an electrically heated surface and insulated from the ambient has been analyzed by Quant and Fink [*134*] and Green and Hunt [*135*].

The analysis showed that in order to obtain a rapid response with a small, steady-state error, it is necessary to use a small junction bead with good surface contact, small diameter wires, and good insulation between the wires and the surroundings.

B. *Surface Heated by Radiation*—Thermocouples mounted on a surface subject to radiant heating at temperature-rise rates up to 30 F per s were investigated by White [*107*]. His results showed that a separated-junction thermocouple produced the least error.

The thermocouple errors increased with increasing plate thickness and with increasing rates of temperature rise. Furthermore, the amount of bare thermocouple wire between the junction and the insulation should be a minimum.

Kovacs and Mesler [*136*] investigated the response of very fast surface thermocouples subject to radiant heating as a function of size and type of junction. Junctions were formed by electroplating or mechanical abrasion. Very thin junctions were subject to an overshoot error for high rates of temperature rise, as heat could not be transmitted back through the thermocouple to subsurface layers fast enough. On the other hand, too thick a junction corresponded to a junction beneath the surface. The junction thickness should be of the same order of magnitude as the distance between the two thermocouple wires to avoid overshooting.

C. *Surface Subject to Aerodynamic Heating*—An analysis and design of a thermocouple installation which can be mounted in a thin-metal skin subjected to aerodynamic heating is given in Ref *137*. A finite difference calculation of the distorted temperature field indicated that errors due to insulation resistance at high temperatures were of the same order of magnitude as those due to the uncertainty of the exact junction location.

9.2.5 *Procedures for Minimizing Error*

The examples quoted in the preceding section have been treated separately. It is generally impossible to extrapolate from one set of conditions to another unless the installation is identical. The analyses show procedures that should be followed to reduce measurement errors. These are:

A. Use the smallest possible installation to avoid perturbation errors.

B. Bring the thermocouple wires away from the junction along an isotherm for at least 20 wire-diameters to reduce conduction errors. The use of thermocouple materials with low thermal conductivity also will reduce this error.

C. Locate the measuring junction as close to the surface as possible rather than above or below it.

D. Design the installation so that it causes a minimum disturbance of any fluid flow or change in the emissivity of the surface, to avoid changes in convective or radiative heat transfer.

E. Design the installation so that the total response is fast enough to cause negligible lag for the transients expected in service.

F. Reduce the thermal resistance between the measuring junction and the surface to as low a value as possible. If the surface has a low thermal conductivity, a heat collecting pad may be used.

9.2.6 Commercial Surface Thermocouples

Many surface installations are custom engineered, but industry does offer several standard surface thermocouples intended for specific applications [*115*, *138*].

9.2.6.1 *Surface Types*—Figure 47 shows several types which are mounted on the surface. Type *a* is a gasket thermocouple which normally is mounted on a stud and Type *b* is a rivet head. The clamp attachment, Type *c*, is used on pipelines and will be reasonably accurate if the pipe is lagged thermally.

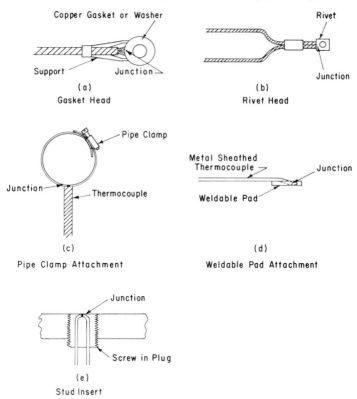

FIG. 47—*Commercially available types of surface thermocouples.*

The weldable pad attachment, Type *d*, is used on boiler or superheater tubes and uses a metal sheathed thermocouple with a grounded junction. The sheath and pad materials are chosen to be compatible with the boiler environment. An accuracy of ± 2 percent is claimed for these thermo- couples (± 1 percent if a correction factor supplied by the manufacturer is used.)

For installations where it is possible to mount the thermocouples in the surface, stud- or rivet-mounted plugs similar to Type *e* are offered by several manufacturers. Variations of this type have been used extensively for heat transfer measurements [122, 139] and for applications requiring very rapid response such as the measurement of surface temperatures in gun barrels and rocket exhaust chambers. The material of the plug must match the mat- erial of the surface, otherwise significant errors can be introduced [140], especially for materials with low thermal conductivity.

9.2.6.2 *Probe Types*—Probe type thermocouples for temporary or spot readings are offered usually as a complete package consisting of a thermo- couple head which contains the measuring junction, a hand probe, and an indicating milliammeter calibrated in degrees. The measuring junctions are normally interchangeable so that one instrument can be used with a variety of heads. The type of head will depend on the surface characteristics. Common types are shown in Fig. 48.

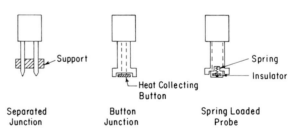

FIG. 48—*Commercial probe thermocouple junctions.*

The separated-junction probe is used on electrically conducting surfaces only. Dirt or oxide layers will introduce a thermal resistance error or even prevent the completion of the circuit. For greatest convenience, the two wires should be separately spring-loaded against the surface. The button type of junction must be held carefully as any deviation from the normal will cause a change in the height of the junction above the surface, and the readings will be inconsistent. The spring-loaded type of junction is available in several forms and has been adapted for measurements on moving surfaces [115].

The accuracy obtainable with these probes is not high. However, the errors can be reduced to 2 or 3 percent for good conducting surfaces in still air cooled by natural convection. If the surface is a poor thermal conductor or the rate of heat transfer is high, the error will be considerably higher than this.

For rotating or moving surfaces, probe instruments utilizing a junction spring-loaded against the surface are used. Heat generated by friction causes an error which can be significant. (Bowden and Ridler [*141*] have shown that the temperatures may reach the melting point of one of the metals.)

A heated thermocouple probe instrument for measuring the temperature of wires or filaments is described by Bensen and Horne [*142*], and a wire temperature meter [*143*] has been marketed recently.

9.3 References

[*78*] Harper, D. R., "Thermometric Lag," NBS Scientific Paper 185, National Bureau of Standards, March 1912.

[*79*] Benedict, R. P., *Fundamentals of Temperature, Pressure, and Flow Measurements*, Wiley, New York, 1969.

[*80*] Wormser, A. F., "Experimental Determination of Thermocouple Time Constants," National AERO Meeting Paper 158D, Society of Automotive Engineers, April 1960.

[*81*] Scadron, M. D. and Warshawsky, I., "Experimental Determination of Time Constants and Nusselt Numbers for Bare-Wire Thermocouples," NACA TN 2599, National Advisory Committee for Aeronautics, Jan. 1952.

[*82*] Coon, G. A., "Response of Temperature-Sensing-Element Analyses," *Transactions*, American Society of Mechanical Engineers, Nov. 1957, p. 1857.

[*83*] Looney, R., "Method for Presenting the Response of Temperature-Measuring Systems," *Transactions*, American Society of Mechanical Engineers, Nov. 1957, p. 1851.

[*84*] Murdock, J. W., Foltz, C. J., and Gregory, C., "A Practical Method of Determining Response Time of Thermometers in Liquid Baths," *Transactions*, American Society of Mechanical Engineers, *Journal of Engineering for Power*, Jan. 1963, p. 27.

[*85*] Caldwell, F. R., Olsen, L. O., and Freeze, P. D., "Intercomparison of Thermocouple Response Data," SAE Paper 158F, Society of Automotive Engineers, April 1960.

[*86*] Hornfeck, A. J., "Response Characteristics of Thermometer Elements," *Transactions*, American Society of Mechanical Engineers, Feb. 1949, p. 121.

[*87*] Benedict, R. P., *Fundamentals of Temperature, Pressure, and Flow Measurements*, Wiley New York, 1969, p. 152.

[*88*] Shepard, C. E. and Warshawsky, I., "Electrical Techniques for Compensation of Thermal Time Lag of Thermocouples and Resistance Thermometer Elements," NACA TN 2703, National Advisory Committee for Aeronautics, Jan. 1952.

[*89*] Benedict, R. P., *Fundamentals of Temperature, Pressure, and Flow Measurements*, Wiley, New York, 1969, p. 126.

[*90*] Roughton, J. E., "Design of Thermometer Pockets for Steam Mains," *Proceedings 1965–66*, Institute of Mechanical Engineers, Vol. 180, Part 1, No. 39.

[*91*] Faul, J. C., "Thermocouple Performance in Gas Streams," *Instruments and Control Systems*, Dec. 1962.

[*92*] Werner, F. D., "Total Temperature Measurement," ASME Paper 58-AU-17, American Society of Mechanical Engineers

[*93*] "Temperature Measurement," ASME PTC 19.3, American Society of Mechanical Engineers, 1961.

[*94*] Benedict, R. P. and Murdock, J. W., "Steady-State Thermal Analysis of a Thermometer Well," *Transactions*, American Society of Mechanical Engineers, *Journal of Engineering for Power*, July 1963, p. 235.

[*95*] Bendersby, D., "A Special Thermocouple for Measuring Transient Temperatures," *Mechanical Engineering*, Vol. 75, No. 2, Feb. 1953, p. 117.

[*96*] Powell, W. B. and Price, T. W., "A Method for the Determination of Local Heat Flux from Transient Temperature Measurements," *Transactions*, Instrument Society of America, Vol. 3, No. 3, July 1964, p. 246.

[97] Moeller, C. E., "Thermocouples for the Measurement of Transient Surface Temperatures," *Temperature, Its Measurement and Control in Science and Industry,* Vol. 3, Part 2, Reinhold, New York, 1962, p. 617.

[98] Ongkiehong, L. and Van Duijn, J., "Construction of a Thermocouple for Measuring Surface Temperatures," *Journal of Scientific Instruments,* Vol. 37, June 1960, p. 221.

[99] Vigor, C. W. and Hornaday, J. R., "A Thermocouple for a Measurement of Temperature Transients in Forging Dies," *Temperature, Its Measurement and Control in Science and Industry,* Vol. 3, Part 2, Reinhold, New York, 1962, p. 625.

[100] Baker, H. D., Ryder, E. A., and Baker, N. H., *Temperature Measurement in Engineering,* Vol. 1, Wiley, New York, 1953, Chapters, 8, 11, 12.

[101] Baker, H. D., Ryder, E. A., and Baker, N. H., *Temperature Measurement in Engineering,* Vol. 2, Wiley, New York, 1961.

[102] Bailey, N. P., "The Response of Thermocouples," *Mechanical Engineering,* Vol. 53, No. 11, Nov. 1931, p. 797.

[103] McCann, J. A., "Temperature Measurement Theory," KAPL-2067-2, 1962, OTS, Department of Commerce, Washington, D. C.

[104] Chapman, A. J., *Heat Transfer,* McMillan, New York, 1960.

[105] Bailey, N. P., "The Measurement of Surface Temperatures, Accuracies Obtainable with Thermocouples," *Mechanical Engineering,* Vol. 54, Aug. 1932, p. 553.

[106] Baker, H. D., Ryder, E. A., and Baker, N. H., *Temperature Measurement in Engineering,* Vol. 1, Wiley, New York, 1961, p. 90.

[107] White, F. J., "Accuracy of Thermocouples in Radiant Heat Testing," *Experimental Mechanics,* Vol. 2, July 1962, p. 204.

[108] Moen, W. K., "Surface Temperature Measurements," *Instruments and Control Systems,* Vol. 33, Jan. 1960, p. 71.

[109] Otter, A. J., "Thermocouples and Surface Temperature Measurement," AECL-3062, March 1968, SDDO, AECL, Chalk River, Ont., Canada.

[110] Roeser, W. F. and Mueller, E. F., "Measurement of Surface Temperatures," *Bureau of Standards Journal of Research,* Vol. 5, No. 4, Oct. 1930, p. 793.

[111] Sasaki, N., "A New Method for Surface Temperature Measurement," *Review of Scientific Instruments,* Vol. 21, No. 1, Jan. 1950, p. 1.

[112] Robertson, D. and Sterbutzel, G. A., "An Accurate Surface Temperature Measuring System," *Proceedings of IEEE Industrial Heating Conference of Philadelphia,* April 1969. See also *Leeds and Northrup Technical Journal,* Issue 7, Summer 1969.

[113] "A Probe for the Instantaneous Measurement of Surface Temperature," RTD-TDR-63-4015, OTS, Department of Commerce, Washington, D. C.

[114] Sasaki, N. and Kamanda, A., "A Recording Device for Surface Temperature Measurements," *Review of Scientific Instruments,* Vol. 23, No. 6, June 1952, p. 261.

[115] "Unusual Thermocouples and Accessories," *Instruments and Control Systems,* Vol. 36, No. 8, Aug. 1963, p. 133.

[116] Shu, H. H., Gaylard, E. W., and Hughes, W. F., "The Relation Between the Rubbing Interface Temperature Distribution and Dynamic Thermocouple Temperature," *Transactions ASME Journal of Basic Engineering,* Vol. 86, Series D, No. 3, Sept. 1964, p. 417.

[117] Dutton, R. and Lee, E. C., "Surface-Temperature Measurement of Current-Carrying Objects," *Journal,* Instrument Society of America, Vol. 6, No. 12, Dec. 1959, p. 49.

[118] Jakob, M., *Heat Transfer,* Vol. 2, Wiley, New York, 1957, p. 183.

[119] Caldwell, W. I., Coon, G. A., and Zoss, L. M., *"Frequency Response for Process Control,"* McGraw-Hill, New York, 1959, p. 295.

[120] "Thermocouples and Thermocouple Extension Wires," *Recommended Practice RPI-7,* Instrument Society of America, July 7, 1959.

[121] Finch, D. I., "General Principles of Thermoelectric Thermometry," *Temperature, Its Measurement and Control in Science and Industry,* Vol. 3, Part 2, Reinhold, New York, p.3.

[122] Jakob, M., *Heat Transfer,* Vol. 2, Wiley, New York, 1957, p. 153.

[123] Chapman, A. J., *Heat Transfer,* McMillan, New York, 1960.
[124] Boelter, L. M. K. et al, "An Investigation of Aircraft Heaters XXVIII-Equations for Steady State Temperature Distribution Caused by Thermal Sources in Flat Plates Applied to Calculation of Thermocouple Errors, Heat Meter Corrections, and Heat Transfer by Pin-Fin Plates," Technical Note 1452, National Advisory Committee for Aeronautics, 1944.
[125] Schneider, P. J., *Conduction Heat Transfer,* Addison-Wesley, Cambridge, Mass., 1955, p. 176.
[126] Fitts, R. L., Flemons, R. S., and Rogers, J. T., "Study of Temperature Distribution in a Finned Nuclear Fuel Sheath by Electrical Analogue and Mathematic Analysis (Revision)," R65CAP1, 19 Jan. 1965, Canadian General Electric, Peterborough, Ont.
[127] Chan, K. S. and Rushton, K. R., "The Simulation of Boundary Conditions in Heat Conduction Problems in a Resistance-Capacitance Electrical Analogue," *Journal of Scientific Instruments,* Vol. 41, No. 9, Sept. 1964, p. 535.
[128] Brindley, J. H., "Calibration of Surface-Attached Thermocouples on a Flat-Plate Fuel Element by Electrical Analogue and Analytical Techniques," *Transactions of the American Nuclear Society,* Vol. 6, No. 2, Nov. 1963, p. 333.
[129] Boelter, L. M. K. and Lockhart, R. W., "Thermocouple Conduction Error Observed in Measuring Surface Temperatures," Technical Note 2427, National Advisory Committee for Aeronautics, 1946.
[130] Stoll, A. M. and Hardy, J. D., "Direct Experimental Comparison of Several Surface Temperature Measuring Devices," *Review of Scientific Instruments,* Vol. 20, No. 9, Sept. 1949, p. 678.
[131] Oetken, E. R., "Evaluation of Surface-Attached Thermocouples During Forced Convection Heat Transfer," IDO-16889, OTS, Department of Commerce, Washington, D. C.
[132] Sudar, S., "Calibration of OMRE Fuel-Element Surface Thermocouple Assembly," *NAA-SR-Memo 3671,* 1959.
[133] Browning, W. E. and Hemphill, H. L., "Thermocouples for Measurement of the Surface Temperature of Nuclear Fuel Elements," *Temperature, Its Measurement and Control in Science and Industry,* Vol. 3, Part 2, Reinhold, New York, p. 723.
[134] Quant, E. R. and Fink, E. W., "Experimental and Theoretical Analysis of the Transient Response of Surface Bonded Thermocouples," *Bettis Technical Review,* WAPD-BT-19, Reactor Technology, June 1960, p. 31.
[135] Green, S. J. and Hunt, T. W., "Accuracy and Response of Thermocouples for Surface and Fluid Temperature Measurements," *Temperature, Its Measurement and Control in Science and Industry,* Vol. 3, Part 2, Reinhold, New York, p. 695.
[136] Kovacs, A. and Mesler, R. B., "Making and Testing Small Surface Thermocouples for Fast Response," *Review of Scientific Instruments,* Vol. 35, No. 4, April 1964, p. 485.
[137] "Design and Construction of a Unit for Measuring Metal Skin Temperatures, Phase I, Theoretical Analysis and Design," SC-4461 (RR), Dec. 1960, OTS, Department of Commerce, Washington, D. C.
[138] See Thermocouple Manufacturers' Catalogs.
[139] Sellers, J. P., "Thermocouple Probes for Evaluating Local Heat Transfer Coefficients in Rocket Motors," *Temperature, Its Measurement and Control in Science and Industry,* Vol. 3, Part 2, Reinhold, New York, p. 673.
[140] Nanigan, J., "Thermal Properties of Thermocouples," *Instruments and Control Systems,* Vol. 36, No. 10, Oct. 1963, p. 87.
[141] Bowden, F. P. and Ridler, K. E. W., "Physical Properties of Surfaces III. The Surface Temperature of Sliding Metals," *Proceedings Royal Society London A,* Vol. 154, 1936, p. 644.
[142] Bensen, J. and Horne, R., "Surface Temperature of Filaments and Thin Sheets," *Instruments and Control Systems,* Vol. 35, No. 10, Part 1, Oct. 1962, p. 115.
[143] Report No. 152, British Iron and Steel Research Organization.

10. REFERENCE TABLES FOR THERMOCOUPLES

The practical use of thermocouples in industrial and laboratory applications requires that the thermocouple conform to an established temperature —electromotive-force relationship within acceptable limits of error. Since the thermocouple in a thermoelectric thermometer system is usually expendable, conformance to established temperature-emf relationships is necessary in order to permit interchangeability.

Section 10.2 consists of reference tables that give temperature-emf relationships for the thermocouple types most commonly used in industry. These are identified as thermocouple Types B, E, J, K, R, S, and T, as defined in ANSI Standard C96.1.

Data in these tables are based upon absolute electrical units, and the International Practical Temperature Scale of 1948.[13] All temperature-emf data contained in Section 10.2 have been extracted from National Bureau of Standards Circular 561, issued 27 April 1955 and Journal of Research Vol. 70C, No. 2, April-June 1966. These tables have been published also as ASTM Standard E 230.

10.1 Thermocouple Types and Limits of Error
10.1.1 *Thermocouple Types*

The letter symbols identifying each reference table are those defined in ANSI Standard C96.1. These symbols which are in common use throughout industry, identify the following thermocouple calibrations:

> Type J —Iron versus constantan (modified 1913 calibration).
> Type K—Originally Chromel-P versus Alumel.[14]
> Type R —Platinum 13 percent rhodium versus platinum.
> Type S —Platinum 10 percent rhodium versus platinum.
> Type T —Copper versus constantan.
> Type E —Originally Chromel-P versus constantan.
> Type B —Platinum 30 percent rhodium versus platinum 6 percent rhodium.

Detailed information covering the advantages and limitations of each of these thermocouple types, their recommended temperature ranges, and detailed physical properties of the thermoelements comprising them are contained in Section 3.1 of this manual.

10.1.2 *Limits of Error*

The limits of error for the common letter designated thermocouple types, as listed in Table 41, are taken from ANSI Standard C96.1. Most manufacturers supply thermocouples and thermocouple wire to these limits of error or better.

[13] New tables will be forthcoming based on IPTS-68.

[14] It should be noted that alloys other than Chromel or Alumel are available which will develop the temperature-emf relationships of the Types K and E calibrations.

TABLE 41—*Limits of error for thermocouples.*

		Limits of Error	
Type	Temperature Range, deg F	Standard	Special
Type J	32 to 530	±4 F	±2 F
	530 to 1400	±¾%	±⅜%
Type K	32 to 530	±4 F	±2 F
	530 to 2300	±¾%	±⅜%
Type R or S	32 to 1000	±5 F	±2½ F
	1000 to 2700	±½%	±¼%
Type T	−300 to −75	...	±1%
	−150 to −75	±2%	±1%
	−75 to +200	±1½ F	±¾ F
	200 to 700	±¾%	±⅜%
Type E	32 to 600	±3 F	±2¼ F
	600 to 1600	±½%	±⅜%
Type B	1600 to 3100	±½%	...

NOTE 1—In this Table the limits of error for each type of thermocouple apply only over the temperature range for which the wire size in question is recommended (see Section 300). These limits of error should be applied only to standard wire sizes. The same limits may not be obtainable in special sizes.

NOTE 2—Where limits of error are given in percent, the percentage applies to the temperature being measured (when expressed in degree Fahrenheit or Celsius).

NOTE 3—Limits of error apply to thermocouples as supplied by the manufacturer. The calibration of a thermocouple may change during use. The magnitude of the change depends upon such factors as temperature and the length of time and conditions of use.

10.2 Thermocouple Reference Tables

Following is a list of the reference tables included in this section:

Table Number	Thermocouple Type	Temperature Range
42	J	−320 to +1600 F
43	J	−195 to + 870 C
44	K	−310 to +2500 F
45	K	−200 to +1371 C
46	R	32 to 3100 F
47	R	0 to 1700 C
48	S	32 to 3215 F
49	S	0 to 1769 C
50	T	−313 to + 752 F
51	T	−192 to + 400 C
52	E	−320 to +1830 F
53	E	−200 to +1000 C
54	B	32 to 3308 F
55	B	0 to 1820 C

TABLE 42—*Type J thermocouples.*
Temperature in Degrees Fahrenheit

EMF in Absolute Millivolts   Reference Junctions at 32 F

°F	0	1	2	3	4	5	6	7	8	9	10	°F
						Millivolts						
−310	−7.66	−7.68	−7.69	−7.70	−7.71	−7.73	−7.74	−7.75	−7.76	−7.78	−7.79	−310
−300	−7.52	−7.54	−7.55	−7.57	−7.58	−7.59	−7.61	−7.62	−7.64	−7.65	−7.66	−300
−290	−7.38	−7.39	−7.40	−7.42	−7.44	−7.45	−7.46	−7.48	−7.49	−7.51	−7.52	−290
−280	−7.22	−7.24	−7.25	−7.27	−7.28	−7.30	−7.31	−7.33	−7.34	−7.36	−7.38	−280
−270	−7.06	−7.07	−7.09	−7.11	−7.12	−7.14	−7.15	−7.17	−7.19	−7.20	−7.22	−270
−260	−6.89	−6.90	−6.92	−6.94	−6.96	−6.97	−6.99	−7.01	−7.02	−7.04	−7.06	−260
−250	−6.71	−6.73	−6.75	−6.77	−6.78	−6.80	−6.82	−6.84	−6.85	−6.87	−6.89	−250
−240	−6.53	−6.55	−6.57	−6.59	−6.61	−6.62	−6.64	−6.66	−6.68	−6.70	−6.71	−240
−230	−6.35	−6.37	−6.38	−6.40	−6.42	−6.44	−6.46	−6.48	−6.50	−6.52	−6.53	−230
−220	−6.16	−6.18	−6.19	−6.21	−6.23	−6.25	−6.27	−6.29	−6.31	−6.33	−6.35	−220
−210	−5.96	−5.98	−6.00	−6.02	−6.04	−6.06	−6.08	−6.10	−6.12	−6.14	−6.16	−210
−200	−5.76	−5.78	−5.80	−5.82	−5.84	−5.86	−5.88	−5.90	−5.92	−5.94	−5.96	−200
−190	−5.55	−5.57	−5.59	−5.61	−5.63	−5.65	−5.67	−5.70	−5.72	−5.74	−5.76	−190
−180	−5.34	−5.36	−5.38	−5.40	−5.42	−5.44	−5.46	−5.49	−5.51	−5.53	−5.55	−180
−170	−5.12	−5.14	−5.16	−5.19	−5.21	−5.23	−5.25	−5.27	−5.30	−5.32	−5.34	−170
−160	−4.90	−4.92	−4.94	−4.97	−4.99	−5.01	−5.03	−5.06	−5.08	−5.10	−5.12	−160
−150	−4.68	−4.70	−4.72	−4.74	−4.76	−4.79	−4.81	−4.83	−4.86	−4.88	−4.90	−150
−140	−4.44	−4.47	−4.49	−4.51	−4.54	−4.56	−4.58	−4.61	−4.63	−4.65	−4.68	−140
−130	−4.21	−4.23	−4.26	−4.28	−4.30	−4.33	−4.35	−4.38	−4.40	−4.42	−4.44	−130
−120	−3.97	−4.00	−4.02	−4.04	−4.07	−4.09	−4.12	−4.14	−4.16	−4.19	−4.21	−120
−110	−3.73	−3.76	−3.78	−3.81	−3.83	−3.85	−3.88	−3.90	−3.93	−3.95	−3.97	−110
−100	−3.49	−3.51	−3.54	−3.56	−3.59	−3.61	−3.64	−3.66	−3.68	−3.71	−3.73	−100
−90	−3.24	−3.27	−3.29	−3.32	−3.34	−3.36	−3.39	−3.41	−3.44	−3.46	−3.49	−90
−80	−2.99	−3.02	−3.04	−3.07	−3.09	−3.12	−3.14	−3.17	−3.19	−3.22	−3.24	−80
−70	−2.74	−2.76	−2.79	−2.81	−2.84	−2.86	−2.89	−2.92	−2.94	−2.97	−2.99	−70
−60	−2.48	−2.51	−2.53	−2.56	−2.58	−2.61	−2.64	−2.66	−2.69	−2.71	−2.74	−60
−50	−2.22	−2.25	−2.27	−2.30	−2.33	−2.35	−2.38	−2.40	−2.43	−2.46	−2.48	−50
−40	−1.96	−1.99	−2.01	−2.04	−2.06	−2.09	−2.12	−2.14	−2.17	−2.20	−2.22	−40
−30	−1.70	−1.72	−1.75	−1.78	−1.80	−1.83	−1.86	−1.88	−1.91	−1.94	−1.96	−30
−20	−1.43	−1.46	−1.48	−1.51	−1.54	−1.56	−1.59	−1.62	−1.64	−1.67	−1.70	−20
−10	−1.16	−1.19	−1.21	−1.24	−1.27	−1.29	−1.32	−1.35	−1.38	−1.40	−1.43	−10
(−)0	−0.89	−0.91	−0.94	−0.97	−1.00	−1.02	−1.05	−1.08	−1.10	−1.13	−1.16	(−)0
(+)0	−0.89	−0.86	−0.83	−0.80	−0.78	−0.75	−0.72	−0.70	−0.67	−0.64	−0.61	(+)0
10	−0.61	−0.58	−0.56	−0.53	−0.50	−0.48	−0.45	−0.42	−0.39	−0.36	−0.34	10
20	−0.34	−0.31	−0.28	−0.25	−0.22	−0.20	−0.17	−0.14	−0.11	−0.09	−0.06	20
30	−0.06	−0.03	0.00	0.03	0.05	0.08	0.11	0.14	0.17	0.19	0.22	30
40	0.22	0.25	0.28	0.31	0.34	0.36	0.39	0.42	0.45	0.48	0.50	40
50	0.50	0.53	0.56	0.59	0.62	0.65	0.67	0.70	0.73	0.76	0.79	50
60	0.79	0.82	0.84	0.87	0.90	0.93	0.96	0.99	1.02	1.04	1.07	60
70	1.07	1.10	1.13	1.16	1.19	1.22	1.25	1.28	1.30	1.33	1.36	70
80	1.36	1.39	1.42	1.45	1.48	1.51	1.54	1.56	1.59	1.62	1.65	80
90	1.65	1.68	1.71	1.74	1.77	1.80	1.83	1.85	1.88	1.91	1.94	90
100	1.94	1.97	2.00	2.03	2.06	2.09	2.12	2.14	2.17	2.20	2.23	100
°F	0	1	2	3	4	5	6	7	8	9	10	°F

TABLE 42—*Type J thermocouples (continued).*
Temperature in Degrees Fahrenheit

EMF in Absolute Millivolts

Reference Junctions at 32 F

°F	0	1	2	3	4	5	6	7	8	9	10	°F
						Millivolts						
100	1. 94	1. 97	2. 00	2. 03	2. 06	2. 09	2. 12	2. 14	2. 17	2. 20	2. 23	100
110	2. 23	2. 26	2. 29	2. 32	2. 35	2. 38	2. 41	2. 44	2. 47	2. 50	2. 52	110
120	2. 52	2. 55	2. 58	2. 61	2. 64	2. 67	2. 70	2. 73	2. 76	2. 79	2. 82	120
130	2. 82	2. 85	2. 88	2. 91	2. 94	2. 97	3. 00	3. 03	3. 06	3. 08	3. 11	130
140	3. 11	3. 14	3. 17	3. 20	3. 23	3. 26	3. 29	3. 32	3. 35	3. 38	3. 41	140
150	3. 41	3. 44	3. 47	3. 50	3. 53	3. 56	3. 59	3. 62	3. 65	3. 68	3. 71	150
160	3. 71	3. 74	3. 77	3. 80	3. 83	3. 86	3. 89	3. 92	3. 95	3. 98	4. 01	160
170	4. 01	4. 04	4. 07	4. 10	4. 13	4. 16	4. 19	4. 22	4. 25	4. 28	4. 31	170
180	4. 31	4. 34	4. 37	4. 40	4. 43	4. 46	4. 49	4. 52	4. 55	4. 58	4. 61	180
190	4. 61	4. 64	4. 67	4. 70	4. 73	4. 76	4. 79	4. 82	4. 85	4. 88	4. 91	190
200	4. 91	4. 94	4. 97	5. 00	5. 03	5. 06	5. 09	5. 12	5. 15	5. 18	5. 21	200
210	5. 21	5. 24	5. 27	5. 30	5. 33	5. 36	5. 39	5. 42	5. 45	5. 48	5. 51	210
220	5. 51	5. 54	5. 57	5. 60	5. 63	5. 66	5. 69	5. 72	5. 75	5. 78	5. 81	220
230	5. 81	5. 84	5. 87	5. 90	5. 93	5. 96	5. 99	6. 02	6. 05	6. 08	6. 11	230
240	6. 11	6. 14	6. 17	6. 20	6. 24	6. 27	6. 30	6. 33	6. 36	6. 39	6. 42	240
250	6. 42	6. 45	6. 48	6. 51	6. 54	6. 57	6. 60	6. 63	6. 66	6. 69	6. 72	250
260	6. 72	6. 75	6. 78	6. 81	6. 84	6. 87	6. 90	6. 93	6. 96	7. 00	7. 03	260
270	7. 03	7. 06	7. 09	7. 12	7. 15	7. 18	7. 21	7. 24	7. 27	7. 30	7. 33	270
280	7. 33	7. 36	7. 39	7. 42	7. 45	7. 48	7. 51	7. 54	7. 58	7. 61	7. 64	280
290	7. 64	7. 67	7. 70	7. 73	7. 76	7. 79	7. 82	7. 85	7. 88	7. 91	7. 94	290
300	7. 94	7. 97	8. 00	8. 04	8. 07	8. 10	8. 13	8. 16	8. 19	8. 22	8. 25	300
310	8. 25	8. 28	8. 31	8. 34	8. 37	8. 40	8. 44	8. 47	8. 50	8. 53	8. 56	310
320	8. 56	8. 59	8. 62	8. 65	8. 68	8. 71	8. 74	8. 77	8. 80	8. 84	8. 87	320
330	8. 87	8. 90	8. 93	8. 96	8. 99	9. 02	9. 05	9. 08	9. 11	9. 14	9. 17	330
340	9. 17	9. 20	9. 24	9. 27	9. 30	9. 33	9. 36	9. 39	9. 42	9. 45	9. 48	340
350	9. 48	9. 51	9. 54	9. 58	9. 61	9. 64	9. 67	9. 70	9. 73	9. 76	9. 79	350
360	9. 79	9. 82	9. 85	9. 88	9. 92	9. 95	9. 98	10. 01	10. 04	10. 07	10. 10	360
370	10. 10	10. 13	10. 16	10. 19	10. 22	10. 25	10. 28	10. 32	10. 35	10. 38	10. 41	370
380	10. 41	10. 44	10. 47	10. 50	10. 53	10. 56	10. 60	10. 63	10. 66	10. 69	10. 72	380
390	10. 72	10. 75	10. 78	10. 81	10. 84	10. 87	10. 90	10. 94	10. 97	11. 00	11. 03	390
400	11. 03	11. 06	11. 09	11. 12	11. 15	11. 18	11. 21	11. 24	11. 28	11. 31	11. 34	400
410	11. 34	11. 37	11. 40	11. 43	11. 46	11. 49	11. 52	11. 55	11. 58	11. 62	11. 65	410
420	11. 65	11. 68	11. 71	11. 74	11. 77	11. 80	11. 83	11. 86	11. 89	11. 92	11. 96	420
430	11. 96	11. 99	12. 02	12. 05	12. 08	12. 11	12. 14	12. 17	12. 20	12. 23	12. 26	430
440	12. 26	12. 30	12. 33	12. 36	12. 39	12. 42	12. 45	12. 48	12. 51	12. 54	12. 57	440
450	12. 57	12. 60	12. 64	12. 67	12. 70	12. 73	12. 76	12. 79	12. 82	12. 85	12. 88	450
460	12. 88	12. 91	12. 94	12. 98	13. 01	13. 04	13. 07	13. 10	13. 13	13. 16	13. 19	460
470	13. 19	13. 22	13. 25	13. 28	13. 31	13. 34	13. 38	13. 41	13. 44	13. 47	13. 50	470
480	13. 50	13. 53	13. 56	13. 59	13. 62	13. 65	13. 68	13. 72	13. 75	13. 78	13. 81	480
490	13. 81	13. 84	13. 87	13. 90	13. 93	13. 96	13. 99	14. 02	14. 05	14. 08	14. 12	490
500	14. 12	14. 15	14. 18	14. 21	14. 24	14. 27	14. 30	14. 33	14. 36	14. 39	14. 42	500
°F	0	1	2	3	4	5	6	7	8	9	10	°F

TABLE 42—*Type J thermocouples (continued)*.
Temperature in Degrees Fahrenheit

EMF in Absolute Millivolts

Reference Junctions at 32 F

°F	0	1	2	3	4	5	6	7	8	9	10	°F
						Millivolts						
500	14. 12	14. 15	14. 18	14. 21	14. 24	14. 27	14. 30	14. 33	14. 36	14. 39	14. 42	500
510	14. 42	14. 45	14. 48	14. 52	14. 55	14. 58	14. 61	14. 64	14. 67	14. 70	14. 73	510
520	14. 73	14. 76	14. 79	14. 82	14. 85	14. 88	14. 91	14. 94	14. 98	15. 01	15. 04	520
530	15. 04	15. 07	15. 10	15. 13	15. 16	15. 19	15. 22	15. 25	15. 28	15. 31	15. 34	530
540	15. 34	15. 37	15. 40	15. 44	15. 47	15. 50	15. 53	15. 56	15. 59	15. 62	15. 65	540
550	15. 65	15. 68	15. 71	15. 74	15. 77	15. 80	15. 84	15. 87	15. 90	15. 93	15. 96	550
560	15. 96	15. 99	16. 02	16. 05	16. 08	16. 11	16. 14	16. 17	16. 20	16. 23	16. 26	560
570	16. 26	16. 30	16. 33	16. 36	16. 39	16. 42	16. 45	16. 48	16. 51	16. 54	16. 57	570
580	16. 57	16. 60	16. 63	16. 66	16. 69	16. 72	16. 75	16. 78	16. 82	16. 85	16. 88	580
590	16. 88	16. 91	16. 94	16. 97	17. 00	17. 03	17. 06	17. 09	17. 12	17. 15	17. 18	590
600	17. 18	17. 21	17. 24	17. 28	17. 31	17. 34	17. 37	17. 40	17. 43	17. 46	17. 49	600
610	17. 49	17. 52	17. 55	17. 58	17. 61	17. 64	17. 68	17. 71	17. 74	17. 77	17. 80	610
620	17. 80	17. 83	17. 86	17. 89	17. 92	17. 95	17. 98	18. 01	18. 04	18. 08	18. 11	620
630	18. 11	18. 14	18. 17	18. 20	18. 23	18. 26	18. 29	18. 32	18. 35	18. 38	18. 41	630
640	18. 41	18. 44	18. 47	18. 50	18. 54	18. 57	18. 60	18. 63	18. 66	18. 69	18. 72	640
650	18. 72	18. 75	18. 78	18. 81	18. 84	18. 87	18. 90	18. 94	18. 97	19. 00	19. 03	650
660	19. 03	19. 06	19. 09	19. 12	19. 15	19. 18	19. 21	19. 24	19. 27	19. 30	19. 34	660
670	19. 34	19. 37	19. 40	19. 43	19. 46	19. 49	19. 52	19. 55	19. 58	19. 61	19. 64	670
680	19. 64	19. 67	19. 70	19. 74	19. 77	19. 80	19. 83	19. 86	19. 89	19. 92	19. 95	680
690	19. 95	19. 98	20. 01	20. 04	20. 07	20. 10	20. 13	20. 16	20. 20	20. 23	20. 26	690
700	20. 26	20. 29	20. 32	20. 35	20. 38	20. 41	20. 44	20. 47	20. 50	20. 53	20. 56	700
710	20. 56	20. 59	20. 62	20. 66	20. 69	20. 72	20. 75	20. 78	20. 81	20. 84	20. 87	710
720	20. 87	20. 90	20. 93	20. 96	20. 99	21. 02	21. 05	21. 08	21. 11	21. 14	21. 18	720
730	21. 18	21. 21	21. 24	21. 27	21. 30	21. 33	21. 36	21. 39	21. 42	21. 45	21. 48	730
740	21. 48	21. 51	21. 54	21. 57	21. 60	21. 64	21. 67	21. 70	21. 73	21. 76	21. 79	740
750	21. 79	21. 82	21. 85	21. 88	21. 91	21. 94	21. 97	22. 00	22. 03	22. 06	22. 10	750
760	22. 10	22.·13	22. 16	22. 19	22. 22	22. 25	22. 28	22. 31	22. 34	22. 37	22. 40	760
770	22. 40	22. 43	22. 46	22. 49	22. 52	22. 55	22. 58	22. 62	22. 65	22. 68	22. 71	770
780	22. 71	22. 74	22. 77	22. 80	22. 83	22. 86	22. 89	22. 92	22. 95	22. 98	23. 01	780
790	23. 01	23. 04	23. 08	23. 11	23. 14	23. 17	23. 20	23. 23	23. 26	23. 29	23. 32	790
800	23. 32	23. 35	23. 38	23. 41	23. 44	23. 47	23. 50	23. 53	23. 56	23. 60	23. 63	800
810	23. 63	23. 66	23. 69	23. 72	23. 75	23. 78	23. 81	23. 84	23. 87	23. 90	23. 93	810.
820	23. 93	23. 96	23. 99	24. 02	24. 06	24. 09	24. 12	24. 15	24. 18	24. 21	24. 24	820
830	24. 24	24. 27	24. 30	24. 33	24. 36	24. 39	24. 42	24. 45	24. 48	24. 52	24. 55	830
840	24. 55	24. 58	24. 61	24. 64	24. 67	24. 70	24. 73	24. 76	24. 79	24. 82	24. 85	840
850	24. 85	24. 88	24. 91	24. 94	24. 98	25. 01	25. 04	25. 07	25. 10	25. 13	25. 16	850
860	25. 16	25. 19	25. 22	25. 25	25. 28	25. 32	25. 35	25. 38	25. 41	25. 44	25. 47	860
870	25. 47	25. 50	25. 53	25. 56	25. 59	25. 62	25. 65	25. 68	25. 72	25. 75	25. 78	870
880	25. 78	25. 81	25. 84	25. 87	25. 90	25. 93	25. 96	25. 99	26. 02	26. 06	26. 09	880
890	26. 09	26. 12	26. 15	26. 18	26. 21	26. 24	26. 27	26. 30	26. 33	26. 36	26. 40	890
900	26. 40	26. 43	26. 46	26. 49	26. 52	26. 55	26. 58	26. 61	26. 64	26. 67	26. 70	900
°F	0	1	2	3	4	5	6	7	8	9	10	°F

TABLE 42—*Type J thermocouples (continued).*
Temperature in Degrees Fahrenheit

EMF in Absolute Millivolts   Reference Junctions at 32 F

°F	0	1	2	3	4	5	6	7	8	9	10	°F
						Millivolts						
900	26.40	26.43	26.46	26.49	26.52	26.55	26.58	26.61	26.64	26.67	26.70	900
910	26.70	26.74	26.77	26.80	26.83	26.86	26.89	26.92	26.95	26.98	27.02	910
920	27.02	27.05	27.08	27.11	27.14	27.17	27.20	27.23	27.26	27.30	27.33	920
930	27.33	27.36	27.39	27.42	27.45	27.48	27.51	27.54	27.58	27.61	27.64	930
940	27.64	27.67	27.70	27.73	27.76	27.80	27.83	27.86	27.89	27.92	27.95	940
950	27.95	27.98	28.02	28.05	28.08	28.11	28.14	28.17	28.20	28.23	28.26	950
960	28.26	28.30	28.33	28.36	28.39	28.42	28.45	28.48	28.52	28.55	28.58	960
970	28.58	28.61	28.64	28.67	28.70	28.74	28.77	28.80	28.83	28.86	28.89	970
980	28.89	28.92	28.96	28.99	29.02	29.05	29.08	29.11	29.14	29.18	29.21	980
990	29.21	29.24	29.27	29.30	29.33	29.37	29.40	29.43	29.46	29.49	29.52	990
1,000	29.52	29.56	29.59	29.62	29.65	29.68	29.71	29.75	29.78	29.81	29.84	1,000
1,010	29.84	29.87	29.90	29.94	29.97	30.00	30.03	30.06	30.10	30.13	30.16	1,010
1,020	30.16	30.19	30.22	30.25	30.28	30.32	30.35	30.38	30.41	30.44	30.48	1,020
1,030	30.48	30.51	30.54	30.57	30.60	30.64	30.67	30.70	30.73	30.76	30.80	1,030
1,040	30.80	30.83	30.86	30.89	30.92	30.96	30.99	31.02	31.05	31.08	31.12	1,040
1,050	31.12	31.15	31.18	31.21	31.24	31.28	31.31	31.34	31.37	31.40	31.44	1,050
1,060	31.44	31.47	31.50	31.53	31.56	31.60	31.63	31.66	31.69	31.72	31.76	1,060
1,070	31.76	31.79	31.82	31.85	31.88	31.92	31.95	31.98	32.01	32.05	32.08	1,070
1,080	32.08	32.11	32.14	32.18	32.21	32.24	32.27	32.30	32.34	32.37	32.40	1,080
1,090	32.40	32.43	32.47	32.50	32.53	32.56	32.60	32.63	32.66	32.69	32.72	1,090
1,100	32.72	32.76	32.79	32.82	32.86	32.89	32.92	32.95	32.98	33.02	33.05	1,100
1,110	33.05	33.08	33.11	33.15	33.18	33.21	33.24	33.28	33.31	33.34	33.37	1,110
1,120	33.37	33.41	33.44	33.47	33.50	33.54	33.57	33.60	33.64	33.67	33.70	1,120
1,130	33.70	33.73	33.76	33.80	33.83	33.86	33.90	33.93	33.96	33.99	34.03	1,130
1,140	34.03	34.06	34.09	34.12	34.16	34.19	34.22	34.26	34.29	34.32	34.36	1,140
1,150	34.36	34.39	34.42	34.45	34.49	34.52	34.55	34.58	34.62	34.65	34.68	1,150
1,160	34.68	34.72	34.75	34.78	34.82	34.85	34.88	34.92	34.95	34.98	35.01	1,160
1,170	35.01	35.05	35.08	35.11	35.15	35.18	35.21	35.25	35.28	35.31	35.35	1,170
1,180	35.35	35.38	35.41	35.45	35.48	35.51	35.54	35.58	35.61	35.64	35.68	1,180
1,190	35.68	35.71	35.74	35.78	35.81	35.84	35.88	35.91	35.94	35.98	36.01	1,190
1,200	36.01	36.05	36.08	36.11	36.15	36.18	36.21	36.25	36.28	36.31	36.35	1,200
1,210	36.35	36.38	36.42	36.45	36.48	36.52	36.55	36.58	36.62	36.65	36.69	1,210
1,220	36.69	36.72	36.75	36.79	36.82	36.86	36.89	36.92	36.96	36.99	37.02	1,220
1,230	37.02	37.06	37.09	37.13	37.16	37.20	37.23	37.26	37.30	37.33	37.36	1,230
1,240	37.36	37.40	37.43	37.47	37.50	37.54	37.57	37.60	37.64	37.67	37.71	1,240
1,250	37.71	37.74	37.78	37.81	37.84	37.88	37.91	37.95	37.98	38.02	38.05	1,250
1,260	38.05	38.08	38.12	38.15	38.19	38.22	38.26	38.29	38.32	38.36	38.39	1,260
1,270	38.39	38.43	38.46	38.50	38.53	38.57	38.60	38.64	38.67	38.70	38.74	1,270
1,280	38.74	38.77	38.81	38.84	38.88	38.91	38.95	38.98	39.02	39.05	39.08	1,280
1,290	39.08	39.12	39.15	39.19	39.22	39.26	39.29	39.33	39.36	39.40	39.43	1,290
1,300	39.43	39.47	39.50	39.54	39.57	39.61	39.64	39.68	39.71	39.75	39.78	1,300
°F	0	1	2	3	4	5	6	7	8	9	10	°F

TABLE 42—*Type J thermocouples (continued)*
Temperature in Degrees Fahrenheit
EMF in Absolute Millivolts   Reference Junctions at 32 F

°F	0	1	2	3	4	5	6	7	8	9	10	°F
						Millivolts						
1,300	39. 43	39. 47	39. 50	39. 54	39. 57	39. 61	39. 64	39. 68	39. 71	39. 75	39. 78	1,300
1,310	39. 78	39. 82	39. 85	39. 89	39. 92	39. 96	39. 99	40. 03	40. 06	40. 10	40. 13	1,310
1,320	40. 13	40. 17	40. 20	40. 24	40. 27	40. 31	40. 34	40. 38	40. 41	40. 45	40. 48	1,320
1,330	40. 48	40. 52	40. 55	40. 59	40. 62	40. 66	40. 69	40. 73	40. 76	40. 80	40. 83	1,330
1,340	40. 83	40. 87	40. 90	40. 94	40. 98	41. 01	41. 05	41. 08	41. 12	41. 15	41. 19	1,340
1,350	41. 19	41. 22	41. 26	41. 29	41. 33	41. 36	41. 40	41. 43	41. 47	41. 50	41. 54	1,350
1,360	41. 54	41. 58	41. 61	41. 65	41. 68	41. 72	41. 75	41. 79	41. 82	41. 86	41. 90	1,360
1,370	41. 90	41. 93	41. 97	42. 00	42. 04	42. 07	42. 11	42. 14	42. 18	42. 22	42. 25	1,370
1,380	42. 25	42. 29	42. 32	42. 36	42. 39	42. 43	42. 46	42. 50	42. 53	42. 57	42. 61	1,380
1,390	42. 61	42. 64	42. 68	42. 71	42. 75	42. 78	42. 82	42. 85	42. 89	42. 92	42. 96	1,390
1,400	42. 96	43. 00	43. 03	43. 07	43. 10	43. 14	43. 18	43. 21	43. 25	43. 28	43. 32	1,400
1,410	43. 32	43. 35	43. 39	43. 43	43. 46	43. 50	43. 53	43. 57	43. 60	43. 64	43. 68	1,410
1,420	43. 68	43. 71	43. 75	43. 78	43. 82	43. 85	43. 89	43. 92	43. 96	44. 00	44. 03	1,420
1,430	44. 03	44. 07	44. 10	44. 14	44. 18	44. 21	44. 25	44. 28	44. 32	44. 35	44. 39	1,430
1,440	44. 39	44. 42	44. 46	44. 50	44. 53	44. 57	44. 60	44. 64	44. 68	44. 71	44. 75	1,440
1,450	44. 75	44. 78	44. 82	44. 85	44. 89	44. 93	44. 96	45. 00	45. 03	45. 07	45. 10	1,450
1,460	45. 10	45. 14	45. 18	45. 21	45. 25	45. 28	45. 32	45. 35	45. 39	45. 42	45. 46	1,460
1,470	45. 46	45. 50	45. 53	45. 57	45. 60	45. 64	45. 68	45. 71	45. 75	45. 78	45. 82	1,470
1,480	45. 82	45. 85	45. 89	45. 92	45. 96	46. 00	46. 03	46. 07	46. 10	46. 14	46. 18	1,480
1,490	46. 18	46. 21	46. 25	46. 28	46. 32	46. 35	46. 39	46. 42	46. 46	46. 50	46. 53	1,490
1,500	46. 53	46. 57	46. 60	46. 64	46. 67	46. 71	46. 74	46. 78	46. 82	46. 85	46. 89	1,500
1,510	46. 89	46. 92	46. 96	47 .00	47. 03	47. 07	47. 10	47. 14	47. 17	47. 21	47. 24	1,510
1,520	47. 24	47. 28	47. 32	47. 35	47. 39	47. 42	47. 46	47. 49	47. 53	47. 56	47. 60	1,520
1,530	47. 60	47. 63	47. 67	47. 70	47. 74	47. 78	47. 81	47. 85	47. 88	47. 92	47. 95	1,530
1,540	47. 95	47. 99	48. 02	48. 06	48. 09	48. 13	48. 16	48. 20	48. 24	48. 27	48. 31	1,540
1,550	48. 31	48. 34	48. 38	48. 41	48. 45	48. 48	48. 52	48. 55	48. 59	48. 62	48. 66	1,550
1,560	48. 66	48. 69	48. 73	48. 76	48. 80	48. 83	48. 87	48. 90	48. 94	48. 97	49. 01	1,560
1,570	49. 01	49. 04	49. 08	49. 11	49. 15	49. 18	49. 22	49. 25	49. 29	49. 32	49. 36	1,570
1,580	49. 36	49. 39	49. 43	49. 46	49. 50	49. 53	49. 56	49. 60	49. 63	49. 67	49. 70	1,580
1,590	49. 70	49. 74	49. 77	49. 81	49. 84	49. 88	49. 91	49. 94	49. 98	50. 01	50. 05	1,590
1,600	50. 05											1,600
°F	0	1	2	3	4	5	6	7	8	9	10	°F

TABLE 43—*Type J thermocouples.*
Temperature in Degrees Celsius

EMF in Absolute Millivolts   Reference Junctions 0 C

°C	0	1	2	3	4	5	6	7	8	9	10	°C
						Millivolts						
−190	−7.66	−7.69	−7.71	−7.73	−7.76	−7.78						−190
−180	−7.40	−7.43	−7.46	−7.49	−7.51	−7.54	−7.56	−7.59	−7.61	−7.64	−7.66	−180
−170	−7.12	−7.15	−7.18	−7.21	−7.24	−7.27	−7.30	−7.32	−7.35	−7.38	−7.40	−170
−160	−6.82	−6.85	−6.88	−6.91	−6.94	−6.97	−7.00	−7.03	−7.06	−7.09	−7.12	−160
−150	−6.50	−6.53	−6.56	−6.60	−6.63	−6.66	−6.69	−6.72	−6.76	−6.79	−6.82	−150
−140	−6.16	−6.19	−6.22	−6.26	−6.29	−6.33	−6.36	−6.40	−6.43	−6.46	−6.50	−140
−130	−5.80	−5.84	−5.87	−5.91	−5.94	−5.98	−6.01	−6.05	−6.08	−6.12	−6.16	−130
−120	−5.42	−5.46	−5.50	−5.54	−5.58	−5.61	−5.65	−5.69	−5.72	−5.76	−5.80	−120
−110	−5.03	−5.07	−5.11	−5.15	−5.19	−5.23	−5.27	−5.31	−5.35	−5.38	−5.42	−110
−100	−4.63	−4.67	−4.71	−4.75	−4.79	−4.83	−4.87	−4.91	−4.95	−4.99	−5.03	−100
−90	−4.21	−4.25	−4.30	−4.34	−4.38	−4.42	−4.46	−4.50	−4.55	−4.59	−4.63	−90
−80	−3.78	−3.82	−3.87	−3.91	−3.96	−4.00	−4.04	−4.08	−4.13	−4.17	−4.21	−80
−70	−3.34	−3.38	−3.43	−3.47	−3.52	−3.56	−3.60	−3.65	−3.69	−3.74	−3.78	−70
−60	−2.89	−2.94	−2.98	−3.03	−3.07	−3.12	−3.16	−3.21	−3.25	−3.30	−3.34	−60
−50	−2.43	−2.48	−2.52	−2.57	−2.62	−2.66	−2.71	−2.75	−2.80	−2.84	−2.89	−50
−40	−1.96	−2.01	−2.06	−2.10	−2.15	−2.20	−2.24	−2.29	−2.34	−2.38	−2.43	−40
−30	−1.48	−1.53	−1.58	−1.63	−1.67	−1.72	−1.77	−1.82	−1.87	−1.91	−1.96	−30
−20	−1.00	−1.04	−1.09	−1.14	−1.19	−1.24	−1.29	−1.34	−1.39	−1.43	−1.48	−20
−10	−0.50	−0.55	−0.60	−0.65	−0.70	−0.75	−0.80	−0.85	−0.90	−0.95	−1.00	−10
(−)0	0.00	−0.05	−0.10	−0.15	−0.20	−0.25	−0.30	−0.35	−0.40	−0.45	−0.50	(−)0
(+)0	0.00	0.05	0.10	0.15	0.20	0.25	0.30	0.35	0.40	0.45	0.50	(+)0
10	0.50	0.56	0.61	0.66	0.71	0.76	0.81	0.86	0.91	0.97	1.02	10
20	1.02	1.07	1.12	1.17	1.22	1.28	1.33	1.38	1.43	1.48	1.54	20
30	1.54	1.59	1.64	1.69	1.74	1.80	1.85	1.90	1.95	2.00	2.06	30
40	2.06	2.11	2.16	2.22	2.27	2.32	2.37	2.42	2.48	2.53	2.58	40
50	2.58	2.64	2.69	2.74	2.80	2.85	2.90	2.96	3.01	3.06	3.11	50
60	3.11	3.17	3.22	3.27	3.33	3.38	3.43	3.49	3.54	3.60	3.65	60
70	3.65	3.70	3.76	3.81	3.86	3.92	3.97	4.02	4.08	4.13	4.19	70
80	4.19	4.24	4.29	4.35	4.40	4.46	4.51	4.56	4.62	4.67	4.73	80
90	4.73	4.78	4.83	4.89	4.94	5.00	5.05	5.10	5.16	5.21	5.27	90
100	5.27	5.32	5.38	5.43	5.48	5.54	5.59	5.65	5.70	5.76	5.81	100
110	5.81	5.86	5.92	5.97	6.03	6.08	6.14	6.19	6.25	6.30	6.36	110
120	6.36	6.41	6.47	6.52	6.58	6.63	6.68	6.74	6.79	6.85	6.90	120
130	6.90	6.96	7.01	7.07	7.12	7.18	7.23	7.29	7.34	7.40	7.45	130
140	7.45	7.51	7.56	7.62	7.67	7.73	7.78	7.84	7.89	7.95	8.00	140
150	8.00	8.06	8.12	8.17	8.23	8.28	8.34	8.39	8.45	8.50	8.56	150
160	8.56	8.61	8.67	8.72	8.78	8.84	8.89	8.95	9.00	9.06	9.11	160
170	9.11	9.17	9.22	9.28	9.33	9.39	9.44	9.50	9.56	9.61	9.67	170
180	9.67	9.72	9.78	9.83	9.89	9.95	10.00	10.06	10.11	10.17	10.22	180
190	10.22	10.28	10.34	10.39	10.45	10.50	10.56	10.61	10.67	10.72	10.78	190
200	10.78	10.84	10.89	10.95	11.00	11.06	11.12	11.17	11.23	11.28	11.34	200
°C	0	1	2	3	4	5	6	7	8	9	10	°C

TABLE 43—*Type J thermocouples (continued).*

Temperature in Degrees Celsius

EMF in Absolute Millivolts   Reference Junctions 0 C

°C	0	1	2	3	4	5	6	7	8	9	10	°C
						Millivolts						
200	10.78	10.84	10.89	10.95	11.00	11.06	11.12	11.17	11.23	11.28	11.34	200
210	11.34	11.39	11.45	11.50	11.56	11.62	11.67	11.73	11.78	11.84	11.89	210
220	11.89	11.95	12.00	12.06	12.12	12.17	12.23	12.28	12.34	12.39	12.45	220
230	12.45	12.50	12.56	12.62	12.67	12.73	12.78	12.84	12.89	12.95	13.01	230
240	13.01	13.06	13.12	13.17	13.23	13.28	13.34	13.40	13.45	13.51	13.56	240
250	13.56	13.62	13.67	13.73	13.78	13.84	13.89	13.95	14.00	14.06	14.12	250
260	14.12	14.17	14.23	14.28	14.34	14.39	14.45	14.50	14.56	14.61	14.67	260
270	14.67	14.72	14.78	14.83	14.89	14.94	15.00	15.06	15.11	15.17	15.22	270
280	15.22	15.28	15.33	15.39	15.44	15.50	15.55	15.61	15.66	15.72	15.77	280
290	15.77	15.83	15.88	15.94	16.00	16.05	16.11	16.16	16.22	16.27	16.33	290
300	16.33	16.38	16.44	16.49	16.55	16.60	16.66	16.71	16.77	16.82	16.88	300
310	16.88	16.93	16.99	17.04	17.10	17.15	17.21	17.26	17.32	17.37	17.43	310
320	17.43	17.48	17.54	17.60	17.65	17.71	17.76	17.82	17.87	17.93	17.98	320
330	17.98	18.04	18.09	18.15	18.20	18.26	18.32	18.37	18.43	18.48	18.54	330
340	18.54	18.59	18.65	18.70	18.76	18.81	18.87	18.92	18.98	19.03	19.09	340
350	19.09	19.14	19.20	19.26	19.31	19.37	19.42	19.48	19.53	19.59	19.64	350
360	19.64	19.70	19.75	19.81	19.86	19.92	19.97	20.03	20.08	20.14	20.20	360
370	20.20	20.25	20.31	20.36	20.42	20.47	20.53	20.58	20.64	20.69	20.75	370
380	20.75	20.80	20.86	20.91	20.97	21.02	21.08	21.13	21.19	21.24	21.30	380
390	21.30	21.35	21.41	21.46	21.52	21.57	21.63	21.68	21.74	21.79	21.85	390
400	21.85	21.90	21.96	22.02	22.07	22.13	22.18	22.24	22.29	22.35	22.40	400
410	22.40	22.46	22.51	22.57	22.62	22.68	22.73	22.79	22.84	22.90	22.95	410
420	22.95	23.01	23.06	23.12	23.17	23.23	23.28	23.34	23.39	23.45	23.50	420
430	23.50	23.56	23.61	23.67	23.72	23.78	23.83	23.89	23.94	24.00	24.06	430
440	24.06	24.11	24.17	24.22	24.28	24.33	24.39	24.44	24.50	24.55	24.61	440
450	24.61	24.66	24.72	24.77	24.83	24.88	24.94	25.00	25.05	25.11	25.16	450
460	25.16	25.22	25.27	25.33	25.38	25.44	25.49	25.55	25.60	25.66	25.72	460
470	25.72	25.77	25.83	25.88	25.94	25.99	26.05	26.10	26.16	26.22	26.27	470
480	26.27	26.33	26.38	26.44	26.49	26.55	26.61	26.66	26.72	26.77	26.83	480
490	26.83	26.89	26.94	27.00	27.05	27.11	27.17	27.22	27.28	27.33	27.39	490
500	27.39	27.45	27.50	27.56	27.61	27.67	27.73	27.78	27.84	27.90	27.95	500
510	27.95	28.01	28.07	28.12	28.18	28.23	28.29	28.35	28.40	28.46	28.52	510
520	28.52	28.57	28.63	28.69	28.74	28.80	28.86	28.91	28.97	29.02	29.08	520
530	29.08	29.14	29.20	29.25	29.31	29.37	29.42	29.48	29.54	29.59	29.65	530
540	29.65	29.71	29.76	29.82	29.88	29.94	29.99	30.05	30.11	30.16	30.22	540
550	30.22	30.28	30.34	30.39	30.45	30.51	30.57	30.62	30.68	30.74	30.80	550
560	30.80	30.85	30.91	30.97	31.02	31.08	31.14	31.20	31.26	31.31	31.37	560
570	31.37	31.43	31.49	31.54	31.60	31.66	31.72	31.78	31.83	31.89	31.95	570
580	31.95	32.01	32.06	32.12	32.18	32.24	32.30	32.36	32.41	32.47	32.53	580
590	32.53	32.59	32.65	32.71	32.76	32.82	32.88	32.94	33.00	33.06	33.11	590
600	33.11	33.17	33.23	33.29	33.35	33.41	33.46	33.52	33.58	33.64	33.70	600
°C	0	1	2	3	4	5	6	7	8	9	10	°C

TABLE 43—*Type J thermocouples (continued)*.
Temperature in Degrees Celsius

EMF in Absolute Millivolts

Reference Junctions 0 C

°C	0	1	2	3	4	5	6	7	8	9	10	°C
						Millivolts						
600	33. 11	33. 17	33. 23	33. 29	33. 35	33. 41	33. 46	33: 52	33. 58	33. 64	33. 70	600
610	33. 70	33. 76	33. 82	33. 88	33. 94	33. 99	34. 05	34. 11	34. 17	34. 23	34. 29	610
620	34. 29	34. 35	34. 41	34. 47	34. 53	34. 58	34. 64	34. 70	34. 76	34. 82	34. 88	620
630	34. 88	34. 94	35. 00	35. 06	35. 12	35. 18	35. 24	35. 30	35. 36	35. 42	35. 48	630
640	35. 48	35. 54	35. 60	35. 66	35. 72	35. 78	35. 84	35. 90	35. 96	36. 02	36. 08	640
650	36. 08	36. 14	36. 20	36. 26	36. 32	36. 38	36. 44	36. 50	36. 56	36. 62	36. 69	650
660	36. 69	36. 75	36. 81	36. 87	36. 93	36. 99	37. 05	37. 11	37. 18	37. 24	37. 30	660
670	37. 30	37. 36	37. 42	37. 48	37. 54	37. 60	37. 66	37. 73	37. 79	37. 85	37. 91	670
680	37. 91	37. 97	38. 04	38. 10	38. 16	38. 22	38. 28	38. 34	38. 41	38. 47	38. 53	680
690	38. 53	38. 59	38. 66	38. 72	38. 78	38. 84	38. 90	38. 97	39. 03	39. 09	39. 15	690
700	39. 15	39. 22	39. 28	39. 34	39. 40	39. 47	39. 53	39. 59	39. 65	39. 72	39. 78	700
710	39. 78	39. 84	39. 91	39. 97	40. 03	40. 10	40. 16	40. 22	40. 28	40. 35	40. 41	710
720	40. 41	40. 48	40. 54	40. 60	40. 66	40. 73	40. 79	40. 86	40. 92	40. 98	41. 05	720
730	41. 05	41. 11	41. 17	41. 24	41. 30	41. 36	41. 43	41. 49	41. 56	41. 62	41. 68	730
740	41. 68	41. 75	41. 81	41. 87	41. 94	42. 00	42. 07	42. 13	42. 19	42. 26	42. 32	740
750	42. 32	42. 38	42. 45	42. 51	42. 58	42. 64	42. 70	42. 77	42. 83	42. 90	42. 96	750
760	42. 96	43. 02	43. 09	43. 15	43. 22	43. 28	43. 35	43. 41	43. 48	43. 54	43. 60	760
770	43. 60	43. 67	43. 73	43. 80	43. 86	43. 92	43. 99	44. 05	44. 12	44. 18	44. 25	770
780	44. 25	44. 31	44. 38	44. 44	44. 50	44. 57	44. 63	44. 70	44. 76	44. 82	44. 89	780
790	44. 89	44. 95	45. 02	45. 08	45. 15	45. 21	45. 28	45. 34	45. 40	45. 47	45. 53	790
800	45. 53	45. 60	45. 66	45. 72	45 79	45. 85	45. 92	45. 98	46. 05	46. 11	46. 18	800
810	46. 18	46. 24	46. 30	46. 37	46. 43	46. 50	46. 56	46. 62	46. 69	46. 75	46. 82	810
820	46. 82	46. 88	46. 94	47. 01	47. 07	47. 14	47. 20	47. 27	47. 33	47. 39	47. 46	820
830	47. 46	47. 52	47. 58	47. 65	47. 71	47. 78	47. 84	47. 90	47. 97	48. 03	48. 09	830
840	48. 09	48. 16	48. 22	48. 28	48. 35	48. 41	48. 48	48. 54	48. 60	48. 66	48. 73	840
850	48. 73	48. 79	48. 85	48. 92	48. 98	49. 04	49. 10	49. 17	49. 23	49 29	49. 36	850
860	49. 36	49. 42	49. 48	49. 54	49. 61	49. 67	49. 73	49. 79	49. 86	49 92	49. 98	860
870	49. 98	50. 04										870
°C.	0	1	2	3	4	5	6	7	8	9	10	°C

TABLE 44—*Type K thermocouples.*
Temperature in Degrees Fahrenheit

EMF in Absolute Millivolts

Reference Junctions at 32 F

°F	0	1	2	3	4	5	6	7	8	9	10	°F
						Millivolts						
−300	−5. 51	−5. 52	−5. 53	−5. 54	−5. 54	−5. 55	−5. 56	−5. 57	−5. 58	−5. 59	−5. 60	−300
−290	−5. 41	−5. 42	−5. 43	−5. 44	−5. 45	−5. 46	−5. 47	−5. 48	−5. 49	−5. 50	−5. 51	−290
−280	−5. 30	−5. 31	−5. 32	−5. 34	−5. 35	−5. 36	−5. 37	−5. 38	−5. 39	−5. 40	−5. 41	−280
−270	−5. 20	−5. 21	−5. 22	−5. 23	−5. 24	−5. 25	−5. 26	−5. 27	−5. 28	−5. 29	−5. 30	−270
−260	−5. 08	−5. 09	−5. 10	−5. 12	−5. 13	−5. 14	−5. 15	−5. 16	−5. 17	−5. 18	−5. 20	−260
−250	−4. 96	−4. 97	−4. 99	−5. 00	−5. 01	−5. 02	−5. 03	−5. 04	−5. 06	−5. 07	−5. 08	−250
−240	−4. 84	−4. 85	−4. 86	−4. 88	−4. 89	−4. 90	−4. 91	−4. 92	−4. 94	−4. 95	−4. 96	−240
−230	−4. 71	−4. 72	−4. 74	−4. 75	−4. 76	−4. 77	−4. 79	−4. 80	−4. 81	−4. 82	−4. 84	−230
−220	−4. 58	−4. 59	−4. 60	−4. 62	−4. 63	−4. 64	−4. 66	−4. 67	−4. 68	−4. 70	−4. 71	−220
−210	−4. 44	−4. 45	−4. 46	−4. 48	−4. 49	−4. 51	−4. 52	−4. 53	−4. 55	−4. 56	−4. 58	−210
−200	−4. 29	−4. 31	−4. 32	−4. 34	−4. 35	−4. 36	−4. 38	−4. 39	−4. 41	−4. 42	−4. 44	−200
−190	−4. 15	−4. 16	−4. 18	−4. 19	−4. 21	−4. 22	−4. 24	−4. 25	−4. 26	−4. 28	−4. 29	−190
−180	−4. 00	−4. 01	−4. 03	−4. 04	−4. 06	−4. 07	−4. 09	−4. 10	−4. 12	−4. 13	−4. 15	−180
−170	−3. 84	−3. 86	−3. 88	−3. 89	−3. 91	−3. 92	−3. 94	−3. 95	−3. 97	−3. 98	−4. 00	−170
−160	−3. 69	−3. 70	−3. 72	−3. 73	−3. 75	−3. 76	−3. 78	−3. 80	−3. 81	−3. 83	−3. 84	−160
−150	−3. 52	−3. 54	−3. 56	−3. 57	−3. 59	−3. 60	−3. 62	−3. 64	−3. 65	−3. 67	−3. 69	−150
−140	−3. 36	−3. 38	−3. 39	−3. 41	−3. 42	−3. 44	−3. 46	−3. 47	−3. 49	−3. 51	−3. 52	−140
−130	−3. 19	−3. 20	−3. 22	−3. 24	−3. 25	−3. 27	−3. 29	−3. 31	−3. 32	−3. 34	−3. 36	−130
−120	−3. 01	−3. 03	−3. 05	−3. 06	−3. 08	−3. 10	−3. 12	−3. 13	−3. 15	−3. 17	−3. 19	−120
−110	−2. 83	−2. 85	−2. 87	−2. 89	−2. 90	−2. 92	−2. 94	−2. 96	−2. 98	−2. 99	−3. 01	−110
−100	−2. 65	−2. 67	−2. 69	−2. 71	−2. 72	−2. 74	−2. 76	−2. 78	−2. 80	−2. 82	−2. 83	−100
−90	−2. 47	−2. 49	−2. 50	−2. 52	−2. 54	−2. 56	−2. 58	−2. 60	−2. 62	−2. 63	−2. 65	−90
−80	−2. 28	−2. 30	−2. 32	−2. 34	−2. 36	−2. 37	−2. 39	−2. 41	−2. 43	−2. 45	−2. 47	−80
−70	−2. 09	−2. 11	−2. 13	−2. 15	−2. 17	−2. 18	−2. 20	−2. 22	−2. 24	−2. 26	−2. 28	−70
−60	−1. 90	−1. 92	−1. 94	−1. 96	−1. 97	−1. 99	−2. 01	−2. 03	−2. 05	−2. 07	−2. 09	−60
−50	−1. 70	−1. 72	−1. 74	−1. 76	−1. 78	−1. 80	−1. 82	−1. 84	−1. 86	−1. 88	−1. 90	−50
−40	−1. 50	−1. 52	−1. 54	−1. 56	−1. 58	−1. 60	−1. 62	−1. 64	−1. 66	−1. 68	−1. 70	−40
−30	−1. 30	−1. 32	−1. 34	−1. 36	−1. 38	−1. 40	−1. 42	−1. 44	−1. 46	−1. 48	−1. 50	−30
−20	−1. 10	−1. 12	−1. 14	−1. 16	−1. 18	−1. 20	−1. 22	−1. 24	−1. 26	−1. 28	−1. 30	−20
−10	−0. 89	−0. 91	−0. 93	−0. 95	−0. 97	−0. 99	−1. 01	−1. 03	−1. 06	−1. 08	−1. 10	−10
(−)0	−0. 68	−0. 70	−0. 72	−0. 75	−0. 77	−0. 79	−0. 81	−0. 83	−0. 85	−0. 87	−0. 89	(−)0
(+)0	−0. 68	−0. 66	−0. 64	−0. 62	−0. 60	−0. 58	−0. 56	−0. 54	−0. 52	−0. 49	−0. 47	(+)0
10	−0. 47	−0. 45	−0. 43	−0. 41	−0. 39	−0. 37	−0. 34	−0. 32	−0. 30	−0. 28	−0. 26	10
20	−0. 26	−0. 24	−0. 22	−0. 19	−0. 17	−0. 15	−0. 13	−0. 11	−0. 09	−0. 07	−0. 04	20
30	−0. 04	−0. 02	0. 00	0. 02	0. 04	0. 07	0. 09	0. 11	0. 13	0. 15	0. 18	30
40	0. 18	0. 20	0. 22	0. 24	0. 26	0. 29	0. 31	0. 33	0. 35	0. 37	0. 40	40
50	0. 40	0. 42	0. 44	0. 46	0. 48	0. 51	0. 53	0. 55	0. 57	0. 60	0. 62	50
60	0. 62	0. 64	0. 66	0. 68	0. 71	0. 73	0. 75	0. 77	0. 80	0. 82	0. 84	60
70	0. 84	0. 86	0. 88	0. 91	0. 93	0. 95	0. 97	1. 00	1. 02	1. 04	1. 06	70
80	1. 06	1. 09	1. 11	1. 13	1. 15	1. 18	1. 20	1. 22	1. 24	1. 27	1. 29	80
90	1. 29	1. 31	1. 33	1. 36	1. 38	1. 40	1. 43	1. 45	1. 47	1. 49	1. 52	90
100	1. 52	1. 54	1. 56	1. 58	1. 61	1. 63	1. 65	1. 68	1. 70	1. 72	1. 74	100
°F	0	1	2	3	4	5	6	7	8	9	10	°F

TABLE 44—*Type K thermocouples (continued).*
Temperature in Degrees Fahrenheit

EMF in Absolute Millivolts

Reference Junctions at 32 F

°F	0	1	2	3	4	5	6	7	8	9	10	°F
						Millivolts						
100	1. 52	1. 54	1. 56	1. 58	1. 61	1. 63	1. 65	1. 68	1. 70	1. 72	1. 74	100
110	1. 74	1. 77	1. 79	1. 81	1. 84	1. 86	1. 88	1. 90	1. 93	1. 95	1. 97	110
120	1. 97	2. 00	2. 02	2. 04	2. 06	2. 09	2. 11	2. 13	2. 16	2. 18	2. 20	120
130	2. 20	2. 23	2. 25	2. 27	2. 29	2. 32	2. 34	2. 36	2. 39	2. 41	2. 43	130
140	2. 43	2. 46	2. 48	2. 50	2. 52	2. 55	2. 57	2. 59	2. 62	2. 64	2. 66	140
150	2. 66	2. 69	2. 71	2. 73	2. 75	2. 78	2. 80	2. 82	2. 85	2. 87	2. 89	150
160	2. 89	2. 92	2. 94	2. 96	2. 98	3. 01	3. 03	3. 05	3. 08	3. 10	3. 12	160
170	3. 12	3. 15	3. 17	3. 19	3. 22	3. 24	3. 26	3. 29	3. 31	3. 33	3. 36	170
180	3. 36	3. 38	3. 40	3. 43	3. 45	3. 47	3. 49	3. 52	3. 54	3. 56	3. 59	180
190	3. 59	3. 61	3. 63	3. 66	3. 68	3. 70	3. 73	3. 75	3. 77	3. 80	3. 82	190
200	3. 82	3. 84	3. 87	3. 89	3. 91	3. 94	3. 96	3. 98	4. 01	4. 03	4. 05	200
210	4. 05	4. 08	4. 10	4. 12	4. 15.	4. 17	4. 19	4. 21	4. 24	4. 26	4. 28	210
220	4. 28	4. 31	4. 33	4. 35	4. 38	4. 40	4. 42	4. 44	4. 47	4. 49	4. 51	220
230	4. 51	4. 54	4. 56	4. 58	4. 61	4. 63	4. 65	4. 67	4. 70	4. 72	4. 74	230
240	4. 74	4. 77	4. 79	4. 81	4. 83	4. 86	4. 88	4. 90	4. 92	4. 95	4. 97	240
250	4. 97	4. 99	5. 02	5. 04	5. 06	5. 08	5. 11	5. 13	5. 15	5. 17	5. 20	250
260	5. 20	5. 22	5. 22	5. 24	5. 26	5. 29	5.31	5. 33	5. 35	5. 38	5. 40	5. 42	260
270	5. 42	5. 44	5. 47	5. 49	5. 51	5. 53	5. 56	5. 58	5. 60	5. 62	5. 65	270
280	5. 65	5. 67	5. 69	5. 71	5. 73	5. 76	5. 78	5. 80	5. 82	5. 85	5. 87	280
290	5. 87	5. 89	5. 91	5. 93	5. 96	5. 98	6. 00	6. 02	6. 05	6. 07	6. 09	290
300	6. 09	6. 11	6. 13	6. 16	6. 18	6. 20	6. 22	6. 25	6. 27	6. 29	6. 31	300
310	6. 31	6. 33	6. 36	6. 38	6. 40	6. 42	6. 45	6. 47	6. 49	6. 51	6. 53	310
320	6. 53	6. 56	6. 58	6. 60	6. 62	6. 65	6. 67	6. 69	6. 71	6. 73	6. 76	320
330	6. 76	6. 78	6. 80	6. 82	6. 84	6. 87	6. 89	6. 91	6. 93	6. 96	6. 98	330
340	6. 98	7. 00	7. 02	7. 04	7. 07	7. 09	7. 11	7. 13	7. 15	7. 18	7. 20	340
350	7. 20	7. 22	7. 24	7. 26	7. 29	7. 31	7. 33	7. 35	7. 38	7. 40	7. 42	350
360	7. 42	7. 44	7. 46	7. 49	7. 51	7. 53	7. 55	7. 58	7. 60	7. 62	7. 64	360
370	7. 64	7. 66	7. 69	7. 71	7. 73	7. 75	7. 78	7. 80	7. 82	7. 84	7. 87	370
380	7. 87	7. 89	7. 91	7. 93	7. 95	7. 98	8. 00	8. 02	8. 04	8. 07	8. 09	380
390	8. 09	8. 11	8. 13	8. 16	8. 18	8. 20	8. 22	8. 24	8. 27	8. 29	8. 31	390
400	8. 31	8. 33	8. 36	8. 38	8. 40	8. 42	8. 45	8. 47	8. 49	8. 51	8. 54	400
410	8. 54	8. 56	8. 58	8. 60	8. 62	8. 65	8. 67	8. 69	8. 71	8. 74	8. 76	410
420	8. 76	8. 78	8. 80	8. 82	8. 85	8. 87	8. 89	8. 91	8. 94	8. 96	8. 98	420
430	8. 98	9. 00	9. 03	9. 05	9. 07	9. 09	9. 12	9. 14	9. 16	9. 18	9. 21	430
440	9. 21	9. 23	9. 25	9. 27	9. 30	9. 32	9. 34	9. 36	9. 39	9. 41	9. 43	440
450	9. 43	9. 45	9. 48	9. 50	9. 52	9. 54	9. 57	9. 59	9. 61	9. 63	9. 66	450
460	9. 66	9. 68	9. 70	9. 73	9. 75	9. 77	9. 79	9. 82	9. 84	9. 86	9. 88	460
470	9. 88	9. 91	9. 93	9. 95	9. 97	10. 00	10. 02	10. 04	10. 06	10. 09	10. 11	470
480	10. 11	10. 13	10. 16	10. 18	10. 20	10. 22	10. 25	10. 27	10. 29	10. 31	10. 34	480
490	10. 34	10. 36	10. 38	10. 40	10. 43	10. 45	10. 47	10. 50	10. 52	10. 54	10. 57	490
500	10. 57	10. 59	10. 61	10. 63	10. 66	10. 68	10. 70	10. 72	10. 75	10. 77	10. 79	500
°F	0	1	2	3	4	5	6	7	8	9	10	°F

TABLE 44—*Type K thermocouples (continued).*
Temperature in Degrees Fahrenheit

EMF in Absolute Millivolts   Reference Junctions at 32 F

°F	0	1	2	3	4	5	6	7	8	9	10	°F
						Millivolts						
500	10.57	10.59	10.61	10.63	10.66	10.68	10.70	10.72	10.75	10.77	10.79	500
510	10.79	10.82	10.84	10.86	10.88	10.91	10.93	10.95	10.98	11.00	11.02	510
520	11.02	11.04	11.07	11.09	11.11	11.13	11.16	11.18	11.20	11.23	11.25	520
530	11.25	11.27	11.29	11.32	11.34	11.36	11.39	11.41	11.43	11.45	11.48	530
540	11.48	11.50	11.52	11.55	11.57	11.59	11.61	11.64	11.66	11.68	11.71	540
550	11.71	11.73	11.75	11.78	11.80	11.82	11.84	11.87	11.89	11.91	11.94	550
560	11.94	11.96	11.98	12.01	12.03	12.05	12.07	12.10	12.12	12.14	12.17	560
570	12.17	12.19	12.21	12.24	12.26	12.28	12.30	12.33	12.35	12.37	12.40	570
580	12.40	12.42	12.44	12.47	12.49	12.51	12.53	12.56	12.58	12.60	12.63	580
590	12.63	12.65	12.67	12.70	12.72	12.74	12.76	12.79	12.81	12.83	12.86	590
600	12.86	12.88	12.90	12.93	12.95	12.97	13.00	13.02	13.04	13.06	13.09	600
610	13.09	13.11	13.13	13.16	13.18	13.20	13.23	13.25	13.27	13.30	13.32	610
620	13.32	13.34	13.36	13.39	13.41	13.44	13.46	13.48	13.50	13.53	13.55	620
630	13.55	13.57	13.60	13.62	13.64	13.67	13.69	13.71	13.74	13.76	13.78	630
640	13.78	13.81	13.83	13.85	13.88	13.90	13.92	13.95	13.97	13.99	14.02	640
650	14.02	14.04	14.06	14.09	14.11	14.13	14.15	14.18	14.20	14.22	14.25	650
660	14.25	14.27	14.29	14.32	14.34	14.36	14.39	14.41	14.43	14.46	14.48	660
670	14.48	14.50	14.53	14.55	14.57	14.60	14.62	14.64	14.67	14.69	14.71	670
680	14.71	14.74	14.76	14.78	14.81	14.83	14.85	14.88	14.90	14.92	14.95	680
690	14.95	14.97	14.99	15.02	15.04	15.06	15.09	15.11	15.13	15.16	15.18	690
700	15.18	15.20	15.23	15.25	15.27	15.30	15.32	15.34	15.37	15.39	15.41	700
710	15.41	15.44	15.46	15.48	15.51	15.53	15.55	15.58	15.60	15.62	15.65	710
720	15.65	15.67	15.69	15.72	15.74	15.76	15.79	15.81	15.83	15.86	15.88	720
730	15.88	15.90	15.93	15.95	15.98	16.00	16.02	16.05	16.07	16.09	16.12	730
740	16.12	16.14	16.16	16.19	16.21	16.23	16.26	16.28	16.30	16.33	16.35	740
750	16.35	16.37	16.40	16.42	16.45	16.47	16.49	16.52	16.54	16.56	16.59	750
760	16.59	16.61	16.63	16.66	16.68	16.70	16.73	16.75	16.77	16.80	16.82	760
770	16.82	16.84	16.87	16.89	16.92	16.94	16.96	16.99	17.01	17.03	17.06	770
780	17.06	17.08	17.10	17.13	17.15	17.17	17.20	17.22	17.24	17.27	17.29	780
790	17.29	17.31	17.34	17.36	17.39	17.41	17.43	17.46	17.48	17.50	17.53	790
800	17.53	17.55	17.57	17.60	17.62	17.64	17.67	17.69	17.71	17.74	17.76	800
810	17.76	17.78	17.81	17.83	17.86	17.88	17.90	17.93	17.95	17.97	18.00	810
820	18.00	18.02	18.04	18.07	18.09	18.11	18.14	18.16	18.18	18.21	18.23	820
830	18.23	18.25	18.28	18.30	18.33	18.35	18.37	18.40	18.42	18.44	18.47	830
840	18.47	18.49	18.51	18.54	18.56	18.58	18.61	18.63	18.65	18.68	18.70	840
850	18.70	18.73	18.75	18.77	18.80	18.82	18.84	18.87	18.89	18.91	18.94	850
860	18.94	18.96	18.99	19.01	19.03	19.06	19.08	19.10	19.13	19.15	19.18	860
870	19.18	19.20	19.22	19.25	19.27	19.29	19.32	19.34	19.36	19.39	19.41	870
880	19.41	19.44	19.46	19.48	19.51	19.53	19.55	19.58	19.60	19.63	19.65	880
890	19.65	19.67	19.70	19.72	19.75	19.77	19.79	19.82	19.84	19.86	19.89	890
900	19.89	19.91	19.94	19.96	19.98	20.01	20.03	20.05	20.08	20.10	20.13	900
°F	0	1	2	3	4	5	6	7	8	9	10	°F

TABLE 44—*Type K thermocouples (continued).*
Temperature in Degrees Fahrenheit

EMF in Absolute Millivolts   Reference Junctions at 32 F

°F	0	1	2	3	4	5	6	7	8	9	10	°F
						Millivolts						
900	19. 89	19. 91	19. 94	19. 96	19. 98	20. 01	20. 03	20. 05	20. 08	20. 10	20. 13	900
910	20. 13	20. 15	20. 17	20. 20	20. 22	20. 24	20. 27	20. 29	20. 32	20. 34	20. 36	910
920	20. 36	20. 39	20. 41	20. 43	20. 46	20. 48	20. 50	20. 53	20. 55	20. 58	20. 60	920
930	20. 60	20. 62	20. 65	20. 67	20. 69	20. 72	20. 74	20. 76	20. 79	20. 81	20. 84	930
940	20. 84	20. 86	20. 88	20. 91	20. 93	20. 95	20. 98	21. 00	21. 03	21. 05	21. 07	940
950	21. 07	21. 10	21. 12	21. 14	21. 17	21. 19	21. 21	21. 24	21. 26	21. 28	21. 31	950
960	21. 31	21. 33	21. 36	21. 38	21. 40	21. 43	21. 45	21. 47	21. 50	21. 52	21. 54	960
970	21. 54	21. 57	21. 59	21. 62	21. 64	21. 66	21. 69	21. 71	21. 73	21. 76	21. 78	970
980	21. 78	21. 81	21. 83	21. 85	21. 88	21. 90	21. 92	21. 95	21. 97	21. 99	22. 02	980
990	22. 02	22. 04	22. 07	22. 09	22. 11	22. 14	22. 16	22. 18	22. 21	22. 23	22. 26	990
1, 000	22. 26	22. 28	22. 30	22. 33	22. 35	22. 37	22. 40	22. 42	22. 44	22. 47	22. 49	1, 000
1, 010	22. 49	22. 52	22. 54	22. 56	22. 59	22. 61	22. 63	22. 66	22. 68	22. 71	22. 73	1, 010
1, 020	22. 73	22. 75	22. 78	22. 80	22. 82	22. 85	22. 87	22. 90	22. 92	22. 94	22. 97	1, 020
1, 030	22. 97	22. 99	23. 01	23. 04	23. 06	23. 08	23. 11	23. 13	23. 16	23. 18	23. 20	1, 030
1, 040	23. 20	23. 23	23. 25	23. 27	23. 30	23. 32	23. 35	23. 37	23. 39	23. 42	23. 44	1, 040
1, 050	23. 44	23. 46	23. 49	23. 51	23. 54	23. 56	23. 58	23. 61	23. 63	23. 65	23. 68	1, 050
1, 060	23. 68	23. 70	23. 72	23. 75	23. 77	23. 80	23. 82	23. 84	23. 87	23. 89	23. 91	1, 060
1, 070	23. 91	23. 94	23. 96	23. 99	24. 01	24. 03	24. 06	24. 08	24. 10	24. 13	24. 15	1, 070
1, 080	24. 15	24. 18	24. 20	24. 22	24. 25	24. 27	24. 29	24. 32	24. 34	24. 36	24. 39	1, 080
1, 090	24. 39	24. 41	24. 44	24. 46	24. 49	24. 51	24. 53	24. 55	2'1. 58	24. 60	24. 63	1, 090
1, 100	24. 63	24. 65	24. 67	24. 70	24. 72	24. 74	24. 77	24. 79	24. 82	24. 84	24. 86	1, 100
1, 110	24. 86	24. 89	24. 91	24. 93	24. 96	24. 98	25. 01	25. 03	25. 05	25. 08	25. 10	1, 110
1, 120	25. 10	25. 12	25. 15	25. 17	25. 20	25. 22	25. 24	25. 27	25. 29	25. 31	25. 34	1, 120
1, 130	25. 34	25. 36	25. 38	25. 41	25. 43	25. 46	25. 48	25. 50	25. 53	25. 55	25. 57	1, 130
1, 140	25. 57	25. 60	25. 62	25. 65	25. 67	25. 69	25. 72	25. 74	25. 76	25. 79	25. 81	1, 140
1, 150	25. 81	25. 83	25. 86	25. 88	25. 91	25. 93	25. 95	25. 98	26. 00	26. 02	26. 05	1, 150
1, 160	26. 05	26. 07	26. 09	26. 12	26. 14	26. 16	26. 19	26. 21	26. 24	26. 26	26. 28	1, 160
1, 170	26. 28	26. 31	26. 33	26. 35	26. 38	26. 40	26. 42	26. 45	26. 47	26. 49	26. 52	1, 170
1, 180	26. 52	26. 54	26. 56	26. 59	26. 61	26. 63	26. 66	26. 68	26. 70	26. 73	26. 75	1, 180
1, 190	26. 75	26. 77	26. 80	26. 82	26. 85	26. 87	26. 89	26. 91	26. 94	26. 96	26. 98	1, 190
1, 200	26. 98	27. 01	27. 03	27. 06	27. 08	27. 10	27. 12	27. 15	27. 17	27. 20	27. 22	1, 200
1, 210	27. 22	27. 24	27. 27	27. 29	27. 31	27. 34	27. 36	27. 38	27. 40	27. 43	27. 45	1, 210
1, 220	27. 45	27. 48	27. 50	27. 52	27. 55	27. 57	27. 59	27. 62	27. 64	27. 66	27. 69	1, 220
1, 230	27. 69	27. 71	27. 73	27. 76	27. 78	27. 80	27. 83	27. 85	27. 87	27. 90	27. 92	1, 230
1, 240	27. 92	27. 94	27. 97	27. 99	28. 01	28. 04	28. 06	28. 08	28. 11	28. 13	28. 15	1, 240
1, 250	28. 15	28. 18	28. 20	28. 22	28. 25	28. 27	28. 29	28. 32	28. 34	28. 37	28. 39	1, 250
1, 260	28. 39	28. 41	28. 44	28. 46	28. 48	28. 50	28. 53	28. 55	28. 58	28. 60	28. 62	1, 260
1, 270	28. 62	28. 65	28. 67	28. 69	28. 72	28. 74	28. 76	28. 79	28. 81	28. 83	28. 86	1, 270
1, 280	28. 86	28. 88	28. 90	28. 93	28. 95	28. 97	29. 00	29. 02	29. 04	29. 07	29. 09	1, 280
1, 290	29. 09	29. 11	29. 14	29. 16	29. 18	29. 21	29. 23	29. 25	29. 28	29. 30	29. 32	1, 290
1, 300	29. 32	29. 35	29. 37	29. 39	29. 42	29. 44	29. 46	29. 49	29. 51	29. 53	29. 56	1, 300
°F	0	1	2	3	4	5	6	7	8	9	10	°F

TABLE 44—*Type K thermocouples (continued).*
Temperature in Degrees Fahrenheit

EMF in Absolute Millivolts   Reference Junctions at 32 F

°F	0	1	2	3	4	5	6	7	8	9	10	°F
						Millivolts						
1,300	29.32	29.35	29.37	29.39	29.42	29.44	29.46	29.49	29.51	29.53	29.56	1,300
1,310	29.56	29.58	29.60	29.63	29.65	29.67	29.70	29.72	29.74	29.77	29.79	1,310
1,320	29.79	29.81	29.84	29.86	29.88	29.91	29.93	29.95	29.97	30.00	30.02	1,320
1,330	30.02	30.05	30.07	30.09	30.11	30.14	30.16	30.18	30.21	30.23	30.25	1,330
1,340	30.25	30.28	30.30	30.32	30.35	30.37	30.39	30.42	30.44	30.46	30.49	1,340
1,350	30.49	30.51	30.53	30.56	30.58	30.60	30.63	30.65	30.67	30.70	30.72	1,350
1,360	30.72	30.74	30.77	30.79	30.81	30.83	30.86	30.88	30.90	30.93	30.95	1,360
1,370	30.95	30.97	31.00	31.02	31.04	31.07	31.09	31.11	31.14	31.16	31.18	1,370
1,380	31.18	31.21	31.23	31.25	31.28	31.30	31.32	31.34	31.37	31.39	31.42	1,380
1,390	31.42	31.44	31.46	31.48	31.51	31.53	31.55	31.58	31.60	31.62	31.65	1,390
1,400	31.65	31.67	31.69	31.72	31.74	31.76	31.78	31.81	31.83	31.85	31.88	1,400
1,410	31.88	31.90	31.92	31.95	31.97	31.99	32.02	32.04	32.06	32.08	32.11	1,410
1,420	32.11	32.13	32.15	32.18	32.20	32.22	32.25	32.27	32.29	32.31	32.34	1,420
1,430	32.34	32.36	32.38	32.41	32.43	32.45	32.48	32.50	32.52	32.54	32.57	1,430
1,440	32.57	32.59	32.61	32.64	32.66	32.68	32.70	32.73	32.75	32.77	32.80	1,440
1,450	32.80	32.82	32.84	32.86	32.89	32.91	32.93	32.96	32.98	33.00	33.02	1,450
1,460	33.02	33.05	33.07	33.09	33.12	33.14	33.16	33.18	33.21	33.23	33.25	1,460
1,470	33.25	33.28	33.30	33.32	33.34	33.37	33.39	33.41	33.43	33.46	33.48	1,470
1,480	33.48	33.50	33.53	33.55	33.57	33.59	33.62	33.64	33.66	33.69	33.71	1,480
1,490	33.71	33.73	33.75	33.78	33.80	33.82	33.84	33.87	33.89	33.91	33.93	1,490
1,500	33.93	33.96	33.98	34.00	34.03	34.05	34.07	34.09	34.12	34.14	34.16	1,500
1,510	34.16	34.18	34.21	34.23	34.25	34.28	34.30	34.32	34.34	34.37	34.39	1,510
1,520	34.39	34.41	34.43	34.46	34.48	34.50	34.53	34.55	34.57	34.59	34.62	1,520
1,530	34.62	34.64	34.66	34.68	34.71	34.73	34.75	34.77	34.80	34.82	34.84	1,530
1,540	34.84	34.87	34.89	34.91	34.93	34.96	34.98	35.00	35.02	35.05	35.07	1,540
1,550	35.07	35.09	35.11	35.14	35.16	35.18	35.21	35.23	35.25	35.27	35.29	1,550
1,560	35.29	35.32	35.34	35.36	35.39	35.41	35.43	35.45	35.48	35.50	35.52	1,560
1,570	35.52	35.54	35.57	35.59	35.61	35.63	35.66	35.68	35.70	35.72	35.75	1,570
1,580	35.75	35.77	35.79	35.81	35.84	35.86	35.88	35.90	35.93	35.95	35.97	1,580
1,590	35.97	35.99	36.02	36.04	36.06	36.08	36.11	36.13	36.15	36.17	36.19	1,590
1,600	36.19	36.22	36.24	36.26	36.29	36.31	36.33	36.35	36.37	36.40	36.42	1,600
1,610	36.42	36.44	36.46	36.49	36.51	36.53	36.55	36.58	36.60	36.62	36.64	1,610
1,620	36.64	36.67	36.69	36.71	36.73	36.76	36.78	36.80	36.82	36.84	36.87	1,620
1,630	36.87	36.89	36.91	36.93	36.96	36.98	37.00	37.02	37.05	37.07	37.09	1,630
1,640	37.09	37.11	37.14	37.16	37.18	37.20	37.23	37.25	37.27	37.29	37.31	1,640
1,650	37.31	37.34	37.36	37.38	37.40	37.43	37.45	37.47	37.49	37.52	37.54	1,650
1,660	37.54	37.56	37.58	37.60	37.63	37.65	37.67	37.69	37.72	37.74	37.76	1,660
1,670	37.76	37.78	37.81	37.83	37.85	37.87	37.89	37.92	37.94	37.96	37.98	1,670
1,680	37.98	38.01	38.03	38.05	38.07	38.09	38.12	38.14	38.16	38.18	38.20	1,680
1,690	38.20	38.23	38.25	38.27	38.29	38.32	38.34	38.36	38.38	38.40	38.43	1,690
1,700	38.43	38.45	38.47	38.49	38.51	38.54	38.56	38.58	38.60	38.62	38.65	1,700
°F	0	1	2	3	4	5	6	7	8	9	10	°F

TABLE 44—*Type K thermocouples (continued).*
Temperature in Degrees Fahrenheit

EMF in Absolute Millivolts

Reference Junctions at 32 F

°F	0	1	2	3	4	5	6	7	8	9	10	°F
						Millivolts						
1,700	38.43	38.45	38.47	38.49	38.51	38.54	38.56	38.58	38.60	38.62	38.65	1,700
1,710	38.65	38.67	38.69	38.71	38.73	38.76	38.78	38.80	38.82	38.84	38.87	1,710
1,720	38.87	38.89	38.91	38.93	38.95	38.98	39.00	39.02	39.04	39.06	39.09	1,720
1,730	39.09	39.11	39.13	39.15	39.17	39.20	39.22	39.24	39.26	39.28	39.31	1,730
1,740	39.31	39.33	39.35	39.37	39.39	39.42	39.44	39.46	39.48	39.50	39.53	1,740
1,750	39.53	39.55	39.57	39.59	39.61	39.64	39.66	39.68	39.70	39.72	39.75	1,750
1,760	39.75	39.77	39.79	39.81	39.83	39.86	39.88	39.90	39.92	39.94	39.96	1,760
1,770	39.96	39.99	40.01	40.03	40.05	40.07	40.10	40.12	40.14	40.16	40.18	1,770
1,780	40.18	40.20	40.23	40.25	40.27	40.29	40.31	40.34	40.36	40.38	40.40	1,780
1,790	40.40	40.42	40.44	40.47	40.49	40.51	40.53	40.55	40.58	40.60	40.62	1,790
1,800	40.62	40.64	40.66	40.68	40.71	40.73	40.75	40.77	40.79	40.82	40.84	1,800
1,810	40.84	40.86	40.88	40.90	40.92	40.95	40.97	40.99	41.01	41.03	41.05	1,810
1,820	41.05	41.08	41.10	41.12	41.14	41.16	41.18	41.21	41.23	41.25	41.27	1,820
1,830	41.27	41.29	41.31	41.34	41.36	41.38	41.40	41.42	41.45	41.47	41.49	1,830
1,840	41.49	41.51	41.53	41.55	41.57	41.60	41.62	41.64	41.66	41.68	41.70	1,840
1,850	41.70	41.73	41.75	41.77	41.79	41.81	41.83	41.85	41.88	41.90	41.92	1,850
1,860	41.92	41.94	41.96	41.99	42.01	42.03	42.05	42.07	42.09	42.11	42.14	1,860
1,870	42.14	42.16	42.18	42.20	42.22	42.24	42.26	42.29	42.31	42.33	42.35	1,870
1,880	42.35	42.37	42.39	42.42	42.44	42.46	42.48	42.50	42.52	42.55	42.57	1,880
1,890	42.57	42.59	42.61	42.63	42.65	42.67	42.69	42.72	42.74	42.76	42.78	1,890
1,900	42.78	42.80	42.82	42.84	42.87	42.89	42.91	42.93	42.95	42.97	42.99	1,900
1,910	42.99	43.01	43.04	43.06	43.08	43.10	43.12	43.14	43.17	43.19	43.21	1,910
1,920	43.21	43.23	43.25	43.27	43.29	43.31	43.34	43.36	43.38	43.40	43.42	1,920
1,930	43.42	43.44	43.47	43.49	43.51	43.53	43.55	43.57	43.59	43.61	43.63	1,930
1,940	43.63	43.66	43.68	43.70	43.72	43.74	43.76	43.78	43.81	43.83	43.85	1,940
1,950	43.85	43.87	43.89	43.91	43.93	43.95	43.98	44.00	44.02	44.04	44.06	1,950
1,960	44.06	44.08	44.10	44.13	44.15	44.17	44.19	44.21	44.23	44.25	44.27	1,960
1,970	44.27	44.30	44.32	44.34	44.36	44.38	44.40	44.42	44.44	44.47	44.49	1,970
1,980	44.49	44.51	44.53	44.55	44.57	44.59	44.61	44.63	44.66	44.68	44.70	1,980
1,990	44.70	44.72	44.74	44.76	44.78	44.80	44.82	44.85	44.87	44.89	44.91	1,990
2,000	44.91	44.93	44.95	44.97	44.99	45.01	45.03	45.06	45.08	45.10	45.12	2,000
2,010	45.12	45.14	45.16	45.18	45.20	45.22	45.24	45.27	45.29	45.31	45.33	2,010
2,020	45.33	45.35	45.37	45.39	45.41	45.43	45.45	45.48	45.50	45.52	45.54	2,020
2,030	45.54	45.56	45.58	45.60	45.62	45.64	45.66	45.69	45.71	45.73	45.75	2,030
2,040	45.75	45.77	45.79	45.81	45.83	45.85	45.87	45.90	45.92	45.94	45.96	2,040
2,050	45.96	45.98	46.00	46.02	46.04	46.06	46.08	46.11	46.13	46.15	46.17	2,050
2,060	46.17	46.19	46.21	46.23	46.25	46.27	46.29	46.31	46.33	46.36	46.38	2,060
2,070	46.38	46.40	46.42	46.44	46.46	46.48	46.50	46.52	46.54	46.56	46.58	2,070
2,080	46.58	46.60	46.63	46.65	46.67	46.69	46.71	46.73	46.75	46.77	46.79	2,080
2,090	46.79	46.81	46.83	46.85	46.87	46.90	46.92	46.94	46.96	46.98	47.00	2,090
2,100	47.00	47.02	47.04	47.06	47.08	47.10	47.12	47.14	47.17	47.19	47.21	2,100
°F	0	1	2	3	4	5	6	7	8	9	10	°F

TABLE 44—*Type K thermocouples (continued).*
Temperature in Degrees Fahrenheit
EMF in Absolute Millivolts | Reference Junctions at 32 F

°F	0	1	2	3	4	5	6	7	8	9	10	°F
						Millivolts						
2,100	47.00	47.02	47.04	47.06	47.08	47.10	47.12	47.14	47.17	47.19	47.21	2,100
2,110	47.21	47.23	47.25	47.27	47.29	47.31	47.33	47.35	47.37	47.39	47.41	2,110
2,120	47.41	47.43	47.45	47.47	47.49	47.52	47.54	47.56	47.58	47.60	47.62	2,120
2,130	47.62	47.64	47.66	47.68	47.70	47.72	47.74	47.76	47.78	47.80	47.82	2,130
2,140	47.82	47.84	47.86	47.89	47.91	47.93	47.95	47.97	47.99	48.01	48.03	2,140
2,150	48.03	48.05	48.07	48.09	48.11	48.13	48.15	48.17	48.19	48.21	48.23	2,150
2,160	48.23	48.25	48.27	48.29	48.32	.48.34	48.36	48.38	48.40	48.42	48.44	2,160
2,170	48.44	48.46	48.48	48.50	48.52	48.54	48.56	48.58	48.60	48.62	48.64	2,170
2,180	48.64	48.66	48.68	48.70	48.72	48.74	48.76	48.79	48.81	48.83	48.85	2,180
2,190	48.85	48.87	48.89	48.91	48.93	48.95	48.97	48.99	49.01	49.03	49.05	2,190
2,200	49.05	49.07	49.09	49.11	49.13	49.15	49.17	49.19	49.21	49.23	49.25	2,200
2,210	49.25	49.27	49.29	49.31	49.33	49.35	49.37	49.39	49.41	49.43	49.45	2,210
2,220	49.45	49.47	49.49	49.51	49.53	49.55	49.57	49.59	49.61	49.63	49.65	2,220
2,230	49.65	49.67	49.69	49.71	49.73	49.76	49.78	49.80	49.82	49.84	49.86	2,230
2,240	49.86	49.88	49.90	49.92	49.94	49.96	49.98	50.00	50.02	50.04	50.06	2,240
2,250	50.06	50.08	50.10	50.12	50.14	50.16	50.18	50.20	50.22	50.24	50.26	2,250
2,260	50.26	50.28	50.30	50.32	50.34	50.36	50.38	50.40	50.42	50.44	50.46	2,260
2,270	50.46	50.48	50.50	50.52	50.54	50.56	50.57	50.59	50.61	50.63	50.65	2,270
2,280	50.65	50.67	50.69	50.71	50.73	50.75	50.77	50.79	50.81	50.83	50.85	2,280
2,290	50.85	50.87	50.89	50.91	50.93	50.95	50.97	50.99	51.01	51.03	51.05	2,290
2,300	51.05	51.07	51.09	51.11	51.13	51.15	51.17	51.19	51.21	51.23	51.25	2,300
2,310	51.25	51.27	51.29	51.31	51.33	51.35	51.37	51.39	51.41	51.43	51.45	2,310
2,320	51.45	51.47	51.48	51.50	51.52	51.54	51.56	51.58	51.60	51.62	51.64	2,320
2,330	51.64	51.66	51.68	51.70	51.72	51.74	51.76	51.78	51.80	51.82	51.84	2,330
2,340	51.84	51.86	51.88	51.90	51.92	51.94	51.96	51.98	52.00	52.01	52.03	2,340
2,350	52.03	52.05	52.07	52.09	52.11	52.13	52.15	52.17	52.19	52.21	52.23	2,350
2,360	52.23	52.25	52.27	52.29	52.31	52.33	52.35	52.37	52.39	52.41	52.42	2,360
2,370	52.42	52.44	52.46	52.48	52.50	52.52	52.54	52.56	52.58	52.60	52.62	2,370
2,380	52.62	52.64	52.66	52.68	52.70	52.72	52.74	52.76	52.77	52.79	52.81	2,380
2,390	52.81	52.83	52.85	52.87	52.89	52.91	52.93	52.95	52.97	52.99	53.01	2,390
2,400	53.01	53.03	53.05	53.07	53.08	53.10	53.12	53.14	53.16	53.18	53.20	2,400
2,410	53.20	53.22	53.24	53.26	53.28	53.30	53.32	53.34	53.35	53.37	53.39	2,410
2,420	53.39	53.41	53.43	53.45	53.47	53.49	53.51	53.53	53.55	53.57	53.59	2,420
2,430	53.59	53.60	53.62	53.64	53.66	53.68	53.70	53.72	53.74	53.76	53.78	2,430
2,440	53.78	53.80	53.82	53.83	53.85	53.87	53.89	53.91	53.93	53.95	53.97	2,440
2,450	53.97	53.99	54.01	54.03	54.04	54.06	54.08	54.10	54.12	54.14	54.16	2,450
2,460	54.16	54.18	54.20	54.22	54.24	54.25	54.27	54.29	54.31	54.33	54.35	2,460
2,470	54.35	54.37	54.39	54.41	54.43	54.44	54.46	54.48	54.50	54.52	54.54	2,470
2,480	54.54	54.56	54.58	54.60	54.62	54.63	54.65	54.67	54.69	54.71	54.73	2,480
2,490	54.73	54.75	54.77	54.79	54.81	54.82	54.84	54.86	54.88	54.90	54.92	2,490
°F	0	1	2	3	4	5	6	7	8	9	10	°F

TABLE 45—*Type K thermocouples.*
Temperature in Degrees Celsius

EMF in Absolute Millivolts   Reference Junctions at 0 C

°C	0	1	2	3	4	5	6	7	8	9	10	°C
						Millivolts						
−190	−5.60	−5.62	−5.63	−5.65	−5.67	−5.68	−5.70	−5.71	−5.73	−5.74	−5.75	−190
−180	−5.43	−5.45	−5.46	−5.48	−5.50	−5.52	−5.53	−5.55	−5.57	−5.58	−5.60	−180
−170	−5.24	−5.26	−5.28	−5.30	−5.32	−5.34	−5.35	−5.37	−5.39	−5.41	−5.43	−170
−160	−5.03	−5.05	−5.08	−5.10	−5.12	−5.14	−5.16	−5.18	−5.20	−5.22	−5.24	−160
−150	−4.81	−4.84	−4.86	−4.88	−4.90	−4.92	−4.95	−4.97	−4.99	−5.01	−5.03	−150
−140	−4.58	−4.60	−4.62	−4.65	−4.67	−4.70	−4.72	−4.74	−4.77	−4.79	−4.81	−140
−130	−4.32	−4.35	−4.37	−4.40	−4.42	−4.45	−4.48	−4.50	−4.52	−4.55	−4.58	−130
−120	−4.06	−4.08	−4.11	−4.14	−4.16	−4.19	−4.22	−4.24	−4.27	−4.30	−4.32	−120
−110	−3.78	−3.81	−3.84	−3.86	−3.89	−3.92	−3.95	−3.98	−4.00	−4.03	−4.06	−110
−100	−3.49	−3.52	−3.55	−3.58	−3.61	−3.64	−3.66	−3.69	−3.72	−3.75	−3.78	−100
−90	−3.19	−3.22	−3.25	−3.28	−3.31	−3.34	−3.37	−3.40	−3.43	−3.46	−3.49	−90
−80	−2.87	−2.90	−2.93	−2.96	−3.00	−3.03	−3.06	−3.09	−3.12	−3.16	−3.19	−80
−70	−2.54	−2.57	−2.61	−2.64	−2.67	−2.71	−2.74	−2.77	−2.80	−2.84	−2.87	−70
−60	−2.20	−2.24	−2.27	−2.30	−2.34	−2.37	−2.41	−2.44	−2.47	−2.51	−2.54	−60
−50	−1.86	−1.89	−1.93	−1.96	−2.00	−2.03	−2.07	−2.10	−2.13	−2.17	−2.20	−50
−40	−1.50	−1.54	−1.57	−1.61	−1.64	−1.68	−1.72	−1.75	−1.79	−1.82	−1.86	−40
−30	−1.14	−1.17	−1.21	−1.25	−1.28	−1.32	−1.36	−1.39	−1.43	−1.47	−1.50	−30
−20	−0.77	−0.80	−0.84	−0.88	−0.92	−0.95	−0.99	−1.03	−1.06	−1.10	−1.14	−20
−10	−0.39	−0.42	−0.46	−0.50	−0.54	−0.58	−0.62	−0.66	−0.69	−0.73	−0.77	−10
(−)0	−0.00	−0.04	−0.08	−0.12	−0.16	−0.19	−0.23	−0.27	−0.31	−0.35	−0.39	(−)0
(+)0	0.00	0.04	0.08	0.12	0.16	0.20	0.24	0.28	0.32	0.36	0.40	(+)0
10	0.40	0.44	0.48	0.52	0.56	0.60	0.64	0.68	0.72	0.76	0.80	10
20	0.80	0.84	0.88	0.92	0.96	1.00	1.04	1.08	1.12	1.16	1.20	20
30	1.20	1.24	1.28	1.32	1.36	1.40	1.44	1.49	1.53	1.57	1.61	30
40	1.61	1.65	1.69	1.73	1.77	1.81	1.85	1.90	1.94	1.98	2.02	40
50	2.02	2.06	2.10	2.14	2.18	2.23	2.27	2.31	2.35	2.39	2.43	50
60	2.43	2.47	2.51	2.56	2.60	2.64	2.68	2.72	2.76	2.80	2.85	60
70	2.85	2.89	2.93	2.97	3.01	3.05	3.10	3.14	3.18	3.22	3.26	70
80	3.26	3.30	3.35	3.39	3.43	3.47	3.51	3.56	3.60	3.64	3.68	80
90	3.68	3.72	3.76	3.81	3.85	3.89	3.93	3.97	4.01	4.06	4.10	90
100	4.10	4.14	4.18	4.22	4.26	4.31	4.35	4.39	4.43	4.47	4.51	100
110	4.51	4.55	4.60	4.64	4.68	4.72	4.76	4.80	4.84	4.88	4.92	110
120	4.92	4.96	5.01	5.05	5.09	5.13	5.17	5.21	5.25	5.29	5.33	120
130	5.33	5.37	5.41	5.45	5.49	5.53	5.57	5.61	5.65	5.69	5.73	130
140	5.73	5.77	5.81	5.85	5.89	5.93	5.97	6.01	6.05	6.09	6.13	140
150	6.13	6.17	6.21	6.25	6.29	6.33	6.37	6.41	6.45	6.49	6.53	150
160	6.53	6.57	6.61	6.65	6.69	6.73	6.77	6.81	6.85	6.89	6.93	160
170	6.93	6.97	7.01	7.05	7.09	7.13	7.17	7.21	7.25	7.29	7.33	170
180	7.33	7.37	7.41	7.45	7.49	7.53	7.57	7.61	7.65	7.69	7.73	180
190	7.73	7.77	7.81	7.85	7.89	7.93	7.97	8.01	8.05	8.09	8.13	190
200	8.13	8.17	8.21	8.25	8.29	8.33	8.37	8.41	8.46	8.50	8.54	200
°C	0	1	2	3	4	5	6	7	8	9	10	°C

TABLE 45—*Type K thermocouples (continued)*.
Temperature in Degrees Celsius

EMF in Absolute Millivolts

Reference Junctions at 0 C

°C	0	1	2	3	4	5	6	7	8	9	10	°C
						Millivolts						
200	8.13	8.17	8.21	8.25	8.29	8.33	8.37	8.41	8.46	8.50	8.54	200
210	8.54	8.58	8.62	8.66	8.70	8.74	8.78	8.82	8.86	8.90	8.94	210
220	8.94	8.98	9.02	9.06	9.10	9.14	9.18	9.22	9.26	9.30	9.34	220
230	9.34	9.38	9.42	9.46	9.50	9.54	9.59	9.63	9.67	9.71	9.75	230
240	9.75	9.79	9.83	9.87	9.91	9.95	9.99	10.03	10.07	10.11	10.16	240
250	10.16	10.20	10.24	10.28	10.32	10.36	10.40	10.44	10.48	10.52	10.57	250
260	10.57	10.61	10.65	10.69	10.73	10.77	10.81	10.85	10.89	10.93	10.98	260
270	10.98	11.02	11.06	11.10	11.14	11.18	11.22	11.26	11.30	11.34	11.39	270
280	11.39	11.43	11.47	11.51	11.55	11.59	11.63	11.67	11.72	11.76	11.80	280
290	11.80	11.84	11.88	11.92	11.96	12.01	12.05	12.09	12.13	12.17	12.21	290
300	12.21	12.25	12.29	12.34	12.38	12.42	12.46	12.50	12.54	12.58	12.63	300
310	12.63	12.67	12.71	12.75	12.79	12.83	12.88	12.92	12.96	13.00	13.04	310
320	13.04	13.08	13.12	13.17	13.21	13.25	13.29	13.33	13.37	13.42	13.46	320
330	13.46	13.50	13.54	13.58	13.62	13.67	13.71	13.75	13.79	13.83	13.88	330
340	13.88	13.92	13.96	14.00	14.04	14.09	14.13	14.17	14.21	14.25	14.29	340
350	14.29	14.34	14.38	14.42	14.46	14.50	14.55	14.59	14.63	14.67	14.71	350
360	14.71	14.76	14.80	14.84	14.88	14.92	14.97	15.01	15.05	15.09	15.13	360
370	15.13	15.18	15.22	15.26	15.30	15.34	15.39	15.43	15.47	15.51	15.55	370
380	15.55	15.60	15.64	15.68	15.72	15.76	15.81	15.85	15.89	15.93	15.98	380
390	15.98	16.02	16.06	16.10	16.14	16.19	16.23	16.27	16.31	16.36	16.40	390
400	16.40	16.44	16.48	16.52	16.57	16.61	16.65	16.69	16.74	16.78	16.82	400
410	16.82	16.86	16.91	16.95	16.99	17.03	17.07	17.12	17.16	17.20	17.24	410
420	17.24	17.29	17.33	17.37	17.41	17.46	17.50	17.54	17.58	17.62	17.67	420
430	17.67	17.71	17.75	17.79	17.84	17.88	17.92	17.96	18.01	18.05	18.09	430
440	18.09	18.13	18.17	18.22	18.26	18.30	18.34	18.39	18.43	18.47	18.51	440
450	18.51	18.56	18.60	18.64	18.68	18.73	18.77	18.81	18.85	18.90	18.94	450
460	18.94	18.98	19.02	19.07	19.11	19.15	19.19	19.24	19.28	19.32	19.36	460
470	19.36	19.41	19.45	19.49	19.54	19.58	19.62	19.66	19.71	19.75	19.79	470
480	19.79	19.84	19.88	19.92	19.96	20.01	20.05	20.09	20.13	20.18	20.22	480
490	20.22	20.26	20.31	20.35	20.39	20.43	20.48	20.52	20.56	20.60	20.65	490
500	20.65	20.69	20.73	20.77	20.82	20.86	20.90	20.94	20.99	21.03	21.07	500
510	21.07	21.11	21.16	21.20	21.24	21.28	21.32	21.37	21.41	21.45	21.50	510
520	21.50	21.54	21.58	21.63	21.67	21.71	21.75	21.80	21.84	21.88	21.92	520
530	21.92	21.97	22.01	22.05	22.09	22.14	22.18	22.22	22.26	22.31	22.35	530
540	22.35	22.39	22.43	22.48	22.52	22.56	22.61	22.65	22.69	22.73	22.78	540
550	22.78	22.82	22.86	22.90	22.95	22.99	23.03	23.07	23.12	23.16	23.20	550
560	23.20	23.25	23.29	23.33	23.38	23.42	23.46	23.50	23.54	23.59	23.63	560
570	23.63	23.67	23.72	23.76	23.80	23.84	23.89	23.93	23.97	24.01	24.06	570
580	24.06	24.10	24.14	24.18	24.23	24.27	24.31	24.36	24.40	24.44	24.49	580
590	24.49	24.53	24.57	24.61	24.65	24.70	24.74	24.78	24.83	24.87	24.91	590
600	24.91	24.95	25.00	25.04	25.08	25.12	25.17	25.21	25.25	25.29	25.34	600
°C	0	1	2	3	4	5	6	7	8	9	10	°C

TABLE 45—*Type K thermocouples (continued).*

Temperature in Degrees Celsius

EMF in Absolute Millivolts

Refere..ce Junctions at 0 C

°C	0	1	2	3	4	5	6	7	8	9	10	°C
						Millivolts						
600	24. 91	24. 95	25. 00	25. 04	25. 08	25. 12	25. 17	25. 21	25. 25	25. 29	25. 34	600
610	25. 34	25. 38	25. 42	25. 47	25. 51	25. 55	25. 59	25. 64	25. 68	25. 72	25. 76	610
620	25. 76	25. 81	25. 85	25. 89	25. 93	25. 98	26. 02	26. 06	26. 10	26. 15	26. 19	620
630	26. 19	26. 23	26. 27	26. 32	26. 36	26. 40	26. 44	26. 48	26. 53	26. 57	26. 61	630
640	26. 61	26. 65	26. 70	26. 74	26. 78	26. 82	26. 86	26. 91	26. 95	26. 99	27. 03	640
650	27. 03	27. 07	27. 12	27. 16	27. 20	27. 24	27. 28	27. 33	27. 37	27. 41	27. 45	650
660	27. 45	27. 49	27. 54	27. 58	27. 62	27. 66	27. 71	27. 75	27. 79	27. 83	27. 87	660
670	27. 87	27. 92	27. 96	28. 00	28. 04	28. 08	28. 13	28. 17	28. 21	28. 25	28. 29	670
680	28. 29	28. 34	28. 38	28. 42	28. 46	28. 50	28. 55	28. 59	28. 63	28. 67	28. 72	680
690	28. 72	28. 76	28. 80	28. 84	28. 88	2S. 93	28. 97	29. 01	29. 05	29. 10	29. 14	690
700	29. 14	29. 18	29. 22	29. 26	29. 30	29. 35	29. 39	29. 43	29. 47	29. 52	29. 56	700
710	29. 56	29. 60	29. 64	29. 68	29. 72	29. 77	29. 81	29. 85	29. 89	29. 93	29. 97	710
720	29. 97	30. 02	30. 06	30. 10	30. 14	30. 18	30. 23	30. 27	30. 31	30. 35	30. 39	720
730	30. 39	30. 44	30. 48	30. 52	30. 56	30. 60	30. 65	30. 69	30. 73	30. 77	30. 81	730
740	30. 81	30. 85	30. 90	30. 94	30. 98	31. 02	31. 06	31. 10	31. 15	31. 19	31. 23	740
750	31. 23	31. 27	31. 31	31. 35	31. 40	31. 44	31. 48	31. 52	31. 56	31. 60	31. 65	750
760	31. 65	31. 69	31. 73	31. 77	31. 81	31. 85	31. 90	31. 94	31. 98	32. 02	32. 06	760
770	32. 06	32. 10	32. 15	32. 19	32. 23	32. 27	32. 31	32. 35	32. 39	32. 43	32. 48	770
780	32. 48	32. 52	32. 56	32. 60	32. 64	32. 68	32. 72	32. 76	32. 81	32. 85	32. 89	780
790	32. 89	32. 93	32. 97	33. 01	33. 05	33. 09	33. 13	33. 18	33. 22	33. 26	33. 30	790
800	33. 30	33. 34	33. 38	33. 42	33. 46	33. 50	33. 54	33. 59	33. 63	33. 67	33. 71	800
810	33. 71	33. 75	33. 79	33. 83	33. 87	33. 91	33. 95	33. 99	34. 04	34. 08	34. 12	810
820	34. 12	34. 16	34. 20	34. 24	34. 28	34. 32	34. 36	34. 40	34. 44	34. 48	34. 53	820
830	34. 53	34. 57	34. 61	34. 65	34. 69	34. 73	34. 77	34. 81	34. 85	34. 89	34. 93	830
840	34. 93	34. 97	35. 02	35. 06	35. 10	35. 14	35. 18	35. 22	35. 26	35. 30	35. 34	840
850	35. 34	35. 38	35. 42	35. 46	35. 50	35. 54	35. 58	35. 63	35. 67	35. 71	35. 75	850
860	35. 75	35. 79	35. 83	35. 87	35. 91	35. 95	35. 99	36. 03	36. 07	36. 11	36. 15	860
870	36. 15	36. 19	36. 23	36. 27	36. 31	36. 35	36. 39	36. 43	36. 47	36. 51	36. 55	870
880	36. 55	36. 59	36. 63	36. 67	36. 72	36. 76	36. 80	36. 84	36. 88	36. 92	36. 96	880
890	36. 96	37. 00	37. 04	37. 08	37. 12	37. 16	37. 20	37. 24	37. 28	37. 32	37. 36	890
900	37. .36	37. 40	37. 44	37. 48	37. 52	37. 56	37. 60	37. 64	37. 68	37. 72	37. 76	900
910	37. 76	37. 80	37. 84	37. 88	37. 92	37. 96	38. 00	38. 04	38. 08	38. 12	38. 16	910
920	38. 16	38. 20	38. 24	38. 28	38. 32	38. 36	38. 40	38. 44	38. 48	38. 52	38. 56	920
930	38. 56	38. 60	38. 64	38. 68	38. 72	38. 76	38. 80	38. 84	38. 88	38. 92	38. 95	930
940	38. 95	38. 99	39. 03	39. 07	39. 11	39. 15	39. 19	39. 23	39. 27	39. 31	39. 35	940
950	39. 35	39. 39	39. 43	39. 47	39. 51	39. 55	39. 59	39. 63	39. 67	39. 71	39. 75	950
960	39. 75	39. 79	39. 83	39. 86	39. 90	39. 94	39. 98	40. 02	40. 06	40. 10	40. 14	960
970	40. 14	40. 18	40. 22	40. 26	40. 30	40. 34	40. 38	40. 41	40. 45	40. 49	40. 53	970
980	40. 53	40. 57	40. 61	40. 65	40. 69	40. 73	40. 77	40. 81	40. 85	40. 89	40. 92	980
990	40. 92	40. 96	41. 00	41. 04	41. 08	41. 12	41. 16	41. 20	41. 24	41. 28	41. 31	990
1,000	41. 31	41. 35	41. 39	41. 43	41. 47	41. 51	41. 55	41. 59	41. 63	41. 67	41. 70	1,000
°C	0	1	2	3	4	5	6	7	8	9	10	°C

TABLE 45—*Type K thermocouples (continued).*
Temperature in Degrees Celsius

EMF in Absolute Millivolts   Reference Junctions at 0 C

°C	0	1	2	3	4	5	6	7	8	9	10	°C
						Millivolts						
1,000	41.31	41.35	41.39	41.43	41.47	41.51	41.55	41.59	41.63	41.67	41.70	1,000
1,010	41.70	41.74	41.78	41.82	41.86	41.90	41.94	41.98	42.02	42.05	42.09	1,010
1,020	42.09	42.13	42.17	42.21	42.25	42.29	42.33	42.36	42.40	42.44	42.48	1,020
1,030	42.48	42.52	42.56	42.60	42.63	42.67	42.71	42.75	42.79	42.83	42.87	1,030
1,040	42.87	42.90	42.94	42.98	43.02	43.06	43.10	43.14	43.17	43.21	43.25	1,040
1,050	43.25	43.29	43.33	43.37	43.41	43.44	43.48	43.52	43.56	43.60	43.63	1,050
1,060	43.63	43.67	43.71	43.75	43.79	43.83	43.87	43.90	43.94	43.98	44.02	1,060
1,070	44.02	44.06	44.10	44.13	44.17	44.21	44.25	44.29	44.33	44.36	44.40	1,070
1,080	44.40	44.44	44.48	44.52	44.55	44.59	44.63	44.67	44.71	44.74	44.78	1,080
1,090	44.78	44.82	44.86	44.90	44.93	44.97	45.01	45.05	45.09	45.12	45.16	1,090
1,100	45.16	45.20	45.24	45.27	45.31	45.35	45.39	45.43	45.46	45.50	45.54	1,100
1,110	45.54	45.58	45.62	45.65	45.69	45.73	45.77	45.80	45.84	45.88	45.92	1,110
1,120	45.92	45.96	45.99	46.03	46.07	46.11	46.14	46.18	46.22	46.26	46.29	1,120
1,130	46.29	46.33	46.37	46.41	46.44	46.48	46.52	46.56	46.59	46.63	46.67	1,130
1,140	46.67	46.70	46.74	46.78	46.82	46.85	46.89	46.93	46.97	47.00	47.04	1,140
1,150	47.04	47.08	47.12	47.15	47.19	47.23	47.26	47.30	47.34	47.38	47.41	1,150
1,160	47.41	47.45	47.49	47.52	47.56	47.60	47.63	47.67	47.71	47.75	47.78	1,160
1,170	47.78	47.82	47.86	47.89	47.93	47.97	48.00	48.04	48.08	48.12	48.15	1,170
1,180	48.15	48.19	48.23	48.26	48.30	48.34	48.37	48.41	48.45	48.48	48.52	1,180
1,190	48.52	48.56	48.59	48.63	48.67	48.70	48.74	48.78	48.81	48.85	48.89	1,190
1,200	48.89	48.92	48.96	49.00	49.03	49.07	49.11	49.14	49.18	49.22	49.25	1,200
1,210	49.25	49.29	49.32	49.36	49.40	49.43	49.47	49.51	49.54	49.58	49.62	1,210
1,220	49.62	49.65	49.69	49.72	49.76	49.80	49.83	49.87	49.90	49.94	49.98	1,220
1,230	49.98	50.01	50.05	50.08	50.12	50.16	50.19	50.23	50.26	50.30	50.34	1,230
1,240	50.34	50.37	50.41	50.44	50.48	50.52	50.55	50.59	50.62	50.66	50.69	1,240
1,250	50.69	50.73	50.77	50.80	50.84	50.87	50.91	50.94	50.98	51.02	51.05	1,250
1,260	51.05	51.09	51.12	51.16	51.19	51.23	51.27	51.30	51.34	51.37	51.41	1,260
1,270	51.41	51.44	51.48	51.51	51.55	51.58	51.62	51.66	51.69	51.73	51.76	1,270
1,280	51.76	51.80	51.83	51.87	51.90	51.94	51.97	52.01	52.04	52.08	52.11	1,280
1,290	52.11	52.15	52.18	52.22	52.25	52.29	52.32	52.36	52.39	52.43	52.46	1,290
1,300	52.46	52.50	52.53	52.57	52.60	52.64	52.67	52.71	52.74	52.78	52.81	1,300
1,310	52.81	52.85	52.88	52.92	52.95	52.99	53.02	53.06	53.09	53.13	53.16	1,310
1,320	53.16	53.20	53.23	53.27	53.30	53.34	53.37	53.41	53.44	53.47	53.51	1,320
1,330	53.51	53.54	53.58	53.61	53.65	53.68	53.72	53.75	53.79	53.82	53.85	1,330
1,340	53.85	53.89	53.92	53.96	53.99	54.03	54.06	54.10	54.13	54.16	54.20	1,340
1,350	54.20	54.23	54.27	54.30	54.34	54.37	54.40	54.44	54.47	54.51	54.54	1,350
1,360	54.54	54.57	54.61	54.64	54.68	54.71	54.74	54.78	54.81	54.85	54.88	1,360
1,370	54.88	54.91	------	------	------	------	------	------	------	------	------	1,370
°C	0	1	2	3	4	5	6	7	8	9	10	°C

TABLE 46—*Type R thermocouples.*
Temperature in Degrees Fahrenheit

EMF in Absolute Millivolts   Reference Junctions at 32 F

°F	0	1	2	3	4	5	6	7	8	9	10	°F
						Millivolts						
30	-----	-----	0.000	0.003	0.006	0.009	0.012	0.015	0.018	0.021	0.024	30
40	0.024	0.027	.030	.033	.036	.039	.042	.045	.048	.052	.055	40
50	.055	.058	.061	.064	.068	.071	.074	.077	.080	.083	.086	50
60	.086	.090	.093	.096	.099	.103	.106	.109	.112	.116	.119	60
70	.119	.122	.126	.129	.132	.135	.139	.142	.145	.149	.152	70
80	.152	.155	.159	.162	.165	.169	.172	.175	.179	.182	.186	80
90	.186	.189	.192	.196	.199	.203	.206	.210	.213	.217	.220	90
100	.220	.224	.227	.230	.234	.237	.241	.244	.248	.251	.255	100
110	.255	.258	.262	.265	.269	.272	.276	.280	.284	.287	.291	110
120	.291	.294	.298	.301	.305	.308	.312	.316	.319	.323	.327	120
130	.327	.330	.334	.337	.341	.345	.349	.352	.356	.359	.363	130
140	.363	.367	.370	.374	.378	.381	.385	.389	.393	.397	.400	140
150	.400	.404	.408	.411	.415	.419	.423	.427	.431	.435	.438	150
160	.438	.442	.446	.450	.453	.457	.461	.465	.469	.473	.476	160
170	.476	.480	.484	.488	.492	.496	.500	.504	.508	.512	.516	170
180	.516	.520	.524	.528	.532	.536	.540	.544	.548	.552	.556	180
190	.556	.560	.564	.568	.572	.576	.580	.584	.588	.592	.596	190
200	.596	.600	.604	.608	.612	.616	.620	.625	.629	.633	.637	200
210	.637	.641	.645	.649	.653	.657	.662	.666	.670	.674	.678	210
220	.678	.683	.687	.691	.695	.700	.704	.708	.712	.716	.721	220
230	.721	.725	.729	.734	.738	.742	.746	.750	.755	.759	.763	230
240	.763	.767	.772	.776	.780	.785	.789	.793	.798	.802	.807	240
250	.807	.811	.815	.820	.824	.828	.833	.837	.842	.846	.850	250
260	.850	.855	.859	.863	.868	.872	.877	.881	.886	.890	.894	260
270	.894	.899	.904	.908	.912	.917	.921	.926	.930	.935	.939	270
280	.939	.944	.948	.953	.957	.962	.966	.971	.975	.980	.984	280
290	.984	.989	.993	.998	1.002	1.007	1.011	1.016	1.020	1.025	1.030	290
300	1.030	1.034	1.039	1.043	1.048	1.052	1.057	1.061	1.066	1.071	1.075	300
310	1.075	1.080	1.084	1.089	1.094	1.098	1.103	1.107	1.112	1.117	1.121	310
320	1.121	1.126	1.130	1.135	1.140	1.144	1.149	1.153	1.158	1.163	1.167	320
330	1.167	1.172	1.176	1.181	1.186	1.191	1.195	1.200	1.205	1.210	1.214	330
340	1.214	1.219	1.223	1.228	1.233	1.238	1.242	1.247	1.252	1.257	1.261	340
350	1.261	1.266	1.271	1.276	1.280	1.285	1.290	1.295	1.300	1.304	1.309	350
360	1.309	1.314	1.319	1.323	1.328	1.333	1.338	1.343	1.348	1.352	1.357	360
370	1.357	1.362	1.367	1.372	1.377	1.381	1.386	1.391	1.396	1.401	1.406	370
380	1.406	1.410	1.415	1.420	1.425	1.430	1.435	1.440	1.445	1.450	1.455	380
390	1.455	1.460	1.465	1.470	1.475	1.480	1.484	1.489	1.494	1.499	1.504	390
400	1.504	1.509	1.514	1.519	1.524	1.529	1.533	1.538	1.543	1.548	1.553	400
°F	0	1	2	3	4	5	6	7	8	9	10	°F

TABLE 46—*Type R thermocouples (continued)*.

Temperature in Degrees Fahrenheit

EMF in Absolute Millivolts

Reference Junctions at 32 F

°F	0	1	2	3	4	5	6	7	8	9	10	°F
						Millivolts						
400	1. 504	1. 509	1. 514	1. 519	1. 524	1. 529	1. 533	1. 538	1. 543	1. 548	1. 553	400
410	1. 553	1. 558	1. 563	1. 568	1. 573	1. 578	1. 583	1. 588	1. 593	1. 598	1. 603	410
420	1. 003	1. 608	1. 613	1. 618	1. 623	1. 628	1. 633	1. 638	1. 643	1. 648	1. 653	420
430	1. 653	1. 658	1. 663	1. 668	1. 673	1. 678	1. 683	1. 688	1. 693	1. 698	1. 703	430
440	1. 703	1. 708	1. 713	1. 719	1. 724	1. 729	1. 734	1. 739	1. 744	1. 749	1. 754	440
450	1. 754	1. 759	1. 764	1. 769	1. 774	1. 779	1. 785	1. 790	1. 795	1. 800	1. 805	450
460	1. 805	1. 811	1. 816	1. 821	1. 826	1. 831	1. 836	1. 841	1. 846	1. 851	1. 856	460
470	1. 856	1. 862	1. 867	1. 872	1. 877	1. 882	1. 887	1. 892	1. 898	1. 903	1. 908	470
480	1. 908	1. 913	1. 918	1. 924	1. 929	1. 934	1. 939	1. 944	1. 950	1. 955	1. 960	480
490	1. 960	1. 965	1. 970	1. 976	1. 981	1. 986	1. 991	1. 996	2. 002	2. 007	2. 012	490
500	2. 012	2. 017	2. 023	2. 028	2. 033	2. 038	2. 044	2. 049	2. 054	2. 059	2. 065	500
510	2. 065	2. 070	2. 075	2. 081	2. 086	2. 091	2. 096	2. 101	2. 107	2. 112	2. 117	510
520	2. 117	2. 123	2. 128	2. 133	2. 139	2. 144	2. 149	2. 154	2. 160	2. 165	2. 170	520
530	2. 170	2. 176	2. 181	2. 186	2. 192	2. 197	2. 202	2. 207	2. 213	2. 218	2. 223	530
540	2. 223	2. 229	2. 234	2. 239	2. 245	2. 250	2. 255	2. 261	2. 266	2. 271	2. 277	540
550	2. 277	2. 282	2. 287	2. 293	2. 298	2. 303	2. 308	2. 314	2. 319	2. 325	2. 330	550
560	2. 330	2. 335	2. 341	2. 346	2. 352	2. 357	2. 363	2. 368	2. 373	2. 379	2. 384	560
570	2. 384	2. 389	2. 395	2. 401	2. 406	2. 412	2. 417	2. 423	2. 428	2. 433	2. 438	570
580	2. 438	2. 444	2. 449	2. 455	2. 460	2. 466	2. 471	2. 477	2. 482	2. 487	2. 493	580
590	2. 493	2. 498	2. 504	2. 509	2. 515	2. 520	2. 526	2. 531	2. 537	2. 542	2. 547	590
600	2. 547	2. 553	2. 558	2. 564	2. 569	2. 575	2. 580	2. 586	2. 591	2. 597	2. 602	600
610	2. 602	2. 608	2. 613	2. 619	2. 624	2. 630	2. 635	2. 641	2. 646	2. 652	2. 657	610
620	2. 657	2. 663	2. 668	2. 674	2. 679	2. 685	2. 690	2. 696	2. 701	2. 707	2. 712	620
630	2. 712	2. 718	2. 723	2. 729	2. 734	2. 740	2. 746	2. 751	2. 757	2. 762	2. 768	630
640	2. 768	2. 773	2. 779	2. 784	2. 790	2. 796	2. 801	2. 807	2. 812	2. 818	2. 823	640
650	2. 823	2. 829	2. 834	2. 840	2. 846	2. 851	2. 857	2. 862	2. 868	2. 873	2. 879	650
660	2. 879	2. 884	2. 890	2. 896	2. 901	2. 907	2. 912	2. 918	2. 923	2. 929	2. 935	660
670	2. 935	2. 940	2. 946	2. 952	2. 957	2. 963	2. 968	2. 974	2. 979	2. 985	2. 991	670
680	2. 991	2. 997	3. 002	3. 008	3. 013	3. 019	3. 024	3. 030	3. 036	3. 041	3. 047	680
690	3. 047	3. 053	3. 058	3. 064	3. 069	3. 075	3. 081	3. 087	3. 092	3. 098	3. 103	690
700	3. 103	3. 109	3. 115	3. 120	3. 126	3. 132	3. 137	3. 143	3. 148	3. 154	3. 160	700
710	3. 160	3. 166	3. 171	3. 177	3. 182	3. 188	3. 194	3. 199	3. 205	3. 211	3. 217	710
720	3. 217	3. 222	3. 228	3. 234	3. 239	3. 245	3. 251	3. 256	3. 262	3. 268	3. 273	720
730	3. 273	3. 279	3. 285	3. 291	3. 296	3. 302	3. 308	3. 313	3. 319	3. 325	3. 330	730
740	3. 330	3. 336	3. 342	3. 348	3. 353	3. 359	3. 365	3. 370	3. 376	3. 382	3. 387	740
750	3. 387	3. 393	3. 399	3. 405	3. 411	3. 416	3. 422	3. 428	3. 433	3. 439	3. 445	750
760	3. 445	3. 451	3. 456	3. 462	3. 468	3. 473	3. 479	3. 485	3. 491	3. 497	3. 502	760
770	3. 502	3. 508	3. 514	3. 519	3. 525	3. 531	3. 537	3. 543	3. 549	3. 554	3. 560	770
780	3. 560	3. 566	3. 572	3. 577	3. 583	3. 589	3. 595	3. 601	3. 607	3. 612	3. 618	780
790	3. 618	3. 624	3. 630	3. 635	3. 641	3. 647	3. 653	3. 659	3. 665	3. 671	3. 677	790
800	3. 677	3. 682	3. 688	3. 694	3. 700	3. 706	3. 712	3. 718	3. 723	3. 729	3. 735	800
°F	0	1	2	3	4	5	6	7	8	9	10	°F

TABLE 46—*Type R thermocouples (continued).*
Temperature in Degrees Fahrenheit

EMF in Absolute Millivolts

Reference Junctions at 32 F

°F	0	1	2	3	4	5	6	7	8	9	10	°F
						Millivolts						
800	3.677	3.682	3.688	3.694	3.700	3.706	3.712	3.718	3.723	3.729	3.735	800
810	3.735	3.741	3.746	3.752	3.758	3.764	3.770	3.776	3.782	3.788	3.794	810
820	3.794	3.799	3.805	3.811	3.817	3.823	3.829	3.835	3.841	3.846	3.852	820
830	3.852	3.858	3.864	3.870	3.876	3.882	3.888	3.894	3.899	3.905	3.911	830
840	3.911	3.917	3.923	3.929	3.935	3.941	3.946	3.952	3.958	3.964	3.970	840
850	3.970	3.976	3.982	3.988	3.994	3.999	4.005	4.011	4.017	4.023	4.029	850
860	4.029	4.035	4.041	4.047	4.052	4.058	4.064	4.070	4.075	4.081	4.087	860
870	4.087	4.093	4.099	4.105	4.111	4.116	4.122	4.128	4.134	4.140	4.146	870
880	4.146	4.152	4.158	4.164	4.169	4.175	4.181	4.187	4.193	4.199	4.205	880
890	4.205	4.211	4.217	4.223	4.229	4.235	4.241	4.246	4.252	4.258	4.264	890
900	4.264	4.270	4.276	4.282	4.288	4.294	4.300	4.306	4.312	4.318	4.324	900
910	4.324	4.330	4.336	4.342	4.348	4.354	4.360	4.366	4.372	4.378	4.384	910
926	4.384	4.389	4.395	4.401	4.407	4.413	4.419	4.425	4.431	4.437	4.443	920
930	4.443	4.449	4.455	4.461	4.467	4.473	4.479	4.485	4.491	4.497	4.503	930
940	4.503	4.509	4.515	4.521	4.527	4.533	4.539	4.545	4.551	4.557	4.563	940
950	4.563	4.569	4.575	4.581	4.587	4.593	4.599	4.605	4.612	4.618	4.624	950
960	4.624	4.630	4.636	4.642	4.648	4.654	4.660	4.666	4.672	4.679	4.685	960
970	4.685	4.691	4.697	4.703	4.709	4.715	4.721	4.727	4.733	4.740	4.746	970
980	4.746	4.752	4.758	4.764	4.770	4.776	4.782	4.788	4.794	4.801	4.807	980
990	4.807	4.813	4.819	4.825	4.831	4.837	4.844	4.850	4.856	4.862	4.868	990
1,000	4.868	4.874	4.881	4.887	4.893	4.899	4.905	4.911	4.917	4.924	4.930	1,000
1,010	4.930	4.936	4.942	4.948	4.954	4.960	4.966	4.972	4.979	4.985	4.991	1,010
1,020	4.991	4.998	5.004	5.010	5.016	5.022	5.028	5.034	5.041	5.047	5.053	1,020
1,030	5.053	5.059	5.066	5.072	5.078	5.084	5.090	5.096	5.102	5.109	5.115	1,030
1,040	5.115	5.121	5.127	5.133	5.139	5.146	5.152	5.158	5.164	5.170	5.176	1,040
1,050	5.176	5.182	5.189	5.195	5.201	5.208	5.214	5.220	5.226	5.232	5.238	1,050
1,060	5.238	5.244	5.251	5.257	5.263	5.270	5.276	5.282	5.288	5.294	5.301	1,060
1,070	5.301	5.307	5.313	5.319	5.326	5.332	5.338	5.344	5.351	5.357	5.363	1,070
1,080	5.363	5.369	5.376	5.382	5.388	5.394	5.401	5.407	5.413	5.419	5.426	1,080
1,090	5.426	5.432	5.438	5.444	5.450	5.457	5.463	5.469	5.476	5.482	5.488	1,090
1,100	5.488	5.494	5.501	5.507	5.513	5.519	5.526	5.532	5.538	5.544	5.551	1,100
1,110	5.551	5.557	5.563	5.570	5.576	5.582	5.589	5.595	5.601	5.607	5.614	1,110
1,120	5.614	5.620	5.626	5.633	5.639	5.645	5.652	5.658	5.664	5.671	5.677	1,120
1,130	5.677	5.684	5.690	5.696	5.703	5.709	5.716	5.722	5.728	5.734	5.741	1,130
1,140	5.741	5.747	5.753	5.760	5.766	5.773	5.779	5.786	5.792	5.798	5.805	1,140
1,150	5.805	5.811	5.817	5.824	5.830	5.837	5.843	5.849	5.856	5.862	5.869	1,150
1,160	5.869	5.875	5.881	5.888	5.894	5.901	5.907	5.913	5.920	5.926	5.933	1,160
1,170	5.933	5.939	5.945	5.952	5.958	5.964	5.971	5.977	5.983	5.990	5.996	1,170
1,180	5.996	6.003	6.009	6.015	6.022	6.028	6.035	6.041	6.047	6.054	6.060	1,180
1,190	6.060	6.067	6.073	6.079	6.086	6.092	6.099	6.105	6.111	6.118	6.125	1,190
1,200	6.125	6.131	6.137	6.143	6.150	6.156	6.163	6.169	6.175	6.182	6.188	1,200
°F	0	1	2	3	4	5	6	7	8	9	10	°F

TABLE 46—*Type R thermocouples (continued)*.
Temperature in Degrees Fahrenheit

EMF in Absolute Millivolts

Reference Junctions at 32 F

°F	0	1	2	3	4	5	6	7	8	9	10	°F
						Millivolts						
1, 200	6. 125	6. 131	6. 137	6. 143	6. 150	6. 156	6. 163	6. 169	6. 175	6. 182	6. 188	1, 200
1, 210	6. 188	6. 195	6. 201	6. 207	6. 214	6. 220	6. 227	6. 233	6. 239	6. 246	6. 252	1, 210
1, 220	6. 252	6. 259	6. 265	6. 272	6. 278	6. 285	6. 291	6. 298	6. 304	6. 310	6. 317	1, 220
1, 230	6. 317	6. 323	6. 329	6. 336	6. 342	6. 349	6. 355	6. 362	6. 368	6. 375	6. 381	1, 230
1, 240	6. 381	6. 388	6. 394	6. 401	6. 407	6. 414	6. 420	6. 427	6. 433	6. 440	6. 446	1, 240
1, 250	6. 446	6. 453	6. 459	6. 466	6. 472	6. 479	6. 485	6. 492	6. 498	6. 505	6. 511	1, 250
1, 260	6. 511	6. 518	6. 524	6. 531	6. 537	6. 544	6. 550	6. 557	6. 563	6. 570	6. 577	1, 260
1, 270	6. 577	6. 583	6. 589	6. 596	6. 602	6. 609	6. 616	6. 622	6. 629	6. 635	6. 642	1, 270
1, 280	6. 642	6. 648	6. 655	6. 661	6. 668	6. 674	6. 681	6. 687	6. 694	6. 701	6. 707	1, 280
1, 290	6. 707	6. 714	6. 720	6. 727	6. 733	6. 740	6. 746	6. 753	6. 759	6. 766	6. 773	1, 290
1, 300	6. 773	6. 779	6. 786	6. 792	6. 799	6. 805	6. 812	6. 818	6. 825	6. 832	6. 838	1, 300
1, 310	6. 838	6. 845	6. 851	6. 858	6. 865	6. 871	6. 877	6. 884	6. 891	6. 898	6. 904	1, 310
1, 320	6. 904	6. 911	6. 917	6. 924	6. 931	6. 937	6. 943	6. 950	6. 957	6. 964	6. 970	1, 320
1, 330	6. 970	6. 977	6. 983	6. 990	6. 997	7. 003	7. 010	7. 017	7. 023	7. 030	7. 037	1, 330
1, 340	7. 037	7. 043	7. 049	7. 056	7. 063	7. 069	7. 076	7. 083	7. 089	7. 096	7. 103	1, 340
1, 350	7. 103	7. 109	7. 116	7. 123	7. 129	7. 136	7. 143	7. 149	7. 155	7. 162	7. 169	1, 350
1, 360	7. 169	7. 175	7. 182	7. 189	7. 195	7. 202	7. 209	7. 215	7. 222	7. 229	7. 235	1, 360
1, 370	7. 235	7. 242	7. 249	7. 255	7. 262	7. 269	7. 275	7. 282	7. 289	7. 295	7. 302	1, 370
1, 380	7. 302	7. 309	7. 315	7. 322	7. 329	7. 336	7. 342	7. 349	7. 356	7. 362	7. 369	1, 380
1, 390	7. 369	7. 376	7. 382	7. 389	7. 396	7. 403	7. 409	7. 416	7. 423	7. 429	7. 436	1, 390
1, 400	7. 436	7. 443	7. 449	7. 456	7. 463	7. 470	7. 477	7. 483	7. 490	7. 497	7. 503	1, 400
1, 410	7. 503	7. 510	7. 517	7. 523	7. 530	7. 537	7. 544	7. 551	7. 557	7. 564	7. 571	1, 410
1, 420	7. 571	7. 578	7. 585	7. 591	7. 598	7. 605	7. 611	7. 618	7. 625	7. 632	7. 639	1, 420
1, 430	7. 639	7. 645	7. 652	7. 659	7. 665	7. 672	7. 679	7. 686	7. 693	7. 699	7. 706	1, 430
1, 440	7. 706	7. 713	7. 720	7. 727	7. 733	7. 740	7. 747	7. 754	7. 761	7. 767	7. 774	1, 440
1, 450	7. 774	7. 781	7. 788	7. 795	7. 801	7. 808	7. 815	7. 822	7. 829	7. 835	7. 842	1, 450
1, 460	7. 842	7. 849	7. 856	7. 863	7. 870	7. 877	7. 884	7. 891	7. 897	7. 904	7. 911	1, 460
1, 470	7. 911	7. 918	7. 924	7. 931	7. 938	7. 945	7. 952	7. 959	7. 965	7. 972	7. 979	1, 470
1, 480	7. 979	7. 986	7. 993	7. 999	8. 006	8. 013	8. 020	8. 027	8. 033	8. 040	8. 047	1, 480
1, 490	8. 047	8. 054	8. 061	8. 068	8. 075	8. 081	8. 089	8. 095	8. 102	8. 109	8. 116	1, 490
1, 500	8. 116	8. 123	8. 129	8. 136	8. 143	8. 150	8. 157	8. 163	8. 170	8. 177	8. 184	1, 500
1, 510	8. 184	8. 191	8. 198	8. 205	8. 212	8. 218	8. 225	8. 232	8. 239	8. 246	8. 253	1, 510
1, 520	8. 253	8. 260	8. 267	8. 274	8. 281	8. 287	8. 294	8. 301	8. 308	8. 315	8. 322	1, 520
1, 530	8. 322	8. 329	8. 336	8. 343	8. 350	8. 356	8. 363	8. 370	8. 377	8. 384	8. 391	1, 530
1, 540	8. 391	8. 398	8. 405	8. 412	8. 419	8. 426	8. 433	8. 439	8. 446	8. 453	8. 460	1, 540
1, 550	8. 460	8. 467	8. 474	8. 481	8. 488	8. 495	8. 502	8. 509	8. 516	8. 523	8. 530	1, 550
1, 560	8. 530	8. 537	8. 544	8. 551	8. 558	8. 565	8. 571	8. 578	8. 585	8. 592	8. 599	1, 560
1, 570	8. 599	8. 606	8. 613	8. 620	8. 627	8. 634	8. 641	8. 648	8. 655	8. 662	8. 669	1, 570
1, 580	8. 669	8. 676	8. 683	8. 690	8. 697	8. 704	8. 711	8. 718	8. 725	8. 732	8. 739	1, 580
1, 590	8. 739	8. 746	8. 753	8. 760	8. 767	8. 774	8. 781	8. 788	8. 795	8. 802	8. 809	1, 590
1, 600	8. 809	8. 816	8. 823	8. 830	8. 837	8. 844	8. 851	8. 858	8. 865	8. 872	8. 879	1, 600
°F	0	1	2	3	4	5	6	7	8	9	10	°F

TABLE 46—*Type R thermocouples (continued)*.

Temperature in Degrees Fahrenheit

EMF in Absolute Millivolts   Reference Junctions at 32 F

°F	0	1	2	3	4	5	6	7	8	9	10	°F
						Millivolts						
1, 600	8. 809	8. 816	8. 823	8. 830	8. 837	8. 844	8. 851	8. 858	8. 865	8. 872	8. 879	1, 600
1, 610	8. 879	8. 886	8. 893	8. 900	8. 907	8. 914	8. 921	8. 928	8. 935	8. 942	8. 949	1, 610
1, 620	8. 949	8. 956	8. 963	8. 970	8. 977	8. 984	8. 991	8. 998	9. 005	9. 012	9. 019	1, 620
1, 630	9. 019	9. 026	9. 033	9. 040	9. 047	9. 054	9. 061	9. 068	9. 075	9. 082	9. 090	1, 630
1, 640	9. 090	9. 097	9. 104	9. 111	9. 118	9. 125	9. 132	9. 139	9. 146	9. 153	9. 161	1, 640
1, 650	9. 161	9. 168	9. 175	9. 182	9. 189	9. 196	9. 203	9. 210	9. 218	9. 225	9. 232	1, 650
1, 660	9. 232	9. 239	9. 246	9. 253	9. 260	9. 267	9. 274	9. 281	9. 289	9. 296	9. 303	1, 660
1, 670	9. 303	9. 310	9. 317	9. 324	9. 331	9. 338	9. 345	9. 353	9. 360	9. 367	9. 374	1, 670
1, 680	9. 374	9. 381	9. 388	9. 395	9. 402	9. 409	9. 416	9. 424	9. 431	9. 438	9. 445	1, 680
1, 690	9. 445	9. 452	9. 459	9. 466	9. 474	9. 481	9. 488	9. 495	9. 502	9. 509	9. 516	1, 690
1, 700	9. 516	9. 523	9. 531	9. 538	9. 545	9. 552	9. 559	9. 566	9. 573	9. 580	9. 587	1, 700
1, 710	9. 587	9. 594	9. 602	9. 609	9. 616	9. 623	9. 630	9. 637	9. 644	9. 651	9. 659	1, 710
1, 720	9. 659	9. 666	9. 673	9. 680	9. 687	9. 694	9. 701	9. 709	9. 716	9. 723	9. 730	1, 720
1, 730	9. 730	9. 737	9. 744	9. 751	9. 759	9. 766	9. 773	9. 780	9. 787	9. 794	9. 802	1, 730
1, 740	9. 802	9. 809	9. 816	9. 823	9. 830	9. 838	9. 845	9. 852	9. 859	9. 866	9. 874	1, 740
1, 750	9. 874	9. 881	9. 888	9. 895	9. 902	9. 910	9. 917	9. 924	9. 931	9. 939	9. 946	1, 750
1, 760	9. 946	9. 953	9. 961	9. 968	9. 975	9. 982	9. 990	9. 997	10. 004	10. 012	10. 019	1, 760
1, 770	10. 019	10. 026	10. 034	10. 041	10. 048	10. 056	10. 063	10. 070	10. 077	10. 084	10. 092	1, 770
1, 780	10. 092	10. 099	10. 106	10. 114	10. 121	10. 129	10. 136	10. 143	10. 150	10. 157	10. 164	1, 780
1, 790	10. 164	10. 172	10. 179	10. 186	10. 194	10. 201	10. 208	10. 215	10. 223	10. 230	10. 237	1, 790
1, 800	10. 237	10. 244	10. 251	10. 259	10. 266	10. 274	10. 281	10. 288	10. 296	10. 303	10. 310	1, 800
1, 810	10. 310	10. 318	10. 325	10. 332	10. 339	10. 347	10. 354	10. 361	10. 369	10. 376	10. 383	1, 810
1, 820	10. 383	10. 391	10. 398	10. 405	10. 412	10. 420	10. 427	10. 434	10. 441	10. 449	10. 456	1, 820
1, 830	10. 456	10. 464	10. 471	10. 478	10. 485	10. 493	10. 500	10. 507	10. 514	10. 522	10. 529	1, 830
1, 840	10. 529	10. 537	10. 544	10. 551	10. 559	10. 566	10. 574	10. 581	10. 588	10. 596	10. 603	1, 840
1, 850	10. 603	10. 610	10. 618	10. 625	10. 632	10. 639	10. 647	10. 654	10. 661	10. 669	10. 676	1, 850
1, 860	10. 676	10. 683	10. 691	10. 698	10. 705	10. 712	10. 720	10. 727	10. 735	10. 742	10. 749	1, 860
1, 870	10. 749	10. 757	10. 764	10. 771	10. 779	10. 786	10. 794	10. 801	10. 809	10. 816	10. 823	1, 870
1, 880	10. 823	10. 831	10. 839	10. 846	10. 854	10. 861	10. 869	10. 876	10. 884	10. 891	10. 898	1, 880
1, 890	10. 898	10. 906	10. 914	10. 921	10. 929	10. 936	10. 944	10. 951	10. 959	10. 966	10. 973	1, 890
1, 900	10. 973	10. 981	10. 988	10. 996	11. 003	11. 011	11. 018	11. 026	11. 033	11. 040	11. 048	1, 900
1, 910	11. 048	11. 055	11. 063	11. 070	11. 078	11. 085	11. 093	11. 100	11. 108	11. 115	11. 122	1, 910
1, 920	11. 122	11. 130	11. 138	11. 145	11. 153	11. 160	11. 168	11. 175	11. 183	11. 190	11. 197	1, 920
1, 930	11. 197	11. 205	11. 213	11. 220	11. 228	11. 235	11. 243	11. 250	11. 258	11. 265	11. 273	1, 930
1, 940	11. 273	11. 280	11. 288	11. 295	11. 303	11. 310	11. 318	11. 325	11. 333	11. 340	11. 348	1, 940
1, 950	11. 348	11. 355	11. 363	11. 371	11. 379	11. 385	11. 393	11. 401	11. 408	11. 416	11. 424	1, 950
1, 960	11. 424	11. 431	11. 439	11. 446	11. 454	11. 461	11. 468	11. 476	11. 484	11. 492	11. 499	1, 960
1, 970	11. 499	11. 507	11. 515	11. 522	11. 529	11. 537	11. 544	11. 552	11. 560	11. 568	11. 575	1, 970
1, 980	11. 575	11. 582	11. 590	11. 598	11. 605	11. 613	11. 620	11. 628	11. 636	11. 643	11. 651	1, 980
1, 990	11. 651	11. 658	11. 666	11. 674	11. 681	11. 689	11. 696	11. 704	11. 712	11. 719	11. 726	1, 990
2, 000	11. 726	11. 734	11. 742	11. 749	11. 757	11. 765	11. 772	11. 779	11. 787	11. 795	11. 802	2, 000
°F	0	1	2	3	4	5	6	7	8	9	10.	°F

TABLE 46—*Type R thermocouples (continued).*
Temperature in Degrees Fahrenheit

EMF in Absolute Millivolts

Reference Junctions at 32 F

°F	0	1	2	3	4	5	6	7	8	9	10	°F
						Millivolts						
2,000	11.726	11.734	11.742	11.749	11.757	11.765	11.772	11.779	11.787	11.795	11.802	2,000
2,010	11.802	11.810	11.817	11.825	11.832	11.840	11.848	11.855	11.863	11.871	11.878	2,010
2,020	11.878	11.885	11.893	11.901	11.908	11.916	11.924	11.931	11.938	11.946	11.954	2,020
2,030	11.954	11.961	11.969	11.976	11.984	11.992	11.999	12.007	12.014	12.022	12.029	2,030
2,040	12.029	12.037	12.045	12.052	12.060	12.068	12.075	12.082	12.090	12.098	12.105	2,040
2,050	12.105	12.113	12.121	12.128	12.136	12.144	12.151	12.159	12.166	12.174	12.182	2,050
2,060	12.182	12.189	12.197	12.205	12.212	12.220	12.227	12.235	12.243	12.250	12.258	2,060
2,070	12.258	12.265	12.273	12.281	12.288	12.296	12.304	12.312	12.319	12.327	12.335	2,070
2,080	12.335	12.342	12.350	12.358	12.365	12.373	12.381	12.388	12.396	12.403	12.411	2,080
2,090	12.411	12.419	12.427	12.434	12.442	12.450	12.458	12.465	12.473	12.480	12.488	2,090
2,100	12.488	12.495	12.503	12.511	12.518	12.526	12.534	12.541	12.549	12.557	12.564	2,100
2,110	12.564	12.572	12.579	12.587	12.595	12.602	12.610	12.618	12.625	12.633	12.641	2,110
2,120	12.641	12.648	12.656	12.664	12.672	12.679	12.687	12.695	12.702	12.710	12.718	2,120
2,130	12.718	12.725	12.733	12.741	12.748	12.756	12.764	12.772	12.779	12.787	12.795	2,130
2,140	12.795	12.802	12.810	12.818	12.825	12.833	12.841	12.848	12.856	12.864	12.871	2,140
2,150	12.871	12.879	12.887	12.894	12.902	12.909	12.917	12.925	12.932	12.940	12.948	2,150
2,160	12.948	12.955	12.963	12.971	12.978	12.986	12.994	13.002	13.009	13.017	13.025	2,160
2,170	13.025	13.032	13.040	13.048	13.055	13.063	13.071	13.078	13.086	13.094	13.102	2,170
2,180	13.102	13.109	13.117	13.125	13.132	13.140	13.148	13.155	13.163	13.170	13.178	2,180
2,190	13.178	13.186	13.193	13.201	13.208	13.216	13.224	13.232	13.239	13.247	13.255	2,190
2,200	13.255	13.263	13.270	13.278	13.285	13.293	13.301	13.309	13.316	13.324	13.332	2,200
2,210	13.332	13.340	13.347	13.355	13.363	13.371	13.378	13.386	13.394	13.402	13.409	2,210
2,220	13.409	13.417	13.425	13.432	13.440	13.448	13.455	13.463	13.471	13.479	13.486	2,220
2,230	13.486	13.494	13.502	13.509	13.517	13.525	13.532	13.540	13.548	13.556	13.564	2,230
2,240	13.564	13.571	13.579	13.587	13.595	13.602	13.610	13.618	13.625	13.633	13.641	2,240
2,250	13.641	13.648	13.656	13.664	13.672	13.679	13.687	13.695	13.702	13.710	13.718	2,250
2,260	13.718	13.725	13.733	13.741	13.749	13.756	13.764	13.772	13.779	13.787	13.795	2,260
2,270	13.795	13.802	13.810	13.818	13.826	13.833	13.841	13.849	13.857	13.865	13.872	2,270
2,280	13.872	13.880	13.888	13.895	13.903	13.911	13.918	13.926	13.934	13.942	13.949	2,280
2,290	13.949	13.957	13.965	13.972	13.980	13.988	13.995	14.003	14.011	14.019	14.027	2,290
2,300	14.027	14.034	14.042	14.050	14.058	14.065	14.073	14.081	14.088	14.096	14.104	2,300
2,310	14.104	14.111	14.119	14.127	14.135	14.142	14.150	14.158	14.165	14.173	14.181	2,310
2,320	14.181	14.188	14.196	14.204	14.212	14.219	14.227	14.235	14.242	14.250	14.258	2,320
2,330	14.258	14.265	14.273	14.281	14.288	14.296	14.304	14.312	14.319	14.327	14.335	2,330
2,340	14.335	14.342	14.350	14.358	14.366	14.374	14.382	14.389	14.397	14.405	14.412	2,340
2,350	14.412	14.420	14.428	14.435	14.443	14.451	14.459	14.467	14.475	14.482	14.490	2,350
2,360	14.490	14.498	14.505	14.513	14.521	14.528	14.536	14.544	14.552	14.560	14.567	2,360
2,370	14.567	14.575	14.583	14.591	14.598	14.606	14.614	14.621	14.629	14.637	14.644	2,370
2,380	14.644	14.652	14.660	14.668	14.675	14.683	14.691	14.698	14.706	14.714	14.721	2,380
2,390	14.721	14.729	14.737	14.745	14.752	14.760	14.768	14.775	14.783	14.791	14.798	2,390
2,400	14.798	14.806	14.814	14.822	14.829	14.837	14.845	14.852	14.860	14.868	14.875	2,400
°F	0	1	2	3	4	5	6	7	8	9	10	°F

TABLE 46—*Type R thermocouples (continued)*.
Temperature in Degrees Fahrenheit

EMF in Absolute Millivolts

Reference Junctions at 32 F

°F	0	1	2	3	4	5	6	7	8	9	10	°F
						Millivolts						
2,400	14.798	14.806	14.814	14.822	14.829	14.837	14.845	14.852	14.860	14.868	14.875	2,400
2,410	14.875	14.883	14.891	14.898	14.906	14.914	14.922	14.929	14.937	14.945	14.952	2,410
2,420	14.952	14.960	14.968	14.975	14.983	14.991	14.999	15.006	15.014	15.022	15.029	2,420
2,430	15.029	15.037	15.045	15.052	15.060	15.068	15.076	15.084	15.091	15.099	15.107	2,430
2,440	15.107	15.115	15.122	15.130	15.138	15.145	15.153	15.161	15.168	15.176	15.184	2,440
2,450	15.184	15.192	15.199	15.207	15.215	15.222	15.230	15.238	15.245	15.253	15.261	2,450
2,460	15.261	15.268	15.276	15.284	15.292	15.299	15.307	15.315	15.322	15.330	15.338	2,460
2,470	15.338	15.345	15.353	15.361	15.369	15.377	15.385	15.392	15.400	15.408	15.415	2,470
2,480	15.415	15.423	15.431	15.438	15.446	15.454	15.462	15.469	15.477	15.484	15.492	2,480
2,490	15.492	15.500	15.508	15.515	15.523	15.531	15.538	15.546	15.553	15.561	15.568	2,490
2,500	15.568	15.576	15.584	15.592	15.599	15.607	15.615	15.623	15.630	15.638	15.645	2,500
2,510	15.645	15.653	15.661	15.668	15.676	15.684	15.692	15.700	15.707	15.715	15.722	2,510
2,520	15.722	15.730	15.738	15.745	15.753	15.761	15.769	15.777	15.785	15.792	15.800	2,520
2,530	15.800	15.808	15.815	15.823	15.831	15.838	15.846	15.854	15.862	15.869	15.877	2,530
2,540	15.877	15.885	15.892	15.900	15.908	15.915	15.923	15.931	15.938	15.946	15.954	2,540
2,550	15.954	15.962	15.969	15.977	15.985	15.992	16.000	16.008	16.015	16.023	16.031	2,550
2,560	16.031	16.039	16.046	16.054	16.062	16.070	16.078	16.085	16.093	16.101	16.108	2,560
2,570	16.108	16.116	16.124	16.132	16.139	16.147	16.155	16.163	16.170	16.178	16.185	2,570
2,580	16.185	16.193	16.201	16.208	16.216	16.224	16.232	16.240	16.247	16.255	16.263	2,580
2,590	16.263	16.271	16.278	16.286	16.294	16.301	16.309	16.317	16.325	16.332	16.340	2,590
2,600	16.340	16.348	16.355	16.363	16.371	16.378	16.386	16.394	16.402	16.409	16.417	2,600
2,610	16.417	16.425	16.432	16.440	16.448	16.455	16.463	16.471	16.478	16.486	16.494	2,610
2,620	16.494	16.502	16.509	16.517	16.524	16.532	16.540	16.548	16.556	16.564	16.571	2,620
2,630	16.571	16.579	16.586	16.594	16.602	16.610	16.618	16.625	16.633	16.641	16.648	2,630
2,640	16.648	16.656	16.663	16.671	16.679	16.687	16.695	16.702	16.710	16.718	16.725	2,640
2,650	16.725	16.733	16.741	16.748	16.756	16.764	16.772	16.780	16.788	16.795	16.802	2,650
2,660	16.802	16.810	16.818	16.826	16.834	16.842	16.849	16.857	16.865	16.872	16.880	2,660
2,670	16.880	16.887	16.895	16.903	16.911	16.918	16.926	16.933	16.941	16.949	16.957	2,670
2,680	16.957	16.964	16.972	16.979	16.987	16.995	17.002	17.010	17.018	17.025	17.033	2,680
2,690	17.033	17.041	17.048	17.056	17.064	17.072	17.079	17.087	17.095	17.102	17.110	2,690
2,700	17.110	17.118	17.125	17.133	17.141	17.148	17.156	17.163	17.171	17.179	17.186	2,700
2,710	17.186	17.194	17.202	17.209	17.217	17.225	17.232	17.240	17.248	17.255	17.263	2,710
2,720	17.263	17.271	17.278	17.286	17.294	17.301	17.309	17.317	17.325	17.332	17.340	2,720
2,730	17.340	17.347	17.355	17.363	17.370	17.378	17.385	17.393	17.401	17.408	17.416	2,730
2,740	17.416	17.424	17.432	17.439	17.447	17.455	17.462	17.470	17.478	17.485	17.493	2,740
2,750	17.493	17.500	17.508	17.516	17.524	17.532	17.539	17.546	17.554	17.562	17.569	2,750
2,760	17.569	17.577	17.585	17.592	17.600	17.608	17.615	17.623	17.631	17.638	17.646	2,760
2,770	17.646	17.654	17.662	17.669	17.677	17.685	17.692	17.700	17.708	17.715	17.723	2,770
2,780	17.723	17.731	17.738	17.746	17.753	17.761	17.768	17.776	17.784	17.792	17.799	2,780
2,790	17.799	17.807	17.814	17.822	17.830	17.837	17.845	17.852	17.860	17.868	17.875	2,790
2,800	17.875	17.882	17.890	17.898	17.906	17.913	17.921	17.928	17.936	17.944	17.951	2,800
°F	0	1	2	3	4	5	6	7	8	9	10	°F

TABLE 46—*Type R thermocouples (continued).*
Temperature in Degrees Fahrenheit

EMF in Absolute Millivolts   Reference Junctions at 32 F

°F	0	1	2	3	4	5	6	7	8	9	10	°F
						Millivolts						
2,800	17.875	17.882	17.890	17.898	17.906	17.913	17.921	17.928	17.936	17.944	17.951	2,800
2,810	17.951	17.958	17.966	17.974	17.982	17.989	17.997	18.004	18.012	18.020	18.027	2,810
2,820	18.027	18.035	18.043	18.050	18.058	18.065	18.073	18.080	18.088	18.096	18.103	2,820
2,830	18.103	18.111	18.119	18.126	18.134	18.141	18.149	18.156	18.164	18.172	18.179	2,830
2,840	18.179	18.187	18.195	18.202	18.210	18.218	18.225	18.233	18.240	18.248	18.255	2,840
2,850	18.255	18.263	18.271	18.278	18.286	18.294	18.301	18.309	18.316	18.324	18.332	2,850
2,860	18.332	18.339	18.347	18.355	18.362	18.370	18.377	18.385	18.392	18.400	18.408	2,860
2,870	18.408	18.415	18.423	18.431	18.438	18.446	18.453	18.461	18.468	18.476	18.484	2,870
2,880	18.484	18.492	18.499	18.507	18.514	18.522	18.529	18.537	18.545	18.552	18.560	2,880
2,890	18.560	18.568	18.575	18.583	18.590	18.598	18.605	18.613	18.621	18.628	18.636	2,890
2,900	18.636	18.644	18.651	18.659	18.666	18.674	18.681	18.689	18.697	18.705	18.712	2,900
2,910	18.712	18.720	18.727	18.735	18.743	18.750	18.758	18.765	18.773	18.781	18.788	2,910
2,920	18.788	18.796	18.803	18.811	18.819	18.826	18.834	18.842	18.849	18.857	18.864	2,920
2,930	18.864	18.872	18.879	18.887	18.895	18.902	18.910	18.918	18.925	18.932	18.940	2,930
2,940	18.940	18.948	18.955	18.963	18.971	18.978	18.986	18.993	19.001	19.008	19.016	2,940
2,950	19.016	19.024	19.031	19.039	19.046	19.054	19.062	19.069	19.077	19.084	19.092	2,950
2,960	19.092	19.099	19.107	19.115	19.122	19.129	19.137	19.145	19.152	19.160	19.168	2,960
2,970	19.168	19.175	19.182	19.190	19.198	19.205	19.213	19.220	19.228	19.235	19.243	2,970
2,980	19.243	19.250	19.258	19.265	19.273	19.281	19.288	19.295	19.303	19.311	19.318	2,980
2,990	19.318	19.326	19.333	19.341	19.348	19.356	19.364	19.371	19.378	19.386	19.394	2,990
3,000	19.394	19.402	19.409	19.417	19.424	19.432	19.439	19.447	19.454	19.462	19.470	3,000
3,010	19.470	19.477	19.485	19.492	19.500	19.508	19.515	19.523	19.530	19.538	19.545	3,010
3,020	19.545	19.553	19.561	19.568	19.576	19.583	19.591	19.598	19.606	19.614	19.621	3,020
3,030	19.621	19.628	19.636	19.644	19.651	19.659	19.667	19.674	19.682	19.689	19.697	3,030
3,040	19.697	19.704	19.712	19.720	19.727	19.735	19.742	19.750	19.758	19.765	19.773	3,040
3,050	19.773	19.780	19.788	19.795	19.803	19.811	19.818	19.826	19.833	19.841	19.848	3,050
3,060	19.848	19.856	19.864	19.871	19.878	19.886	19.894	19.902	19.909	19.916	19.924	3,060
3,070	19.924	19.932	19.939	19.947	19.954	19.962	19.969	19.977	19.984	19.992	19.999	3,070
3,080	19.999	20.007	20.014	20.022	20.029	20.037	20.044	20.052	20.059	20.067	20.075	3,080
3,090	20.075	20.082	20.090	20.097	20.105	20.112	20.120	20.127	20.135	20.142	20.150	3,090
°F	0	1	2	3	4	5	6	7	8	9	10	°F

TABLE 47—*Type R thermocouples.*
Temperature in Degrees Celsius

EMF in Absolute Millivolts

Reference Junctions at 0 C

° C	0	1	2	3	4	5	6	7	8	9	10	° C
						Millivolts						
0	0. 000	0. 005	0. 011	0. 016	0. 022	0. 027	0. 033	0. 038	0. 043	0. 049	0. 055	0
10	. 055	. 061	. 066	. 072	. 078	. 083	. 089	. 095	. 101	. 106	. 112	10
20	. 112	. 118	. 124	. 130	. 136	. 142	. 148	. 154	. 160	. 166	. 172	20
30	. 172	. 178	. 184	. 190	. 196	. 203	. 209	. 215	. 221	. 228	. 234	30
40	. 234	. 240	. 246	. 252	. 259	. 265	. 272	. 278	. 285	. 291	. 298	40
50	. 298	. 304	. 311	. 317	. 324	. 330	. 337	. 343	. 350	. 357	. 363	50
60	. 363	. 370	. 377	. 383	. 390	. 397	. 403	. 410	. 417	. 424	. 431	60
70	. 431	. 438	. 445	. 451	. 458	. 465	. 472	. 479	. 486	. 493	. 500	70
80	. 500	. 507	. 514	. 521	. 528	. 536	. 543	. 550	. 557	. 565	. 572	80
90	. 572	. 579	. 586	. 594	. 601	. 609	. 616	. 623	. 631	. 638	. 645	90
100	. 645	. 653	. 660	. 668	. 675	. 683	. 690	698	. 705	. 713	. 721	100
110	. 721	. 728	. 736	. 744	. 752	. 759	. 767	. 775	. 782	. 790	. 798	110
120	. 798	. 805	. 813	. 821	. 829	. 837	. 845	. 853	. 861	. 869	. 877	120
130	. 877	. 885	. 893	. 901	. 909	. 917	. 925	. 933	. 941	. 949	. 957	130
140	. 957	. 966	. 974	. 982	. 990	. 998	1. 006	1. 014	1. 022	1. 031	1. 039	140
150	1. 039	1. 047	1. 055	1. 063	1. 072	1. 080	1. 088	1. 096	1. 104	1. 112	1. 121	150
160	1. 121	1. 129	1. 138	1. 146	1. 154	1. 163	1. 171	1. 179	1. 188	1. 196	1. 205	160
170	1. 205	1. 213	1. 222	1. 231	1. 239	1. 247	1. 256	1. 265	1. 273	1. 282	1. 290	170
180	1. 290	1. 298	1. 307	1. 316	1. 324	1. 333	1. 342	1. 351	1. 359	1. 368	1. 377	180
190	1. 377	1. 386	1. 395	1. 403	1. 412	1. 420	1. 429	1. 438	1. 447	1. 456	1. 465	190
200	1. 465	1. 473	1. 482	1. 491	1. 500	1. 509	1. 517	1. 526	1. 535	1. 544	1. 553	200
210	1. 553	1. 562	1. 571	1. 580	1. 589	1. 598	1. 607	1. 616	1. 625	1. 634	1. 643	210
220	1. 643	1. 652	1. 661	1. 670	1. 679	1. 688	1. 697	1. 706	1. 715	1. 725	1. 734	220
230	1. 734	1. 743	1. 752	1. 761	1. 770	1. 779	1. 788	1. 798	1. 807	1. 816	1. 826	230
240	1. 826	1. 835	1. 844	1. 853	1. 863	1. 872	1. 881	1. 890	1. 900	1. 909	1. 918	240
250	1. 918	1. 928	1. 937	1. 946	1. 956	1. 965	1. 974	1. 984	1. 993	2. 002	2. 012	250
260	2. 012	2. 021	2. 031	2. 040	2. 050	2. 059	2. 068	2. 078	2. 087	2. 097	2. 107	260
270	2. 107	2. 116	2. 126	2. 135	2. 145	2. 154	2. 164	2. 173	2. 183	2. 192	2. 202	270
280	2. 202	2. 211	2. 221	2. 231	2. 240	2. 250	2. 259	2. 269	2. 279	2. 288	2. 298	280
290	2. 298	2. 308	2. 317	2. 327	2. 337	2. 346	2. 356	2. 366	2. 375	2. 385	2. 395	290
300	2. 395	2. 405	2. 415	2. 424	2. 434	2. 444	2. 454	2. 464	2. 473	2. 483	2. 493	300
310	2. 493	2. 503	2. 513	2. 522	2. 532	2. 542	2. 552	2. 562	2. 572	2. 581	2. 591	310
320	2. 591	2. 601	2. 611	2. 621	2. 631	2. 641	2. 650	2. 660	2. 670	2. 680	2. 690	320
330	2. 690	2. 700	2. 710	2. 720	2. 730	2. 740	2. 750	2. 760	2. 770	2. 780	2. 790	330
340	2. 790	2. 800	2. 810	2. 820	2. 830	2. 840	2. 850	2. 860	2. 870	2. 880	2. 890	340
350	2. 890	2. 900	2. 910	2. 920	2. 930	2. 940	2. 950	2. 961	2. 971	2. 981	2. 991	350
360	2. 991	3. 001	3. 011	3. 021	3. 031	3. 041	3. 051	3. 062	3. 072	3. 082	3. 092	360
370	3. 092	3. 102	3. 112	3. 122	3. 133	3. 143	3. 153	3. 163	3. 173	3. 183	3. 194	370
380	3. 194	3. 204	3. 214	3. 224	3. 234	3. 245	3. 255	3. 265	3. 276	3. 286	3. 296	380
390	3. 296	3. 306	3. 317	3. 327	3. 337	3. 347	3. 358	3. 368	3. 378	3. 389	3. 399	390
400	3. 399	3. 409	3. 420	3. 430	3. 440	3. 451	3. 461	3. 471	3. 481	3. 492	3. 502	400
° C	0	1	2	3	4	5	6	7	8	9	10	° C

TABLE 47—*Type R thermocouples (continued)*.
Temperature in Degrees Celsius
EMF in Absolute Millivolts

Reference Junctions at 0 C

°C	0	1	2	3	4	5	6	7	8	9	10	°C
						Millivolts						
400	3. 399	3. 409	3. 420	3. 430	3. 440	3. 451	3. 461	3. 471	3. 481	3. 492	3. 502	400
410	3. 502	3. 512	3. 523	3. 533	3. 544	3. 554	3. 565	3. 575	3. 586	3. 596	3. 607	410
420	3. 607	3. 617	3. 627	3. 638	3. 648	3. 659	3. 669	3. 680	3. 690	3. 701	3. 712	420
430	3. 712	3. 722	3. 732	3. 743	3. 753	3. 764	3. 774	3. 785	3. 796	3. 806	3. 817	430
440	3. 817	3. 827	3. 838	3. 848	3. 859	3. 870	3. 880	3. 891	3. 901	3. 912	3. 923	440
450	3. 923	3. 933	3. 944	3. 954	3. 965	3. 976	3. 987	3. 997	4. 008	4. 018	4. 029	450
460	4. 029	4. 039	4. 050	4. 060	4. 071	4. 081	4. 092	4. 102	4. 113	4. 123	4. 134	460
470	4. 134	4. 145	4. 156	4. 166	4. 177	4. 187	4. 198	4. 209	4. 219	4. 230	4. 241	470
480	4. 241	4. 251	4. 262	4. 273	4. 283	4. 294	4. 305	4. 315	4. 326	4. 337	4. 348	480
490	4. 348	4. 358	4. 369	4. 380	4. 390	4. 401	4. 412	4. 422	4. 433	4. 444	4. 455	490
500	4. 455	4. 466	4. 477	4. 488	4. 498	4. 509	4. 520	4. 531	4. 542	4. 552	4. 563	500
510	4. 563	4. 574	4. 585	4. 596	4. 607	4. 618	4. 629	4. 640	4. 651	4. 662	4. 672	510
520	4. 672	4. 683	4. 694	4. 705	4. 716	4. 727	4. 738	4. 749	4. 760	4. 771	4. 782	520
530	4. 782	4. 793	4. 804	4. 815	4. 826	4. 837	4. 848	4. 859	4. 870	4. 881	4. 893	530
540	4. 893	4. 904	4. 915	4. 926	4. 937	4. 948	4. 959	4. 970	4. 981	4. 992	5. 004	540
550	5. 004	5. 015	5. 026	5. 037	5. 048	5. 059	5. 070	5. 081	5. 092	5. 104	5. 115	550
560	5. 115	5. 126	5. 137	5. 148	5. 159	5. 170	5. 182	5. 193	5. 204	5. 215	5. 226	560
570	5. 226	5. 238	5. 249	5. 260	5. 271	5. 282	5. 293	5. 304	5. 316	5. 327	5. 338	570
580	5. 338	5. 349	5. 360	5. 371	5. 383	5. 394	5. 405	5. 416	5. 428	5. 439	5. 450	580
590	5. 450	5. 461	5. 472	5. 484	5. 495	5. 507	5. 518	5. 529	5. 540	5. 551	5. 563	590
600	5. 563	5. 574	5. 586	5. 597	5. 609	5. 620	5. 631	5. 642	5. 654	5. 665	5. 677	600
610	5. 677	5. 688	5. 700	5. 711	5. 723	5. 734	5. 746	5. 757	5. 769	5. 780	5. 792	610
620	5. 792	5. 803	5. 814	5. 826	5. 837	5. 849	5. 861	5. 872	5. 883	5. 895	5. 907	620
630	5. 907	5. 918	5. 930	5. 941	5. 952	5. 964	5. 976	5. 987	5. 999	6. 010	6. 022	630
640	6. 022	6. 033	6. 044	6. 056	6. 068	6. 079	6. 091	6. 102	6. 114	6. 126	6. 137	640
650	6. 137	6. 149	6. 160	6. 171	6. 183	6. 194	6. 206	6. 218	6. 229	6. 240	6. 252	650
660	6. 252	6. 264	6. 275	6. 287	6. 299	6. 310	6. 321	6. 333	6. 344	6. 356	6. 368	660
670	6. 368	6. 380	6. 391	6. 403	6. 415	6. 427	6. 438	6. 450	6. 461	6. 473	6. 485	670
680	6. 485	6. 497	6. 508	6. 520	6. 532	6. 544	6. 555	6. 567	6. 579	6. 590	6. 602	680
690	6. 602	6. 614	6. 626	6. 637	6. 649	6. 661	6. 672	6. 684	6. 696	6. 708	6. 720	690
700	6. 720	6. 732	6. 744	6. 756	6. 768	6. 779	6. 791	6. 803	6. 815	6. 827	6. 838	700
710	6. 838	6. 850	6. 862	6. 874	6. 886	6. 898	6. 910	6. 922	6. 934	6. 946	6. 957	710
720	6. 957	6. 969	6. 981	6. 993	7. 005	7. 017	7. 029	7. 040	7. 052	7. 064	7. 076	720
730	7. 076	7. 088	7. 100	7. 112	7. 124	7. 136	7. 147	7. 159	7. 171	7. 183	7. 195	730
740	7. 195	7. 207	7. 219	7. 231	7. 243	7. 255	7. 267	7. 279	7. 291	7. 303	7. 315	740
750	7. 315	7. 327	7. 339	7. 351	7. 364	7. 376	7. 388	7. 400	7. 412	7. 424	7. 436	750
760	7. 436	7. 448	7. 460	7. 472	7. 485	7. 497	7. 509	7. 521	7. 533	7. 545	7. 557	760
770	7. 557	7. 570	7. 582	7. 594	7. 606	7. 618	7. 631	7. 643	7. 655	7. 667	7. 679	770
780	7. 679	7. 692	7. 704	7. 716	7. 728	7. 740	7. 752	7. 765	7. 777	7. 789	7. 801	780
790	7. 801	7. 814	7. 826	7. 838	7. 850	7. 863	7. 875	7. 888	7. 900	7. 912	7. 924	790
800	7. 924	7. 936	7. 949	7. 961	7. 973	7. 986	7. 998	8. 010	8. 022	8. 035	8. 047	800
°C	0	1	2	3	4	5	6	7	8	9	10	°C

TABLE 47—*Type R thermocouples (continued)*.
Temperature in Degrees Celsius

EMF in Absolute Millivolts

Reference Junctions at 0 C

°C	0	1	2	3	4	5	6	7	8	9	10	°C
						Millivolts						
800	7. 924	7. 936	7. 949	7. 961	7. 973	7. 986	7. 998	8. 010	8. 022	8. 035	8. 047	800
810	8. 047	8. 059	8. 071	8. 084	8. 096	8. 109	8. 121	8. 134	8. 146	8. 158	8. 170	810
820	8. 170	8. 182	8. 195	8. 208	8. 220	8. 232	8. 245	8. 257	8. 269	8. 281	8. 294	820
830	8. 294	8. 306	8. 319	8. 331	8. 343	8. 356	8. 369	8. 381	8. 394	8. 406	8. 419	830
840	8. 419	8. 431	8. 444	8. 456	8. 469	8. 481	8. 494	8. 506	8. 519	8. 531	8. 544	840
850	8. 544	8. 556	8. 569	8. 581	8. 594	8. 606	8. 619	8. 631	8. 644	8. 656	8. 669	850
860	8. 669	8. 681	8. 694	8. 706	8. 719	8. 732	8. 744	8. 757	8. 769	8. 782	8. 795	860
870	8. 795	8. 807	8. 820	8. 832	8. 845	8. 858	8. 870	8. 883	8. 895	8. 908	8. 921	870
880	8. 921	8. 933	8. 946	8. 959	8. 971	8. 984	8. 996	9. 009	9. 021	9. 034	9. 047	880
890	9. 047	9. 060	9. 072	9. 085	9. 098	9. 111	9. 123	9. 136	9. 149	9. 161	9. 175	890
900	9. 175	9. 188	9. 200	9. 213	9. 226	9. 239	9. 251	9. 264	9. 277	9. 290	9. 303	900
910	9. 303	9. 316	9. 328	9. 341	9. 354	9. 367	9. 379	9. 392	9. 405	9. 418	9. 431	910
920	9. 431	9. 444	9. 456	9. 469	9. 482	9. 495	9. 508	9. 520	9. 533	9. 546	9. 559	920
930	9. 559	9. 572	9. 585	9. 598	9. 610	9. 623	9. 636	9. 649	9. 661	9. 674	9. 687	930
940	9. 687	9. 700	9. 713	9. 726	9. 739	9. 752	9. 765	9. 778	9. 790	9. 803	9. 816	940
950	9. 816	9. 829	9. 842	9. 855	9. 868	9. 881	9. 894	9. 907	9. 920	9. 933	9. 946	950
960	9. 946	9. 960	9. 973	9. 986	9. 999	10. 012	10. 025	10. 038	10. 051	10. 064	10. 077	960
970	10. 077	10. 090	10. 103	10. 116	10. 130	10. 143	10. 156	10. 169	10. 182	10. 195	10. 208	970
980	10. 208	10. 221	10. 234	10. 247	10. 260	10. 274	10. 287	10. 300	10. 313	10. 326	10. 339	980
990	10. 339	10. 352	10. 366	10. 379	10. 392	10. 405	10. 419	10. 432	10. 445	10. 458	10. 471	990
1, 000	10. 471	10. 484	10. 497	10. 510	10. 523	10. 537	10. 550	10. 563	10. 576	10. 589	10. 603	1, 000
1, 010	10. 603	10. 616	10. 629	10. 642	10. 655	10. 669	10. 682	10. 695	10. 709	10. 722	10. 735	1, 010
1, 020	10. 735	10. 748	10. 761	10. 775	10. 788	10. 801	10. 815	10. 828	10. 841	10. 855	10. 869	1, 020
1, 030	10. 869	10. 882	10. 895	10. 909	10. 922	10. 936	10. 949	10. 963	10. 976	10. 989	11. 003	1, 030
1, 040	11. 003	11. 016	11. 030	11. 043	11. 057	11. 070	11. 084	11. 097	11. 111	11. 124	11. 138	1, 040
1, 050	11. 138	11. 151	11. 165	11. 178	11. 191	11. 205	11. 219	11. 232	11. 246	11. 259	11. 273	1, 050
1, 060	11. 273	11. 286	11. 300	11. 313	11. 327	11. 340	11. 354	11. 367	11. 381	11. 394	11. 408	1, 060
1, 070	11. 408	11. 421	11. 435	11. 449	11. 463	11. 476	11. 490	11. 504	11. 517	11. 531	11. 544	1, 070
1, 080	11. 544	11. 558	11. 571	11. 585	11. 599	11. 613	11. 626	11. 640	11. 654	11. 667	11. 681	1, 080
1, 090	11. 681	11. 694	11. 708	11. 722	11. 736	11. 749	11. 763	11. 776	11. 790	11. 803	11. 817	1, 090
1, 100	11. 817	11. 830	11. 844	11. 858	11. 871	11. 885	11. 899	11. 913	11. 926	11. 940	11. 954	1, 100
1, 110	11. 954	11. 967	11. 981	11. 994	12. 008	12. 022	12. 035	12. 049	12. 063	12. 077	12. 090	1, 110
1, 120	12. 090	12. 104	12. 118	12. 131	12. 145	12. 159	12. 173	12. 186	12. 200	12. 214	12. 227	1, 120
1, 130	12. 227	12. 241	12. 254	12. 268	12. 282	12. 296	12. 310	12. 323	12. 337	12. 351	12. 365	1, 130
1, 140	12. 365	12. 378	12. 392	12. 406	12. 420	12. 434	12. 447	12. 461	12. 475	12. 489	12. 503	1, 140
1, 150	12. 503	12. 516	12. 530	12. 544	12. 558	12. 572	12. 585	12. 599	12. 613	12. 627	12. 641	1, 150
1, 160	12. 641	12. 654	12. 668	12. 682	12. 696	12. 710	12. 723	12. 737	12. 751	12. 765	12. 779	1, 160
1, 170	12. 779	12. 792	12. 806	12. 820	12. 834	12. 848	12. 861	12. 875	12. 889	12. 903	12. 917	1, 170
1, 180	12. 917	12. 931	12. 944	12. 958	12. 972	12. 986	13. 000	13. 014	13. 028	13. 042	13. 055	1, 180
1, 190	13. 055	13. 069	13. 083	13. 097	13. 111	13. 125	13. 139	13. 152	13. 166	13. 180	13. 193	1, 190
1, 200	13. 193	13. 207	13. 221	13. 235	13. 249	13. 263	13. 277	13. 291	13. 305	13. 319	13. 333	1, 200
°C	0	1	2	3	4	5	6	7	8	9	10	°C

TABLE 47—*Type R thermocouples (continued)*.

Temperature in Degrees Celsius

EMF in Absolute Millivolts   Reference Junctions at 0 C

°C	0	1	2	3	4	5	6	7	8	9	10	°C
						Millivolts						
1,200	13.193	13.207	13.221	13.235	13.249	13.263	13.277	13.291	13.305	13.319	13.332	1,200
1,210	13.332	13.346	13.360	13.374	13.388	13.402	13.416	13.429	13.443	13.457	13.471	1,210
1,220	13.471	13.485	13.499	13.513	13.526	13.540	13.554	13.568	13.582	13.596	13.610	1,220
1,230	13.610	13.624	13.638	13.652	13.666	13.679	13.693	13.707	13.721	13.735	13.749	1,230
1,240	13.749	13.763	13.777	13.791	13.805	13.818	13.832	13.846	13.860	13.874	13.888	1,240
1,250	13.888	13.902	13.916	13.930	13.943	13.957	13.971	13.985	13.999	14.013	14.027	1,250
1,260	14.027	14.041	14.055	14.069	14.082	14.096	14.110	14.124	14.138	14.152	14.165	1,260
1,270	14.165	14.179	14.193	14.207	14.221	14.235	14.249	14.263	14.277	14.291	14.304	1,270
1,280	14.304	14.318	14.332	14.346	14.360	14.374	14.388	14.402	14.416	14.430	14.443	1,280
1,290	14.443	14.457	14.471	14.485	14.499	14.513	14.527	14.541	14.555	14.569	14.582	1,290
1,300	14.582	14.596	14.610	14.624	14.638	14.652	14.666	14.680	14.694	14.707	14.721	1,300
1,310	14.721	14.735	14.749	14.763	14.777	14.791	14.804	14.818	14.832	14.846	14.860	1,310
1,320	14.860	14.874	14.888	14.901	14.915	14.929	14.943	14.957	14.971	14.985	14.999	1,320
1,330	14.999	15.013	15.026	15.040	15.054	15.068	15.082	15.096	15.110	15.124	15.138	1,330
1,340	15.138	15.151	15.165	15.179	15.193	15.207	15.221	15.234	15.248	15.262	15.276	1,340
1,350	15.276	15.290	15.304	15.318	15.331	15.345	15.359	15.373	15.387	15.401	15.415	1,350
1,360	15.415	15.429	15.443	15.456	15.470	15.484	15.498	15.512	15.526	15.540	15.553	1,360
1,370	15.553	15.567	15.581	15.595	15.609	15.623	15.637	15.651	15.665	15.679	15.692	1,370
1,380	15.692	15.706	15.720	15.734	15.748	15.761	15.775	15.789	15.803	15.817	15.831	1,380
1,390	15.831	15.845	15.859	15.873	15.886	15.900	15.914	15.928	15.942	15.956	15.969	1,390
1,400	15.969	15.983	15.997	16.011	16.025	16.039	16.053	16.067	16.081	16.095	16.108	1,400
1,410	16.108	16.122	16.136	16.150	16.164	16.178	16.192	16.206	16.219	16.233	16.247	1,410
1,420	16.247	16.261	16.275	16.289	16.303	16.317	16.330	16.344	16.358	16.372	16.386	1,420
1,430	16.386	16.400	16.414	16.427	16.441	16.455	16.469	16.483	16.497	16.511	16.524	1,430
1,440	16.524	16.538	16.552	16.566	16.580	16.594	16.608	16.621	16.635	16.649	16.663	1,440
1,450	16.663	16.677	16.691	16.705	16.719	16.733	16.746	16.760	16.774	16.788	16.802	1,450
1,460	16.802	16.816	16.830	16.844	16.858	16.872	16.885	16.899	16.913	16.927	16.940	1,460
1,470	16.940	16.954	16.968	16.982	16.996	17.010	17.024	17.037	17.051	17.065	17.079	1,470
1,480	17.079	17.092	17.106	17.120	17.134	17.148	17.161	17.175	17.189	17.203	17.217	1,480
1,490	17.217	17.230	17.244	17.258	17.272	17.286	17.299	17.313	17.327	17.341	17.355	1,490
1,500	17.355	17.368	17.382	17.396	17.410	17.424	17.437	17.451	17.465	17.479	17.493	1,500
1,510	17.493	17.506	17.520	17.534	17.547	17.561	17.575	17.589	17.603	17.617	17.631	1,510
1,520	17.631	17.644	17.658	17.672	17.686	17.699	17.713	17.726	17.740	17.754	17.768	1,520
1,530	17.768	17.781	17.795	17.809	17.823	17.837	17.850	17.864	17.878	17.892	17.906	1,530
1,540	17.906	17.919	17.933	17.947	17.960	17.974	17.988	18.002	18.016	18.029	18.043	1,540
1,550	18.043	18.056	18.070	18.084	18.098	18.111	18.125	18.139	18.152	18.166	18.179	1,550
1,560	18.179	18.193	18.207	18.220	18.234	18.248	18.261	18.275	18.289	18.303	18.316	1,560
1,570	18.316	18.330	18.344	18.357	18.371	18.385	18.399	18.412	18.426	18.440	18.453	1,570
1,580	18.453	18.467	18.481	18.494	18.508	18.522	18.536	18.549	18.563	18.576	18.590	1,580
1,590	18.590	18.604	18.618	18.631	18.645	18.659	18.672	18.686	18.700	18.714	18.727	1,590
1,600	18.727	18.741	18.754	18.768	18.782	18.796	18.810	18.823	18.836	18.850	18.864	1,600
°C	0	1	2	3	4	5	6	7	8	9	10	°C

TABLE 47—*Type R thermocouples (continued)*.

Temperature in Degrees Celsius

EMF in Absolute Millivolts

Reference Junctions at 0 C

°C	0	1	2	3	4	5	6	7	8	9	10	°C
						Millivolts						
1,600	18.727	18.741	18.754	18.768	18.782	18.796	18.810	18.823	18.836	18.850	18.864	1,600
1,610	18.864	18.878	18.891	18.905	18.919	18.932	18.946	18.960	18.973	18.987	19.001	1,610
1,620	19.001	19.014	19.028	19.042	19.056	19.069	19.083	19.096	19.110	19.124	19.137	1,620
1,630	19.137	19.150	19.164	19.178	19.191	19.205	19.219	19.232	19.246	19.260	19.273	1,630
1,640	19.273	19.287	19.300	19.314	19.328	19.341	19.355	19.369	19.382	19.396	19.409	1,640
1,650	19.409	19.423	19.437	19.450	19.464	19.477	19.491	19.504	19.518	19.531	19.545	1,650
1,660	19.545	19.559	19.573	19.586	19.600	19.614	19.627	19.641	19.654	19.668	19.682	1,660
1,670	19.682	19.695	19.709	19.722	19.736	19.750	19.763	19.777	19.790	19.804	19.818	1,670
1,680	19.818	19.831	19.845	19.859	19.873	19.886	19.900	19.913	19.927	19.940	19.954	1,680
1,690	19.954	19.967	19.981	19.994	20.008	20.022	20.035	20.049	20.062	20.076	20.090	1,690
°C	0	1	2	3	4	5	6	7	8	9	10	°C

TABLE 48—*Type S thermocouples*
Temperature in Degrees Fahrenheit

EMF in Absolute Millivolts

Reference Junctions at 32 F

°F	0	1	2	3	4	5	6	7	8	9	10	°F
						Millivolts						
30	-------	-------	0. 000	0. 003	0. 006	0. 009	0. 012	0. 015	0. 018	0. 021	0. 024	30
40	0. 024	0. 028	. 031	. 034	. 037	. 040	. 043	. 046	. 049	. 052	. 056	40
50	. 056	. 059	. 062	. 065	. 068	. 071	. 075	. 078	. 081	. 084	. 087	50
60	. 087	. 091	. 094	. 097	. 100	. 104	. 107	. 110	. 113	. 117	. 120	60
70	. 120	. 123	. 126	. 130	. 133	. 136	. 140	. 143	. 146	. 150	. 153	70
80	. 153	. 156	. 160	. 163	. 166	. 170	. 173	. 176	. 180	. 183	. 187	80
90	. 187	. 190	. 193	. 197	. 200	. 204	. 207	. 211	. 214	. 218	. 221	90
100	. 221	. 224	. 228	. 231	. 235	. 238	. 242	. 245	. 249	. 252	. 256	100
110	. 256	. 259	. 263	. 266	. 270	. 274	. 277	. 281	. 284	. 288	. 291	110
120	. 291	. 295	. 299	. 302	. 306	. 309	. 313	. 317	. 320	. 324	. 327	120
130	. 327	. 331	. 335	. 338	. 342	. 346	. 349	. 353	. 357	. 360	. 364	130
140	. 364	. 368	. 371	. 375	. 379	. 383	. 386	. 390	. 394	. 397	. 401	140
150	. 401	. 405	. 409	. 412	. 416	. 420	. 424	. 428	. 431	. 435	. 439	150
160	. 439	. 443	. 447	. 450	. 454	. 458	. 462	. 466	. 469	. 473	. 477	160
170	. 477	. 481	. 485	. 489	. 493	. 496	. 500	. 504	. 508	. 512	. 516	170
180	. 516	. 520	. 524	. 528	. 532	. 535	. 539	. 543	. 547	. 551	. 555	180
190	. 555	. 559	. 563	. 567	. 571	. 575	. 579	. 583	. 587	. 591	. 595	190
200	. 595	. 599	. 603	. 607	. 611	. 615	. 619	. 623	. 627	. 631	. 635	200
210	. 635	. 639	. 643	. 647	. 651	. 655	. 659	. 664	. 668	. 672	. 676	210
220	. 676	. 680	. 684	. 688	. 692	. 696	. 700	. 705	. 709	. 713	. 717	220
230	. 717	. 721	. 725	. 729	. 734	. 738	. 742	. 746	. 750	. 754	. 758	230
240	. 758	. 763	. 767	. 771	. 775	. 779	. 784	. 788	. 792	. 796	. 800	240
250	. 800	. 805	. 809	. 813	. 817	. 822	. 826	. 830	. 834	. 839	. 843	250
260	. 843	. 847	. 851	. 856	. 860	. 864	. 869	. 873	. 877	. 881	. 886	260
270	. 886	. 890	. 894	. 899	. 903	. 907	. 912	. 916	. 920	. 925	. 929	270
280	. 929	. 933	. 938	. 942	. 946	. 951	. 955	. 959	. 964	. 968	. 973	280
290	. 973	. 977	. 981	. 986	. 990	. 994	. 999	1. 003	1. 008	1. 012	1. 017	290
300	1. 017	1. 021	1. 025	1. 030	1. 034	1. 039	1. 043	1. 048	1. 052	1. 056	1. 061	300
310	1. 061	1. 065	1. 070	1. 074	1. 079	1. 083	1. 088	1. 092	1. 097	1. 101	1. 106	310
320	1. 106	1. 110	1. 115	1. 119	1. 124	1. 128	1. 132	1. 137	1. 142	1. 146	1. 151	320
330	1. 151	1. 155	1. 160	1. 164	1. 169	1. 173	1. 178	1. 182	1. 187	1. 191	1. 196	330
340	1. 196	1. 200	1. 205	1. 210	1. 214	1. 219	1. 223	1. 228	1. 232	1. 237	1. 242	340
350	1. 242	1. 246	1. 251	1. 255	1. 260	1. 264	1. 269	1. 274	1. 278	1. 283	1. 287	350
360	1. 287	1. 292	1. 297	1. 301	1. 306	1. 311	1. 315	1. 320	1. 324	1. 329	1. 334	360
370	1. 334	1. 338	1. 343	1. 348	1. 352	1. 357	1. 362	1. 366	1. 371	1. 376	1. 380	370
380	1. 380	1. 385	1. 390	1. 394	1. 399	1. 404	1. 408	1. 413	1. 418	1. 422	1. 427	380
390	1. 427	1. 432	1. 436	1. 441	1. 446	1. 450	1. 455	1. 460	1. 465	1. 469	1. 474	390
400	1. 474	1. 479	1. 483	1. 488	1. 493	1. 498	1. 502	1. 507	1. 512	1. 516	1. 521	400
°F	0	1	2	3	4	5	6	7	8	9	10	°F

TABLE 48—*Type S thermocouples (continued).*
Temperature in Degrees Fahrenheit

EMF in Absolute Millivolts

Reference Junctions at 32 F

°F	0	1	2	3	4	5	6	7	8	9	10	°F
						Millivolts						
400	1. 474	1. 479	1. 483	1. 488	1. 493	1. 498	1. 502	1. 507	1. 512	1. 516	1. 521	400
410	1. 521	1. 526	1. 531	1. 535	1. 540	1. 545	1. 550	1. 554	1. 559	1. 564	1. 569	410
420	1. 569	1. 573	1. 578	1. 583	1. 588	1. 593	1. 597	1. 602	1. 607	1. 612	1. 616	420
430	1. 616	1. 621	1. 626	1. 631	1. 636	1. 640	1. 645	1. 650	1. 655	1. 660	1. 664	430
440	1. 664	1. 669	1. 674	1. 679	1. 684	1. 688	1. 693	1. 698	1. 703	1. 708	1. 712	440
450	1. 712	1. 717	1. 722	1. 727	1. 732	1. 736	1. 741	1. 746	1. 751	1. 756	1. 761	450
460	1. 761	1. 765	1. 770	1. 775	1. 780	1. 785	1. 790	1. 795	1. 799	1. 804	1. 809	460
470	1. 809	1. 814	1. 819	1. 824	1. 829	1. 833	1. 838	1. 843	1. 848	1. 853	1. 858	470
480	1. 858	1. 863	1. 868	1. 873	1. 877	1. 882	1. 887	1. 892	1. 897	1. 902	1. 907	480
490	1. 907	1. 912	1. 917	1. 922	1. 927	1. 931	1. 936	1. 941	1. 946	1. 951	1. 956	490
500	1. 956	1. 961	1. 966	1. 971	1. 976	1. 981	1. 986	1. 991	1. 996	2. 000	2. 005	500
510	2. 005	2. 010	2. 015	2. 020	2. 025	2. 030	2. 035	2. 040	2. 045	2. 050	2. 055	510
520	2. 055	2. 060	2. 065	2. 070	2. 075	2. 080	2. 085	2. 090	2. 095	2. 100	2. 105	520
530	2. 105	2. 110	2. 115	2. 120	2. 125	2. 130	2. 135	2. 140	2. 145	2. 150	2. 155	530
540	2. 155	2. 160	2. 165	2. 170	2. 175	2. 180	2. 185	2. 190	2. 195	2. 200	2. 205	540
550	2. 205	2. 210	2. 215	2. 220	2. 225	2. 230	2. 235	2. 240	2. 245	2. 250	2. 255	550
560	2. 255	2. 260	2. 265	2. 270	2. 276	2. 281	2. 286	2. 291	2. 296	2. 301	2. 306	560
570	2. 306	2. 311	2. 316	2. 321	2. 326	2. 331	2. 336	2. 341	2. 346	2. 351	2. 357	570
580	2. 357	2. 362	2. 367	2. 372	2. 377	2. 382	2. 387	2. 392	2. 397	2. 402	2. 407	580
590	2. 407	2. 413	2. 418	2. 423	2. 428	2. 433	2. 438	2. 443	2. 448	2. 453	2. 458	590
600	2. 458	2. 464	2. 469	2. 474	2. 479	2. 484	2. 489	2. 494	2. 499	2. 505	2. 510	600
610	2. 510	2. 515	2. 520	2. 525	2. 530	2. 535	2. 540	2. 546	2. 551	2. 556	2. 561	610
620	2. 561	2. 566	2. 571	2. 576	2. 582	2. 587	2. 592	2. 597	2. 602	2. 607	2. 613	620
630	2. 613	2. 618	2. 623	2. 628	2. 633	2. 638	2. 644	2. 649	2. 654	2. 659	2. 664	630
640	2. 664	2. 669	2. 675	2. 680	2. 685	2. 690	2. 695	2. 700	2. 706	2. 711	2. 716	640
650	2. 716	2. 721	2. 726	2. 731	2. 737	2. 742	2. 747	2. 752	2. 757	2. 763	2. 768	650
660	2. 768	2. 773	2. 778	2. 783	2. 789	2. 794	2. 799	2. 804	2. 809	2. 815	2. 820	660
670	2. 820	2. 825	2. 830	2. 836	2. 841	2. 846	2. 851	2. 856	2. 862	2. 867	2. 872	670
680	2. 872	2. 877	2. 882	2. 888	2. 893	2. 898	2. 903	2. 909	2. 914	2. 919	2. 924	680
690	2. 924	2. 930	2. 935	2. 940	2. 945	2. 951	2. 956	2. 961	2. 966	2. 972	2. 977	690
700	2. 977	2. 982	2. 987	2. 992	2. 998	3. 003	3. 008	3. 014	3. 019	3. 024	3. 029	700
710	3. 029	3. 035	3. 040	3. 045	3. 050	3. 056	3. 061	3. 066	3. 071	3. 077	3. 082	710
720	3. 082	3. 087	3. 092	3. 098	3. 102	3. 108	3. 114	3. 119	3. 124	3. 129	3. 135	720
730	3. 135	3. 140	3. 145	3. 150	3. 156	3. 161	3. 166	3. 172	3. 177	3. 182	3. 188	730
740	3. 188	3. 193	3. 198	3. 203	3. 209	3. 214	3. 219	3. 225	3. 230	3. 235	3. 240	740
750	3. 240	3. 246	3. 251	3. 256	3. 262	3. 267	3. 272	3. 278	3. 283	3. 288	3. 293	750
760	3. 293	3. 299	3. 304	3. 309	3. 315	3. 320	3. 325	3. 331	3. 336	3. 341	3. 347	760
770	3. 347	3. 352	3. 357	3. 363	3. 368	3. 373	3. 378	3. 384	3. 389	3. 394	3. 400	770
780	3. 400	3. 405	3. 410	3. 416	3. 421	3. 426	3. 432	3. 437	3. 442	3. 448	3. 453	780
790	3. 453	3. 458	3. 464	3. 469	3. 474	3. 480	3. 485	3. 490	3. 496	3. 501	3. 506	790
800	3. 506	3. 512	3. 517	3. 522	3. 528	3. 533	3. 539	3. 544	3. 549	3. 555	3. 560	800
°F	0	1	2	3	4	5	6	7	8	9	10	°F

TABLE 48—*Type S thermocouples (continued)*.
Temperature in Degrees Fahrenheit

EMF in Absolute Millivolts   Reference Junctions at 32 F

°F	0	1	2	3	4	5	6	7	8	9	10	°F
						Millivolts						
800	3. 506	3. 512	3. 517	3. 522	3. 528	3. 533	3. 539	3. 544	3. 549	3. 555	3. 560	800
810	3. 560	3. 565	3. 571	3. 576	3. 581	3. 587	3. 592	3. 597	3. 603	3. 608	3. 614	810
820	3. 614	3. 619	3. 624	3. 630	3. 635	3. 640	3. 646	3. 651	3. 656	3. 662	3. 667	820
830	3. 667	3. 673	3. 678	3. 683	3. 689	3. 694	3. 699	3. 705	3. 710	3. 716	3. 721	830
840	3. 721	3. 726	3. 732	3. 737	3. 743	3. 748	3. 753	3. 759	3. 764	3. 769	3. 775	840
850	3. 775	3. 780	3. 786	3. 791	3. 796	3. 802	3. 807	3. 813	3. 818	3. 823	3. 829	850
860	3. 829	3. 834	3. 840	3. 845	3. 850	3. 856	3. 861	3. 867	3. 872	3. 878	3. 883	860
870	3. 883	3. 888	3. 894	3. 899	3. 905	3. 910	3. 915	3. 921	3. 926	3. 932	3. 937	870
880	3. 937	3. 943	3. 948	3. 953	3. 959	3. 964	3. 970	3. 975	3. 981	3. 986	3. 991	880
890	3. 991	3. 997	4. 002	4. 008	4. 013	4. 019	4. 024	4. 030	4. 035	4. 040	4. 046	890
900	4. 046	4. 051	4. 057	4. 062	4. 068	4. 073	4. 079	4. 084	4. 089	4. 095	4. 100	900
910	4. 100	4. 106	4. 111	4. 117	4. 122	4. 128	4. 133	4. 139	4. 144	4. 149	4. 155	910
920	4. 155	4. 160	4. 166	4. 171	4. 177	4. 182	4. 188	4. 193	4. 199	4. 204	4. 210	920
930	4. 210	4. 215	4. 221	4. 226	4. 232	4. 237	4. 243	4. 248	4. 254	4. 259	4. 264	930
940	4. 264	4. 270	4. 275	4. 281	4. 286	4. 292	4. 297	4. 303	4. 308	4. 314	4. 319	940
950	4. 319	4. 325	4. 330	4. 336	4. 341	4. 347	4. 352	4. 358	4. 363	4. 369	4. 374	950
960	4. 374	4. 380	4. 385	4. 391	4. 396	4. 402	4. 408	4. 413	4. 419	4. 424	4. 430	960
970	4. 430	4. 435	4. 441	4. 446	4. 452	4. 457	4. 463	4. 468	4. 474	4. 479	4. 485	970
980	4. 485	4. 490	4. 496	4. 501	4. 507	4. 512	4. 518	4. 524	4. 529	4. 535	4. 540	980
990	4. 540	4. 546	4. 551	4. 557	4. 562	4. 568	4. 573	4. 579	4. 584	4. 590	4. 596	990
1, 000	4. 596	4. 601	4. 607	4. 612	4. 618	4. 623	4. 629	4. 634	4. 640	4. 646	4. 651	1, 000
1, 010	4. 651	4. 657	4. 662	4. 668	4. 673	4. 679	4. 685	4. 690	4. 696	4. 701	4. 707	1, 010
1, 020	4. 707	4. 712	4. 718	4. 724	4. 729	4. 735	4. 740	4. 746	4. 751	4. 757	4. 763	1, 020
1, 030	4. 763	4. 768	4. 774	4. 779	4. 785	4. 790	4. 796	4. 802	4. 807	4. 813	4. 818	1, 030
1, 040	4. 818	4. 824	4. 830	4. 835	4. 841	4. 846	4. 852	4. 858	4. 863	4. 869	4. 874	1, 040
1, 050	4. 874	4. 880	4. 886	4. 891	4. 897	4. 902	4. 908	4. 914	4. 919	4. 925	4. 930	1, 050
1, 060	4. 930	4. 936	4. 942	4. 947	4. 953	4. 959	4. 964	4. 970	4. 975	4. 981	4. 987	1, 060
1, 070	4. 987	4. 992	4. 998	5. 004	5. 009	5. 015	5. 020	5. 026	5. 032	5. 037	5. 043	1, 070
1, 080	5. 043	5. 049	5. 054	5. 060	5. 066	5. 071	5. 077	5. 082	5. 088	5. 094	5. 099	1, 080
1, 090	5. 099	5. 105	5. 111	5. 116	5. 122	5. 128	5. 133	5. 139	5. 145	5. 150	5. 156	1, 090
1, 100	5. 156	5. 162	5. 167	5. 173	5. 178	5. 184	5. 190	5. 195	5. 201	5. 207	5. 212	1, 100
1, 110	5. 212	5. 218	5. 224	5. 229	5. 235	5. 241	5. 246	5. 252	5. 258	5. 264	5. 269	1, 110
1, 120	5. 269	5. 275	5. 281	5. 286	5. 292	5. 298	5. 303	5. 309	5. 315	5. 320	5. 326	1, 120
1, 130	5. 326	5. 332	5. 337	5. 343	5. 349	5. 354	5. 360	5. 366	5. 372	5. 377	5. 383	1, 130
1, 140	5. 383	5. 389	5. 394	5. 400	5. 406	5. 411	5. 417	5. 423	5. 429	5. 434	5. 440	1, 140
1, 150	5. 440	5. 446	5. 451	5. 457	5. 463	5. 469	5. 474	5. 480	5. 486	5. 491	5. 497	1, 150
1, 160	5. 497	5. 503	5. 509	5. 514	5. 520	5. 526	5. 532	5. 537	5. 543	5. 549	5. 555	1, 160
1, 170	5. 555	5. 560	5. 566	5. 572	5. 577	5. 583	5. 589	5. 595	5. 600	5. 606	5. 612	1, 170
1, 180	5. 612	5. 617	5. 623	5. 629	5. 635	5. 640	5. 646	5. 652	5. 658	5. 663	5. 669	1, 180
1, 190	5. 669	5. 675	5. 681	5. 686	5. 692	5. 698	5. 704	5. 709	5. 715	5. 721	5. 726	1, 190
1, 200	5. 726	5. 732	5. 738	5. 744	5. 749	5. 755	5. 761	5. 767	5. 773	5. 778	5. 784	1, 200
°F	0	1	2	3	4	5	6	7	8	9	10	°F

TABLE 48—*Type S thermocouples (continued).*
Temperature in Degrees Fahrenheit

EMF in Absolute Millivolts   Reference Junctions at 32 F

°F	0	1	2	3	4	5	6	7	8	9	10	°F
						Millivolts						
1,200	5.726	5.732	5.738	5.744	5.749	5.755	5.761	5.767	5.773	5.778	5,784	1,200
1,210	5.784	5.790	5.796	5.801	5.807	5.813	5.819	5.824	5.830	5.836	5.842	1,210
1,220	5.842	5.847	5.853	5.859	5.865	5.871	5.876	5.882	5.888	5.894	5.899	1,220
1,230	5.899	5.905	5.911	5.917	5.923	5.928	5.934	5.940	5.946	5.951	5.957	1,230
1,240	5.957	5.963	5.969	5.975	5.980	5.986	5.992	5.998	6.004	6.009	6.015	1,240
1,250	6.015	6.021	6.027	6.033	6.038	6.044	6.050	6.056	6.062	6.067	6.073	1,250
1,260	6.073	6.079	6.085	6.091	6.096	6.102	6.108	6.114	6.120	6.126	6.131	1,260
1,270	6.131	6.137	6.143	6.149	6.155	6.161	6.166	6.172	6.178	6.184	6.190	1,270
1,280	6.190	6.196	6.201	6.207	6.213	6.219	6.225	6.231	6.236	6.242	6.248	1,280
1,290	6.248	6.254	6.260	6.266	6.271	6.277	6.283	6.289	6.295	6.301	6.307	1,290
1,300	6.307	6.312	6.318	6.324	6.330	6.336	6.342	6.348	6.353	6.359	6.365	1,300
1,310	6.365	6.371	6.377	6.383	6.389	6.394	6.400	6.406	6.412	6.418	6.424	1,310
1,320	6.424	6.430	6.436	6.441	6.447	6.453	6.459	6.465	6.471	6.477	6.483	1,320
1,330	6.483	6.488	6.494	6.500	6.506	6.512	6.518	6.524	6.530	6.536	6.542	1,330
1,340	6.542	6.547	6.553	6.559	6.565	6.571	6.577	6.583	6.589	6.595	6.601	1,340
1,350	6.601	6.606	6.612	6.618	6.624	6.630	6.636	6.642	6.648	6.654	6.660	1,350
1,360	6.660	6.666	6.671	6.677	6.683	6.689	6.695	6.701	6.707	6.713	6.719	1,360
1,370	6.719	6.725	6.731	6.737	6.743	6.749	6.754	6.760	6.766	6.772	6.778	1,370
1,380	6.778	6.784	6.790	6.796	6.802	6.808	6.814	6.820	6.826	6.832	6.838	1,380
1,390	6.838	6.844	6.850	6.855	6.861	6.867	6.873	6.879	6.885	6.891	6.897	1,390
1,400	6.897	6.903	6.909	6.915	6.921	6.927	6.933	6.939	6.945	6.951	6.957	1,400
1,410	6.957	6.963	6.969	6.975	6.981	6.987	6.993	6.999	7.005	7.011	7.017	1,410
1,420	7.017	7.023	7.029	7.034	7.040	7.046	7.052	7.058	7.064	7.070	7.076	1,420
1,430	7.076	7.082	7.088	7.094	7.100	7.106	7.112	7.118	7.124	7.130	7.136	1,430
1,440	7.136	7.142	7.148	7.154	7.160	7.166	7.172	7.178	7.184	7.190	7.196	1,440
1,450	7.196	7.202	7.208	7.214	7.220	7.226	7.233	7.239	7.245	7.251	7.257	1,450
1,460	7.257	7.263	7.269	7.275	7.281	7.287	7.293	7.299	7.305	7.311	7.317	1,460
1,470	7.317	7.323	7.329	7.335	7.341	7.347	7.353	7.359	7.365	7.371	7.377	1,470
1,480	7.377	7.383	7.389	7.395	7.401	7.407	7.414	7.420	7.426	7.432	7.438	1,480
1,490	7.438	7.444	7.450	7.456	7.462	7.468	7.474	7.480	7.486	7.492	7.498	1,490
1,500	7.498	7.504	7.510	7.517	7.523	7.529	7.535	7.541	7.547	7.553	7.559	1,500
1,510	7.559	7.565	7.571	7.577	7.583	7.589	7.596	7.602	7.608	7.614	7.620	1,510
1,520	7.620	7.626	7.632	7.638	7.644	7.650	7.656	7.662	7.669	7.675	7.681	1,520
1,530	7.681	7.687	7.693	7.699	7.705	7.711	7.717	7.723	7.730	7.736	7.742	1,530
1,540	7.742	7.748	7.754	7.760	7.766	7.772	7.778	7.785	7.791	7.797	7.803	1,540
1,550	7.803	7.809	7.815	7.821	7.827	7.834	7.840	7.846	7.852	7.858	7.864	1,550
1,560	7.864	7.870	7.876	7.882	7.889	7.895	7.901	7.907	7.913	7.919	7.925	1,560
1,570	7.925	7.932	7.938	7.944	7.950	7.956	7.962	7.968	7.975	7.981	7.987	1,570
1,580	7.987	7.993	7.999	8.005	8.012	8.018	8.024	8.030	8.036	8.042	8.048	1,580
1,590	8.048	8.055	8.061	8.067	8.073	8.079	8.085	8.092	8.098	8.104	8.110	1,590
1,600	8.110	8.116	8.122	8.129	8.135	8.141	8.147	8.153	8.159	8.166	8.172	1,600
°F	0	1	2	3	4	5	6	7	8	9	10	°F

TABLE 48—*Type S thermocouples (continued)*.
Temperature in Degrees Fahrenheit

EMF in Absolute Millivolts   Reference Junctions at 32 F

°F	0	1	2	3	4	5	6	7	8	9	10	°F
						Millivolts						
1,600	8. 110	8. 116	8. 122	8. 129	8. 135	8. 141	8. 147	8. 153	8. 159	8. 166	8. 172	1,600
1,610	8. 172	8. 178	8. 184	8. 190	8. 197	8. 203	8. 209	8. 215	8. 221	8. 228	8. 234	1,610
1,620	8. 234	8. 240	8. 246	8. 252	8. 258	8. 265	8. 271	8. 277	8. 283	8. 289	8. 296	1,620
1,630	8. 296	8. 302	8. 308	8. 314	8. 320	8. 327	8. 333	8. 339	8. 345	8. 352	8. 358	1,630
1,640	8. 358	8. 364	8. 370	8. 376	8. 383	8. 389	8. 395	8. 401	8. 407	8. 414	8. 420	1,640
1,650	8. 420	8. 426	8. 432	8. 439	8. 445	8. 451	8. 457	8. 464	8. 470	8. 476	8. 482	1,650
1,660	8. 482	8. 488	8. 495	8. 501	8. 507	8. 513	8. 520	8. 526	8. 532	8. 538	8. 545	1,660
1,670	8. 545	8. 551	8. 557	8. 563	8. 570	8. 576	8. 582	8. 588	8. 595	8. 601	8. 607	1,670
1,680	8. 607	8. 613	8. 620	8. 626	8. 632	8. 638	8. 645	8. 651	8. 657	8. 663	8. 670	1,680
1,690	8. 670	8. 676	8. 682	8. 689	8. 695	8. 701	8. 707	8. 714	8. 720	8. 726	8. 732	1,690
1,700	8. 732	8. 739	8. 745	8. 751	8. 758	8. 764	8. 770	8. 776	8. 783	8. 789	8. 795	1,700
1,710	8. 795	8. 802	8. 808	8. 814	8. 820	8. 827	8. 833	8. 839	8. 846	8. 852	8. 858	1,710
1,720	8. 858	8. 864	8. 871	8. 877	8. 883	8. 890	8. 896	8. 902	8. 909	8. 915	8. 921	1,720
1,730	8. 921	8. 927	8. 934	8. 940	8. 946	8. 953	8. 959	8. 965	8. 972	8. 978	8. 984	1,730
1,740	8. 984	8. 991	8. 997	9. 003	9. 010	9. 016	9. 022	9. 029	9. 035	9. 041	9. 048	1,740
1,750	9. 048	9. 054	9. 060	9. 067	9. 073	9. 079	9. 086	9. 092	9. 098	9. 105	9. 111	1,750
1,760	9. 111	9. 117	9. 124	9. 130	9. 136	9. 143	9. 149	9. 155	9. 162	9. 168	9. 174	1,760
1,770	9. 174	9. 181	9. 187	9. 193	9. 200	9. 206	9. 212	9. 219	9. 225	9. 232	9. 238	1,770
1,780	9. 238	9. 244	9. 251	9. 257	9. 263	9. 270	9. 276	9. 282	9. 289	9. 295	9. 302	1,780
1,790	9. 302	9. 308	9. 314	9. 321	9. 327	9. 333	9. 340	9. 346	9. 353	9. 359	9. 365	1,790
1,800	9. 365	9. 372	9. 378	9. 384	9. 391	9. 397	9. 404	9. 410	9. 416	9. 423	9. 429	1,800
1,810	9. 429	9. 436	9. 442	9. 448	9. 455	9. 461	9. 468	9. 474	9. 480	9. 487	9. 493	1,810
1,820	9. 493	9. 500	9. 506	9. 512	9. 519	9. 525	9. 532	9. 538	9. 544	9. 551	9. 557	1,820
1,830	9. 557	9. 564	9. 570	9. 576	9. 583	9. 589	9. 596	9. 602	9. 609	9. 615	9. 621	1,830
1,840	9. 621	9. 628	9. 634	9. 641	9. 647	9. 654	9. 660	9. 666	9. 673	9. 679	9. 686	1,840
1,850	9. 686	9. 692	9. 699	9. 705	9. 711	9. 718	9. 724	9. 731	9. 737	9. 744	9. 750	1,850
1,860	9. 750	9. 757	9. 763	9. 769	9. 776	9. 782	9. 789	9. 795	9. 802	9. 808	9. 815	1,860
1,870	9. 815	9. 821	9. 828	9. 834	9. 840	9. 847	9. 853	9. 860	9. 866	9. 873	9. 879	1,870
1,880	9. 879	9. 886	9. 892	9. 899	9. 905	9. 912	9. 918	9. 925	9. 931	9. 937	9. 944	1,880
1,890	9. 944	9. 950	9. 957	9. 963	9. 970	9. 976	9. 983	9. 989	9. 996	10. 002	10. 009	1,890
1,900	10. 009	10. 015	10. 022	10. 028	10. 035	10. 041	10. 048	10. 054	10. 061	10. 067	10. 074	1,900
1,910	10. 074	10. 080	10. 087	10. 093	10. 100	10. 106	10. 113	10. 119	10. 126	10. 132	10. 139	1,910
1,920	10. 139	10. 145	10. 152	10. 158	10. 165	10. 171	10. 178	10. 184	10. 191	10. 197	10. 204	1,920
1,930	10. 204	10. 210	10. 217	10 223	10. 230	10. 237	10. 243	10. 250	10. 256	10. 263	10. 269	1,930
1,940	10. 269	10. 276	10. 282	10. 289	10. 295	10. 302	10. 308	10. 315	10. 321	10. 328	10. 334	1,940
1,950	10. 334	10. 341	10. 348	10. 354	10. 361	10. 367	10. 374	10. 380	10. 387	10. 393	10. 400	1,950
1,960	10. 400	10. 406	10. 413	10. 420	10. 426	10. 433	10. 439	10. 446	10. 452	10. 459	10. 465	1,960
1,970	10. 465	10. 472	10. 478	10. 485	10. 492	10. 498	10. 505	10. 511	10. 518	10. 524	10. 531	1,970
1,980	10. 531	10. 538	10. 544	10. 551	10. 557	10. 564	10. 570	10. 577	10. 583	10. 590	10. 597	1,980
1,990	10. 597	10. 603	10. 610	10. 616	10. 623	10. 629	10. 636	10. 643	10. 649	10. 656	10. 662	1,990
2,000	10. 662	10. 669	10. 675	10. 682	10. 689	10. 695	10. 702	10. 708	10. 715	10. 722	10. 728	2,000
°F	0	1	2	3	4	5	6	7	8	9	10	°F

TABLE 48—*Type S thermocouples (continued)*.
Temperature in Degrees Fahrenheit

EMF in Absolute Millivolts

Reference Junctions at 32 F

°F	0	1	2	3	4	5	6	7	8	9	10	°F
						Millivolts						
2,000	10.662	10.669	10.675	10.682	10.689	10.695	10.702	10.708	10.715	10.722	10.728	2,000
2,010	10.728	10.735	10.741	10.748	10.754	10.761	10.768	10.774	10.781	10.787	10.794	2,010
2,020	10.794	10.801	10.807	10.814	10.820	10.827	10.834	10.840	10.847	10.853	10.860	2,020
2,030	10.860	10.866	10.873	10.880	10.886	10.893	10.899	10.906	10.913	10.919	10.926	2,030
2,040	10.926	10.932	10.939	10.946	10.952	10.959	10.966	10.972	10.979	10.985	10.992	2,040
2,050	10.992	10.999	11.005	11.012	11.018	11.025	11.032	11.038	11.045	11.051	11.058	2,050
2,060	11.058	11.065	11.071	11.078	11.085	11.091	11.098	11.104	11.111	11.118	11.124	2,060
2,070	11.124	11.131	11.137	11.144	11.151	11.157	11.164	11.171	11.177	11.184	11.190	2,070
2,080	11.190	11.197	11.204	11.210	11.217	11.224	11.230	11.237	11.243	11.250	11.257	2,080
2,090	11.257	11.263	11.270	11.277	11.283	11.290	11.296	11.303	11.310	11.316	11.323	2,090
2,100	11.323	11.330	11.336	11.343	11.350	11.356	11.363	11.369	11.376	11.383	11.389	2,100
2,110	11.389	11.396	11.403	11.409	11.416	11.423	11.429	11.436	11.443	11.449	11.456	2,110
2,120	11.456	11.462	11.469	11.476	11.482	11.489	11.496	11.502	11.509	11.516	11.522	2,120
2,130	11.522	11.529	11.536	11.542	11.549	11.556	11.562	11.569	11.575	11.582	11.589	2,130
2,140	11.589	11.595	11.602	11.609	11.615	11.622	11.629	11.635	11.642	11.649	11.655	2,140
2,150	11.655	11.662	11.669	11.675	11.682	11.689	11.695	11.702	11.709	11.715	11.722	2,150
2,160	11.722	11.729	11.735	11.742	11.749	11.755	11.762	11.769	11.775	11.782	11.789	2,160
2,170	11.789	11.795	11.802	11.809	11.815	11.822	11.829	11.835	11.842	11.848	11.855	2,170
2,180	11.855	11.862	11.868	11.875	11.882	11.888	11.895	11.902	11.908	11.915	11.922	2,180
2,190	11.922	11.928	11.935	11.942	11.949	11.955	11.962	11.969	11.975	11.982	11.989	2,190
2,200	11.989	11.995	12.002	12.009	12.015	12.022	12.029	12.035	12.042	12.049	12.055	2,200
2,210	12.055	12.062	12.069	12.075	12.082	12.089	12.095	12.102	12.109	12.115	12.122	2,210
2,220	12.122	12.129	12.135	12.142	12.149	12.155	12.162	12.169	12.175	12.182	12.189	2,220
2,230	12.189	12.196	12.202	12.209	12.216	12.222	12.229	12.236	12.242	12.249	12.256	2,230
2,240	12.256	12.262	12.269	12.276	12.282	12.289	12.296	12.302	12.309	12.316	12.322	2,240
2,250	12.322	12.329	12.336	12.342	12.349	12.356	12.363	12.369	12.376	12.383	12.389	2,250
2,260	12.389	12.396	12.403	12.409	12.416	12.423	12.429	12.436	12.443	12.449	12.456	2,260
2,270	12.456	12.463	12.470	12.476	12.483	12.490	12.496	12.503	12.510	12.516	12.523	2,270
2,280	12.523	12.530	12.536	12.543	12.550	12.556	12.563	12.570	12.577	12.583	12.590	2,280
2,290	12.590	12.597	12.603	12.610	12.617	12.623	12.630	12.637	12.643	12.650	12.657	2,290
2,300	12.657	12.663	12.670	12.677	12.684	12.690	12.697	12.704	12.710	12.717	12.724	2,300
2,310	12.724	12.730	12.737	12.744	12.750	12.757	12.764	12.770	12.777	12.784	12.790	2,310
2,320	12.790	12.797	12.804	12.810	12.817	12.824	12.830	12.837	12.844	12.851	12.857	2,320
2,330	12.857	12.864	12.871	12.877	12.884	12.891	12.897	12.904	12.911	12.917	12.924	2,330
2,340	12.924	12.931	12.937	12.944	12.951	12.957	12.964	12.971	12.977	12.984	12.991	2,340
2,350	12.991	12.997	13.004	13.011	13.018	13.024	13.031	13.038	13.044	13.051	13.058	2,350
2,360	13.058	13.064	13.071	13.078	13.084	13.091	13.098	13.104	13.111	13.118	13.124	2,360
2,370	13.124	13.131	13.138	13.144	13.151	13.158	13.164	13.171	13.178	13.184	13.191	2,370
2,380	13.191	13.198	13.204	13.211	13.218	13.224	13.231	13.238	13.244	13.251	13.258	2,380
2,390	13.258	13.265	13.271	13.278	13.285	13.291	13.298	13.305	13.311	13.318	13.325	2,390
2,400	13.325	13.331	13.338	13.345	13.351	13.358	13.365	13.371	13.378	13.385	13.391	2,400
°F	0	1	2	3	4	5	6	7	8	9	10	°F

TABLE 48—*Type S thermocouples (continued)*.

Temperature in Degrees Fahrenheit

EMF in Absolute Millivolts

Reference Junctions at 32 F

°F	0	1	2	3	4	5	6	7	8	9	10	°F
						Millivolts						
2,400	13.325	13.331	13.338	13.345	13.351	13.358	13.365	13.371	13.378	13.385	13.391	2,400
2,410	13.391	13.398	13.405	13.411	13.418	13.425	13.431	13.438	13.445	13.451	13.458	2,410
2,420	13.458	13.465	13.471	13.478	13.485	13.491	13.498	13.505	13.511	13.518	13.525	2,420
2,430	13.525	13.531	13.538	13.545	13.551	13.558	13.565	13.571	13.578	13.585	13.591	2,430
2,440	13.591	13.598	13.605	13.611	13.618	13.625	13.631	13.638	13.645	13.651	13.658	2,440
2,450	13.658	13.665	13.671	13.678	13.685	13.691	13.698	13.705	13.711	13.718	13.725	2,450
2,460	13.725	13.731	13.738	13.745	13.751	13.758	13.765	13.771	13.778	13.785	13.791	2,460
2,470	13.791	13.798	13.805	13.811	13.818	13.825	13.831	13.838	13.845	13.851	13.858	2,470
2,480	13.858	13.865	13.871	13.878	13.885	13.891	13.898	13.905	13.911	13.918	13.924	2,480
2,490	13.924	13.931	13.938	13.944	13.951	13.958	13.964	13.971	13.978	13.984	13.991	2,490
2,500	13.991	13.998	14.004	14.011	14.018	14.024	14.031	14.038	14.044	14.051	14.058	2,500
2,510	14.058	14.064	14.071	14.078	14.084	14.091	14.098	14.104	14.111	14.118	14.124	2,510
2,520	14.124	14.131	14.137	14.144	14.151	14.157	14.164	14.171	14.177	14.184	14.191	2,520
2,530	14.191	14.197	14.204	14.211	14.217	14.224	14.231	14.237	14.244	14.251	14.257	2,530
2,540	14.257	14.264	14.271	14.277	14.284	14.290	14.297	14.304	14.310	14.317	14.324	2,540
2,550	14.324	14.330	14.337	14.344	14.350	14.357	14.364	14.370	14.377	14.384	14.390	2,550
2,560	14.390	14.397	14.403	14.410	14.417	14.423	14.430	14.437	14.443	14.450	14.457	2,560
2,570	14.457	14.463	14.470	14.477	14.483	14.490	14.497	14.503	14.510	14.516	14.523	2,570
2,580	14.523	14.530	14.536	14.543	14.550	14.556	14.563	14.570	14.576	14.583	14.589	2,580
2,590	14.589	14.596	14.603	14.609	14.616	14.623	14.629	14.636	14.643	14.649	14.656	2,590
2,600	14.656	14.663	14.669	14.676	14.682	14.689	14.696	14.702	14.709	14.716	14.722	2,600
2,610	14.722	14.729	14.736	14.742	14.749	14.755	14.762	14.769	14.775	14.782	14.789	2,610
2,620	14.789	14.795	14.802	14.809	14.815	14.822	14.828	14.835	14.842	14.848	14.855	2,620
2,630	14.855	14.862	14.868	14.875	14.881	14.888	14.895	14.901	14.908	14.915	14.921	2,630
2,640	14.921	14.928	14.935	14.941	14.948	14.954	14.961	14.968	14.974	14.981	14.988	2,640
2,650	14.988	14.994	15.001	15.007	15.014	15.021	15.027	15.034	15.041	15.047	15.054	2,650
2,660	15.054	15.060	15.067	15.074	15.080	15.087	15.094	15.100	15.107	15.113	15.120	2,660
2,670	15.120	15.127	15.133	15.140	15.147	15.153	15.160	15.166	15.173	15.180	15.186	2,670
2,680	15.186	15.193	15.200	15.206	15.213	15.219	15.226	15.233	15.239	15.246	15.253	2,680
2,690	15.253	15.259	15.266	15.272	15.279	15.286	15.292	15.299	15.305	15.312	15.319	2,690
2,700	15.319	15.325	15.332	15.339	15.345	15.352	15.358	15.365	15.372	15.378	15.385	2,700
2,710	15.385	15.391	15.398	15.405	15.411	15.418	15.425	15.431	15.438	15.444	15.451	2,710
2,720	15.451	15.458	15.464	15.471	15.477	15.484	15.491	15.497	15.504	15.510	15.517	2,720
2,730	15.517	15.524	15.530	15.537	15.544	15.550	15.557	15.563	15.570	15.577	15.583	2,730
2,740	15.583	15.590	15.596	15.603	15.610	15.616	15.623	15.629	15.636	15.643	15.649	2,740
2,750	15.649	15.656	15.662	15.669	15.676	15.682	15.689	15.695	15.702	15.709	15.715	2,750
2,760	15.715	15.722	15.728	15.735	15.742	15.748	15.755	15.761	15.768	15.775	15.781	2,760
2,770	15.781	15.788	15.794	15.801	15.808	15.814	15.821	15.827	15.834	15.841	15.847	2,770
2,780	15.847	15.854	15.860	15.867	15.874	15.880	15.887	15.893	15.900	15.907	15.913	2,780
2,790	15.913	15.920	15.926	15.933	15.940	15.946	15.953	15.959	15.966	15.973	15.979	2,790
2,800	15.979	15.986	15.992	15.999	16.006	16.012	16.019	16.025	16.032	16.038	16.045	2,800
°F	0	1	2	3	4	5	6	7	8	9	10	°F

TABLE 48—*Type S thermocouples (continued).*
Temperature in Degrees Fahrenheit

EMF in Absolute Millivolts   Reference Junctions at 32 F

°F	0	1	2	3	4	5	6	7	8	9	10	°F
						Millivolts						
2,800	15.979	15.986	15.992	15.999	16.006	16.012	16.019	16.025	16.032	16.038	16.045	2,800
2,810	16.045	16.052	16.058	16.065	16.071	16.078	16.085	16.091	16.098	16.104	16.111	2,810
2,820	16.111	16.117	16.124	16.131	16.137	16.144	16.150	16.157	16.164	16.170	16.177	2,820
2,830	16.177	16.183	16.190	16.196	16.203	16.210	16.216	16.223	16.229	16.236	16.243	2,830
2,840	16.243	16.249	16.256	16.262	16.269	16.275	16.282	16.289	16.295	16.302	16.308	2,840
2,850	16.308	16.315	16.322	16.328	16.335	16.341	16.348	16.354	16.361	16.368	16.374	2,850
2,860	16.374	16.381	16.387	16.394	16.400	16.407	16.414	16.420	16.427	16.433	16.440	2,860
2,870	16.440	16.446	16.453	16.460	16.466	16.473	16.479	16.486	16.492	16.499	16.506	2,870
2,880	16.506	16.512	16.519	16.525	16.532	16.538	16.545	16.552	16.558	16.565	16.571	2,880
2,890	16.571	16.578	16.584	16.591	16.597	16.604	16.611	16.617	16.624	16.630	16.637	2,890
2,900	16.637	16.643	16.650	16.657	16.663	16.670	16.676	16.683	16.689	16.696	16.702	2,900
2,910	16.702	16.709	16.716	16.722	16.729	16.735	16.742	16.748	16.755	16.761	16.768	2,910
2,920	16.768	16.775	16.781	16.788	16.794	16.801	16.807	16.814	16.820	16.827	16.834	2,920
2,930	16.834	16.840	16.847	16.853	16.860	16.866	16.873	16.879	16.886	16.893	16.899	2,930
2,940	16.899	16.906	16.912	16.919	16.925	16.932	16.938	16.945	16.952	16.958	16.965	2,940
2,950	16.965	16.971	16.978	16.984	16.991	16.997	17.004	17.010	17.017	17.023	17.030	2,950
2,960	17.030	17.037	17.043	17.050	17.056	17.063	17.069	17.076	17.082	17.089	17.095	2,960
2,970	17.095	17.102	17.109	17.115	17.122	17.128	17.135	17.141	17.148	17.154	17.161	2,970
2,980	17.161	17.167	17.174	17.180	17.187	17.194	17.200	17.207	17.213	17.220	17.226	2,980
2,990	17.226	17.233	17.239	17.246	17.252	17.259	17.265	17.272	17.278	17.285	17.292	2,990
3,000	17.292	17.298	17.305	17.311	17.318	17.324	17.331	17.337	17.344	17.350	17.357	3,000
3,010	17.357	17.363	17.370	17.376	17.383	17.389	17.396	17.402	17.409	17.416	17.422	3,010
3,020	17.422	17.429	17.435	17.442	17.448	17.455	17.461	17.468	17.474	17.481	17.487	3,020
3,030	17.487	17.494	17.500	17.507	17.513	17.520	17.526	17.533	17.539	17.546	17.552	3,030
3,040	17.552	17.559	17.565	17.572	17.578	17.585	17.592	17.598	17.605	17.611	17.618	3,040
3,050	17.618	17.624	17.631	17.637	17.644	17.650	17.657	17.663	17.670	17.676	17.683	3,050
3,060	17.683	17.689	17.696	17.702	17.709	17.715	17.722	17.728	17.735	17.741	17.748	3,060
3,070	17.748	17.754	17.761	17.767	17.774	17.780	17.787	17.793	17.800	17.806	17.813	3,070
3,080	17.813	17.819	17.826	17.832	17.839	17.845	17.852	17.858	17.865	17.871	17.878	3,080
3,090	17.878	17.884	17.891	17.897	17.904	17.910	17.917	17.923	17.930	17.936	17.943	3,090
3,100	17.943	17.949	17.956	17.962	17.969	17.975	17.982	17.988	17.995	18.001	18.008	3,100
3,110	18.008	18.014	18.021	18.027	18.034	18.040	18.047	18.053	18.060	18.066	18.073	3,110
3,120	18.073	18.079	18.086	18.092	18.098	18.105	18.111	18.118	18.124	18.131	18.137	3,120
3,130	18.137	18.144	18.150	18.157	18.163	18.170	18.176	18.183	18.189	18.196	18.202	3,130
3,140	18.202	18.209	18.215	18.222	18.228	18.235	18.241	18.248	18.254	18.260	18.267	3,140
3,150	18.267	18.273	18.280	18.286	18.293	18.299	18.306	18.312	18.319	18.325	18.332	3,150
3,160	18.332	18.338	18.345	18.351	18.358	18.364	18.371	18.377	18.383	18.390	18.396	3,160
3,170	18.396	18.403	18.409	18.416	18.422	18.429	18.435	18.442	18.448	18.455	18.461	3,170
3,180	18.461	18.468	18.474	18.480	18.487	18.493	18.500	18.506	18.513	18.519	18.526	3,180
3,190	18.526	18.532	18.539	18.545	18.551	18.558	18.564	18.571	18.577	18.584	18.590	3,190
3,200	18.590	18.597	18.603	18.610	18.616	18.622	18.629	18.635	18.642	18.648	18.655	3,200
3,210	18.655	18.661	18.668	18.674	18.681	18.687	------	------	------	------	------	3,210
°F	0	1	2	3	4	5	6	7	8	9	10	°F

TABLE 49—*Type S thermocouples.*
Temperature in Degrees Celsius

EMF in Absolute Millivolts

Reference Junctions at 0 C

°C	0	1	2	3	4	5	6	7	8	9	10	°C
						Millivolts						
0	0.000	0.005	0.011	0.016	0.022	0.028	0.033	0.039	0.044	0.050	0.056	0
10	.056	.061	.067	.073	.078	.084	.090	.096	.102	.107	.113	10
20	.113	.119	.125	.131	.137	.143	.149	.155	.161	.167	.173	20
30	.173	.179	.185	.191	.198	.204	.210	.216	.222	.229	.235	30
40	.235	.241	.247	.254	.260	.266	.273	.279	.286	.292	.299	40
50	.299	.305	.312	.318	.325	.331	.338	.344	.351	.357	.364	50
60	.364	.371	.377	.384	.391	.397	.404	.411	.418	.425	.431	60
70	.431	.438	.445	.452	.459	.466	.473	.479	.486	.493	.500	70
80	.500	.507	.514	.521	.528	.535	.543	.550	.557	.564	.571	80
90	.571	.578	.585	.593	.600	.607	.614	.621	.629	.636	.643	90
100	.643	.651	.658	.665	.673	.680	.687	.694	.702	.709	.717	100
110	.717	.724	.732	.739	.747	.754	.762	.769	.777	.784	.792	110
120	.792	.800	.807	.815	.823	.830	.838	.845	.853	.861	.869	120
130	.869	.876	.884	.892	.900	.907	.915	.923	.931	.939	.946	130
140	.946	.954	.962	.970	.978	.986	.994	1.002	1.009	1.017	1.025	140
150	1.025	1.033	1.041	1.049	1.057	1.065	1.073	1.081	1.089	1.097	1.106	150
160	1.106	1.114	1.122	1.130	1.138	1.146	1.154	1.162	1.170	1.179	1.187	160
170	1.187	1.195	1.203	1.211	1.220	1.228	1.236	1.244	1.253	1.261	1.269	170
180	1.269	1.277	1.286	1.294	1.302	1.311	1.319	1.327	1.336	1.344	1.352	180
190	1.352	1.361	1.369	1.377	1.386	1.394	1.403	1.411	1.419	1.428	1.436	190
200	1.436	1.445	1.453	1.462	1.470	1.479	1.487	1.496	1.504	1.513	1.521	200
210	1.521	1.530	1.538	1.547	1.555	1.564	1.573	1.581	1.590	1.598	1.607	210
220	1.607	1.615	1.624	1.633	1.641	1.650	1.659	1.667	1.676	1.685	1.693	220
230	1.693	1.702	1.710	1.719	1.728	1.736	1.745	1.754	1.763	1.771	1.780	230
240	1.780	1.789	1.798	1.806	1.815	1.824	1.833	1.841	1.850	1.859	1.868	240
250	1.868	1.877	1.885	1.894	1.903	1.912	1.921	1.930	1.938	1.947	1.956	250
260	1.956	1.965	1.974	1.983	1.992	2.001	2.009	2.018	2.027	2.036	2.045	260
270	2.045	2.054	2.063	2.072	2.081	2.090	2.099	2.108	2.117	2.126	2.135	270
280	2.135	2.144	2.153	2.162	2.171	2.180	2.189	2.198	2.207	2.216	2.225	280
290	2.225	2.234	2.243	2.252	2.261	2.271	2.280	2.289	2.298	2.307	2.316	290
300	2.316	2.325	2.334	2.343	2.353	2.362	2.371	2.380	2.389	2.398	2.408	300
310	2.408	2.417	2.426	2.435	2.444	2.453	2.463	2.472	2.481	2.490	2.499	310
320	2.499	2.509	2.518	2.527	2.536	2.546	2.555	2.564	2.573	2.583	2.592	320
330	2.592	2.601	2.610	2.620	2.629	2.638	2.648	2.657	2.666	2.676	2.685	330
340	2.685	2.694	2.704	2.713	2.722	2.731	2.741	2.750	2.760	2.769	2.778	340
350	2.778	2.788	2.797	2.806	2.816	2.825	2.834	2.844	2.853	2.863	2.872	350
360	2.872	2.881	2.891	2.900	2.910	2.919	2.929	2.938	2.947	2.957	2.966	360
370	2.966	2.976	2.985	2.995	3.004	3.014	3.023	3.032	3.042	3.051	3.061	370
380	3.061	3.070	3.080	3.089	3.099	3.108	3.118	3.127	3.137	3.146	3.156	380
390	3.156	3.165	3.175	3.184	3.194	3.203	3.213	3.222	3.232	3.241	3.251	390
400	3.251	3.261	3.270	3.280	3.289	3.299	3.308	3.318	3.327	3.337	3.347	400
°C	0	1	2	3	4	5	6	7	8	9	10	°C

TABLE 49— *Type S thermocouples (continued)*.
Temperature in Degrees Celsius

EMF in Absolute Millivolts   Reference Junctions at 0 C

°C	0	1	2	3	4	5	6	7	8	9	10	°C
						Millivolts						
400	3. 251	3. 261	3. 270	3. 280	3. 289	3. 299	3. 308	3. 318	3. 327	3. 337	3. 347	400
410	3. 347	3. 356	3. 366	3. 375	3. 385	3. 394	3. 404	3. 414	3. 423	3. 433	3. 442	410
420	3. 442	3. 452	3. 462	3. 471	3. 481	3. 490	3. 500	3. 510	3. 519	3. 529	3. 539	420
430	3. 539	3. 548	3. 558	3. 567	3. 577	3. 587	3. 596	3. 606	3. 616	3. 625	3. 635	430
440	3. 635	3. 645	3. 654	3. 664	3. 674	3. 683	3. 693	3. 703	3. 712	3. 722	3. 732	440
450	3. 732	3. 741	3. 751	3. 761	3. 771	3. 780	3. 790	3. 800	3. 809	3. 819	3. 829	450
460	3. 829	3. 839	3. 848	3. 858	3. 868	3. 878	3. 887	3. 897	3. 907	3. 917	3. 926	460
470	3. 926	3. 936	3. 946	3. 956	3. 965	3. 975	3. 985	3. 995	4. 004	4. 014	4. 024	470
480	4. 024	4. 034	4. 044	4. 053	4. 063	4. 073	4. 083	4. 093	4. 103	4. 112	4. 122	480
490	4. 122	4. 132	4. 142	4. 152	4. 162	4. 171	4. 181	4. 191	4. 201	4. 211	4. 221	490
500	4. 221	4. 230	4. 240	4. 250	4. 260	4. 270	4. 280	4. 290	4. 300	4. 310	4. 319	500
510	4. 319	4. 329	4. 339	4. 349	4. 359	4. 369	4. 379	4. 389	4. 399	4. 409	4. 419	510
520	4. 419	4. 428	4. 438	4. 448	4. 458	4. 468	4. 478	4. 488	4. 498	4. 508	4. 518	520
530	4. 518	4. 528	4. 538	4. 548	4. 558	4. 568	4. 578	4. 588	4. 598	4. 608	4. 618	530
540	4. 618	4. 628	4. 638	4. 648	4. 658	4. 668	4. 678	4. 688	4. 698	4. 708	4. 718	540
550	4. 718	4. 728	4. 738	4. 748	4. 758	4. 768	4. 778	4. 788	4. 798	4. 808	4. 818	550
560	4. 818	4. 828	4. 839	4. 849	4. 859	4. 869	4. 879	4. 889	4. 899	4. 909	4. 919	560
570	4. 919	4. 929	4. 939	4. 950	4. 960	4. 970	4. 980	4. 990	5. 000	5. 010	5. 020	576
580	5. 020	5. 031	5. 041	5. 051	5. 061	5. 071	5. 081	5. 091	5. 102	5. 112	5. 122	580
590	5. 122	5. 132	5. 142	5. 152	5. 163	5. 173	5. 183	5. 193	5. 203	5. 214	5. 224	590
600	5. 224	5. 234	5. 244	5. 254	5. 265	5. 275	5. 285	5. 295	5. 306	5. 316	5. 326	600
610	5. 326	5. 336	5. 346	5. 357	5. 367	5. 377	5. 388	5. 398	5. 408	5. 418	5. 429	610
620	5. 429	5. 439	5. 449	5. 459	5. 470	5. 480	5. 490	5. 501	5. 511	5. 521	5. 532	620
630	5. 532	5. 542	5. 552	5. 563	5. 573	5. 583	5. 593	5. 604	5. 614	5. 624	5. 635	630
640	5. 635	5. 645	5. 655	5. 666	5. 676	5. 686	5. 697	5. 707	5. 717	5. 728	5. 738	640
650	5. 738	5. 748	5. 759	5. 769	5. 779	5. 790	5. 800	5. 811	5. 821	5. 831	5. 842	650
660	5. 842	5. 852	5. 862	5. 873	5. 883	5. 894	5. 904	5. 914	5. 925	5. 935	5. 946	660
670	5. 946	5. 956	5. 967	5. 977	5. 987	5. 998	6. 008	6. 019	6. 029	6. 040	6. 050	670
680	6. 050	6. 060	6. 071	6. 081	6. 092	6. 102	6. 113	6. 123	6. 134	6. 144	6. 155	680
690	6. 155	6. 165	6. 176	6. 186	6. 197	6. 207	6. 218	6. 228	6. 239	6. 249	6. 260	690
700	6. 260	6. 270	6. 281	6. 291	6. 302	6. 312	6. 323	6. 333	6. 344	6. 355	6. 365	700
710	6. 365	6. 376	6. 386	6. 397	6. 407	6. 418	6. 429	6. 439	6. 450	6. 460	6. 471	710
720	6. 471	6. 481	6. 492	6. 503	6. 513	6. 524	6. 534	6. 545	6. 556	6. 566	6. 577	720
730	6. 577	6. 588	6. 598	6. 609	6. 619	6. 630	6. 641	6. 651	6. 662	6. 673	6. 683	730
740	6. 683	6. 694	6. 705	6. 715	6. 726	6. 737	6. 747	6. 758	6. 769	6. 779	6. 790	740
750	6. 790	6. 801	6. 811	6. 822	6. 833	6. 844	6. 854	6. 865	6. 876	6. 886	6. 897	750
760	6. 897	6. 908	6. 919	6. 929	6. 940	6. 951	6. 962	6. 972	6. 983	6. 994	7. 005	760
770	7. 005	7. 015	7. 026	7. 037	7. 047	7. 058	7. 069	7. 080	7. 091	7. 102	7. 112	770
780	7. 112	7. 123	7. 134	7. 145	7. 156	7. 166	7. 177	7. 188	7. 199	7. 210	7. 220	780
790	7. 220	7. 231	7. 242	7. 253	7. 264	7. 275	7. 286	7. 296	7. 307	7. 318	7. 329	790
800	7. 329	7. 340	7. 351	7. 362	7. 372	7. 383	7. 394	7. 405	7. 416	7. 427	7. 438	800
°C	0	1	2	3	4	5	6	7	8	9	10	°C

TABLE 49 — *Type S thermocouples (continued)*.

Temperature in Degrees Celsius

EMF in Absolute Millivolts   Reference Junctions at 0 C

°C	0	1	2	3	4	5	6	7	8	9	10	°C
						Millivolts						
800	7.329	7.340	7.351	7.362	7.372	7.383	7.394	7.405	7.416	7.427	7.438	800
810	7.438	7.449	7.460	7.470	7.481	7.492	7.503	7.514	7.525	7.536	7.547	810
820	7.547	7.558	7.569	7.580	7.591	7.602	7.613	7.623	7.634	7.645	7.656	820
830	7.656	7.667	7.678	7.689	7.700	7.711	7.722	7.733	7.744	7.755	7.766	830
840	7.766	7.777	7.788	7.799	7.810	7.821	7.832	7.843	7.854	7.865	7.876	840
850	7.876	7.887	7.898	7.910	7.921	7.932	7.943	7.954	7.965	7.976	7.987	850
860	7.987	7.998	8.009	8.020	8.031	8.042	8.053	8.064	8.076	8.087	8.098	860
870	8.098	8.109	8.120	8.131	8.142	8.153	8.164	8.176	8.187	8.198	8.209	870
880	8.209	8.220	8.231	8.242	8.254	8.265	8.276	8.287	8.298	8.309	8.320	880
890	8.320	8.332	8.343	8.354	8.365	8.376	8.388	8.399	8.410	8.421	8.432	890
900	8.432	8.444	8.455	8.466	8.477	8.488	8.500	8.511	8.522	8.533	8.545	900
910	8.545	8.556	8.567	8.578	8.590	8.601	8.612	8.623	8.635	8.646	8.657	910
920	8.657	8.668	8.680	8.691	8.702	8.714	8.725	8.736	8.747	8.759	8.770	920
930	8.770	8.781	8.793	8.804	8.815	8.827	8.838	8.849	8.861	8.872	8.883	930
940	8.883	8.895	8.906	8.917	8.929	8.940	8.951	8.963	8.974	8.986	8.997	940
950	8.997	9.008	9.020	9.031	9.042	9.054	9.065	9.077	9.088	9.099	9.111	950
960	9.111	9.122	9.134	9.145	9.157	9.168	9.179	9.191	9.202	9.214	9.225	960
970	9.225	9.236	9.248	9.260	9.271	9.282	9.294	9.305	9.317	9.328	9.340	970
980	9.340	9.351	9.363	9.374	9.386	9.397	9.409	9.420	9.432	9.443	9.455	980
990	9.455	9.466	9.478	9.489	9.501	9.512	9.524	9.535	9.547	9.559	9.570	990
1,000	9.570	9.582	9.593	9.605	9.616	9.628	9.639	9.651	9.663	9.674	9.686	1,000
1,010	9.686	9.697	9.709	9.720	9.732	9.744	9.755	9.767	9.779	9.790	9.802	1,010
1,020	9.802	9.813	9.825	9.837	9.848	9.860	9.871	9.883	9.895	9.906	9.918	1,020
1,030	9.918	9.930	9.941	9.953	9.965	9.976	9.988	10.000	10.011	10.023	10.035	1,030
1,040	10.035	10.046	10.058	10.070	10.082	10.093	10.105	10.117	10.128	10.140	10.152	1,040
1,050	10.152	10.163	10.175	10.187	10.199	10.210	10.222	10.234	10.246	10.257	10.269	1,050
1,060	10.269	10.281	10.293	10.304	10.316	10.328	10.340	10.351	10.363	10.375	10.387	1,060
1,070	10.387	10.399	10.410	10.422	10.434	10.446	10.458	10.469	10.481	10.493	10.505	1,070
1,080	10.505	10.517	10.528	10.540	10.552	10.564	10.576	10.587	10.599	10.611	10.623	1,080
1,090	10.623	10.635	10.647	10.658	10.670	10.682	10.694	10.706	10.718	10.729	10.741	1,090
1,100	10.741	10.753	10.765	10.777	10.789	10.801	10.812	10.824	10.836	10.848	10.860	1,100
1,110	10.860	10.872	10.884	10.896	10.907	10.919	10.931	10.943	10.955	10.967	10.979	1,110
1,120	10.979	10.991	11.003	11.014	11.026	11.038	11.050	11.062	11.074	11.086	11.098	1,120
1,130	11.098	11.110	11.122	11.133	11.145	11.157	11.169	11.181	11.193	11.205	11.217	1,130
1,140	11.217	11.229	11.241	11.253	11.265	11.277	11.289	11.300	11.312	11.324	11.336	1,140
1,150	11.336	11.348	11.360	11.372	11.384	11.396	11.408	11.420	11.432	11.444	11.456	1,150
1,160	11.456	11.468	11.480	11.492	11.504	11.516	11.528	11.540	11.552	11.564	11.575	1,160
1,170	11.575	11.587	11.599	11.611	11.623	11.635	11.647	11.659	11.671	11.683	11.695	1,170
1,180	11.695	11.707	11.719	11.731	11.743	11.755	11.767	11.779	11.791	11.803	11.815	1,180
1,190	11.815	11.827	11.839	11.851	11.863	11.875	11.887	11.899	11.911	11.923	11.935	1,190
1,200	11.935	11.947	11.959	11.971	11.983	11.995	12.007	12.019	12.031	12.043	12.055	1,200
°C	0	1	2	3	4	5	6	7	8	9	10	°C

TABLE 49—*Type S thermocouples (continued).*

Temperature in Degrees Celsius

EMF in Absolute Millivolts

Reference Junctions at 0 C

°C	0	1	2	3	4	5	6	7	8	9	10	°C
						Millivolts						
1,200	11.935	11.947	11.959	11.971	11.983	11.995	12.007	12.019	12.031	12.043	12.055	1,200
1,210	12.055	12.067	12.079	12.091	12.103	12.115	12.127	12.139	12.151	12.163	12.175	1,210
1,220	12.175	12.187	12.200	12.212	12.224	12.236	12.248	12.260	12.272	12.284	12.296	1,220
1,230	12.296	12.308	12.320	12.332	12.344	12.356	12.368	12.380	12.392	12.404	12.416	1,230
1,240	12.416	12.428	12.440	12.452	12.464	12.476	12.488	12.500	12.512	12.524	12.536	1,240
1,250	12.536	12.548	12.560	12.573	12.585	12.597	12.609	12.621	12.633	12.645	12.657	1,250
1,260	12.657	12.669	12.681	12.693	12.705	12.717	12.729	12.741	12.753	12.765	12.777	1,260
1,270	12.777	12.789	12.801	12.813	12.825	12.837	12.849	12.861	12.873	12.885	12.897	1,270
1,280	12.897	12.909	12.921	12.933	12.945	12.957	12.969	12.981	12.993	13.005	13.018	1,280
1,290	13.018	13.030	13.042	13.054	13.066	13.078	13.090	13.102	13.114	13.126	13.138	1,290
1,300	13.138	13.150	13.162	13.174	13.186	13.198	13.210	13.222	13.234	13.246	13.258	1,300
1,310	13.258	13.270	13.282	13.294	13.306	13.318	13.330	13.342	13.354	13.366	13.378	1,310
1,320	13.378	13.390	13.402	13.414	13.426	13.438	13.450	13.462	13.474	13.486	13.498	1,320
1,330	13.498	13.510	13.522	13.534	13.546	13.558	13.570	13.582	13.594	13.606	13.618	1,330
1,340	13.618	13.630	13.642	13.654	13.666	13.678	13.690	13.702	13.714	13.726	13.738	1,340
1,350	13.738	13.750	13.762	13.774	13.786	13.798	13.810	13.822	13.834	13.846	13.858	1,350
1,360	13.858	13.870	13.882	13.894	13.906	13.918	13.930	13.942	13.954	13.966	13.978	1,360
1,370	13.978	13.990	14.002	14.014	14.026	14.058	14.050	14.062	14.074	14.086	14.098	1,370
1,380	14.098	14.110	14.122	14.133	14.145	14.157	14.169	14.181	14.193	14.205	14.217	1,380
1,390	14.217	14.229	14.241	14.253	14.265	14.277	14.289	14.301	14.313	14.325	14.337	1,390
1,400	14.337	14.349	14.361	14.373	14.385	14.397	14.409	14.421	14.433	14.445	14.457	1,400
1,410	14.457	14.469	14.481	14.493	14.504	14.516	14.528	14.540	14.552	14.564	14.576	1,410
1,420	14.576	14.588	14.600	14.612	14.624	14.636	14.648	14.660	14.672	14.684	14.696	1,420
1,430	14.696	14.708	14.720	14.732	14.744	14.755	14.767	14.779	14.791	14.803	14.815	1,430
1,440	14.815	14.827	14.839	14.851	14.863	14.875	14.887	14.899	14.911	14.923	14.935	1,440
1,450	14.935	14.946	14.958	14.970	14.982	14.994	15.006	15.018	15.030	15.042	15.054	1,450
1,460	15.054	15.066	15.078	15.090	15.102	15.113	15.125	15.137	15.149	15.161	15.173	1,460
1,470	15.173	15.185	15.197	15.209	15.221	15.233	15.245	15.256	15.268	15.280	15.292	1,470
1,480	15.292	15.304	15.316	15.328	15.340	15.352	15.364	15.376	15.387	15.399	15.411	1,480
1,490	15.411	15.423	15.435	15.447	15.459	15.471	15.483	15.495	15.507	15.518	15.530	1,490
1,500	15.530	15.542	15.554	15.566	15.578	15.590	15.602	15.614	15.625	15.637	15.649	1,500
1,510	15.649	15.661	15.673	15.685	15.697	15.709	15.721	15.732	15.744	15.756	15.768	1,510
1,520	15.768	15.780	15.792	15.804	15.816	15.827	15.839	15.851	15.863	15.875	15.887	1,520
1,530	15.887	15.899	15.911	15.922	15.934	15.946	15.958	15.970	15.982	15.994	16.006	1,530
1,540	16.006	16.017	16.029	16.041	16.053	16.065	16.077	16.089	16.100	16.112	16.124	1,540
1,550	16.124	16.136	16.148	16.160	16.171	16.183	16.195	16.207	16.219	16.231	16.243	1,550
1,560	16.243	16.254	16.266	16.278	16.290	16.302	16.314	16.325	16.337	16.349	16.361	1,560
1,570	16.361	16.373	16.385	16.396	16.408	16.420	16.432	16.444	16.456	16.467	16.479	1,570
1,580	16.479	16.491	16.503	16.515	16.527	16.538	16.550	16.562	16.574	16.586	16.597	1,580
1,590	16.597	16.609	16.621	16.633	16.645	16.657	16.668	16.680	16.692	16.704	16.716	1,590
1,600	16.716	16.727	16.739	16.751	16.763	16.775	16.786	6.798	16.810	16.822	16.834	1,600
°C	0	1	2	3	4	5	6	7	8	9	10	°C

TABLE 49—*Type S thermocouples (continued)*.
Temperature in Degrees Celsius

EMF in Absolute Millivolts

Reference Junctions at 0 C

° C	0	1	2	3	4	5	6	7	8	9	10	° C
	Millivolts											
1,600	16.716	16.727	16.739	16.751	16.763	16.775	16.786	16.798	16.810	16.822	16.834	1,600
1,610	16.834	16.845	16.857	16.869	16.881	16.893	16.904	16.916	16.928	16.940	16.952	1,610
1,620	16.952	16.963	16.975	16.987	16.999	17.010	17.022	17.034	17.046	17.058	17.069	1,620
1,630	17.069	17.081	17.093	17.105	17.116	17.128	17.140	17.152	17.163	17.175	17.187	1,630
1,640	17.187	17.199	17.211	17.222	17.234	17.246	17.258	17.269	17.281	17.293	17.305	1,640
1,650	17.305	17.316	17.328	17.340	17.352	17.363	17.375	17.387	17.398	17.410	17.422	1,650
1,660	17.422	17.434	17.446	17.457	17.469	17.481	17.492	17.504	17.516	17.528	17.539	1,660
1,670	17.539	17.551	17.563	17.575	17.586	17.598	17.610	17.621	17.633	17.645	17.657	1,670
1,680	17.657	17.668	17.680	17.692	17.704	17.715	17.727	17.739	17.750	17.762	17.774	1,680
1,690	17.774	17.785	17.797	17.809	17.821	17.832	17.844	17.856	17.867	17.879	17.891	1,690
1,700	17.891	17.902	17.914	17.926	17.938	17.949	17.961	17.973	17.984	17.996	18.008	1,700
1,710	18.008	18.019	18.031	18.043	18.054	18.066	18.078	18.089	18.101	18.113	18.124	1,710
1,720	18.124	18.136	18.148	18.159	18.171	18.183	18.194	18.206	18.218	18.229	18.241	1,720
1,730	18.241	18.253	18.264	18.276	18.288	18.299	18.311	18.323	18.334	18.346	18.358	1,730
1,740	18.358	18.369	18.381	18.393	18.404	18.416	18.427	18.439	18.451	18.462	18.474	1,740
1,750	18.474	18.486	18.497	18.509	18.520	18.532	18.544	18.555	18.567	18.579	18.590	1,750
1,760	18.590	18.602	18.613	18.625	18.637	18.648	18.660	18.672	18.683	18.695	-------	1,760
° C	0	1	2	3	4	5	6	7	8	9	10	° C

TABLE 50—*Type T thermocouples.*
Temperature in Degrees Fahrenheit

EMF in Absolute Millivolts   Reference Junctions at 32 F

°F	0	1	2	3	4	5	6	7	8	9	10	°F
						Millivolts						
—310	—5.379	—5.388	—5.397	—5.406	-------	-------	-------	-------	-------	-------	-------	**—310**
—300	—5.284	—5.294	—5.303	—5.313	—5.322	—5.332	—5.341	—5.351	—5.360	—5.370	—5.379	**—300**
—290	—5.185	—5.195	—5.205	—5.215	—5.225	—5.235	—5.245	—5.254	—5.264	—5.274	—5.284	**—290**
—280	—5.081	—5.092	—5.102	—5.113	—5.124	—5.134	—5.144	—5.154	—5.165	—5.175	—5.185	**—280**
—270	—4.974	—4.985	—4.996	—5.007	—5.018	—5.029	—5.039	—5.050	—5.060	—5.071	—5.081	**—270**
—260	—4.863	—4.874	—4.885	—4.897	—4.908	—4.919	—4.930	—4.941	—4.952	—4.963	—4.974	**—260**
—250	—4.747	—4.759	—4.770	—4.782	—4.794	—4.805	—4.817	—4.829	—4.840	—4.851	—4.863	**—250**
—240	—4.627	—4.640	—4.652	—4.664	—4.676	—4.688	—4.700	—4.712	—4.724	—4.735	—4.747	**—240**
—230	—4.504	—4.517	—4.529	—4.542	—4.554	—4.566	—4.579	—4.591	—4.603	—4.615	—4.627	**—230**
—220	—4.377	—4.390	—4.403	—4.415	—4.428	—4.441	—4.454	—4.466	—4.479	—4.492	—4.504	**—220**
—210	—4.246	—4.259	—4.272	—4.286	—4.299	—4.312	—4.325	—4.338	—4.351	—4.364	—4.377	**—210**
—200	—4.111	—4.125	—4.138	—4.151	—4.165	—4.179	—4.192	—4.206	—4.219	—4.232	—4.246	**—200**
—190	—3.972	—3.986	—4.000	—4.014	—4.028	—4.042	—4.056	—4.069	—4.083	—4.097	—4.111	**—190**
—180	—3.829	—3.844	—3.858	—3.873	—3.887	—3.901	—3.915	—3.929	—3.944	—3.958	—3.972	**—180**
—170	—3.684	—3.698	—3.713	—3.727	—3.742	—3.757	—3.771	—3.786	—3.800	—3.815	—3.829	**—170**
—160	—3.533	—3.548	—3.564	—3.579	—3.594	—3.609	—3.624	—3.639	—3.654	—3.669	—3.684	**—160**
—150	—3.380	—3.396	—3.411	—3.426	—3.441	—3.457	—3.472	—3.488	—3.503	—3.518	—3.533	**—150**
—140	—3.223	—3.238	—3.254	—3.270	—3.286	—3.301	—3.317	—3.333	—3.349	—3.365	—3.380	**—140**
—130	—3.062	—3.078	—3.094	—3.110	—3.127	—3.143	—3.159	—3.175	—3.191	—3.207	—3.223	**—130**
—120	—2.897	—2.914	—2.931	—2.947	—2.964	—2.980	—2.997	—3.013	—3.030	—3.046	—3.062	**—120**
—110	—2.730	—2.747	—2.764	—2.781	—2.797	—2.814	—2.831	—2.847	—2.864	—2.881	—2.897	**—110**
—100	—2.559	—2.577	—2.594	—2.611	—2.628	—2.645	—2.662	—2.679	—2.696	—2.713	—2.730	**—100**
—90	—2.385	—2.402	—2.420	—2.437	—2.455	—2.472	—2.490	—2.507	—2.525	—2.542	—2.559	**—90**
—80	—2.207	—2.225	—2.243	—2.260	—2.278	—2.296	—2.314	—2.332	—2.349	—2.367	—2.385	**—80**
—70	—2.026	—2.044	—2.063	—2.081	—2.099	—2.117	—2.135	—2.153	—2.171	—2.189	—2.207	**—70**
—60	—1.842	—1.860	—1.879	—1.897	—1.916	—1.934	—1.953	—1.971	—1.989	—2.008	—2.026	**—60**
—50	—1.654	—1.673	—1.692	—1.711	—1.729	—1.748	—1.767	—1.786	—1.804	—1.823	—1.842	**—50**
—40	—1.463	—1.482	—1.502	—1.521	—1.540	—1.559	—1.578	—1.597	—1.616	—1.635	—1.654	**—40**
—30	—1.270	—1.289	—1.308	—1.328	—1.347	—1.367	—1.386	—1.406	—1.425	—1.444	—1.463	**—30**
—20	—1.072	—1.092	—1.112	—1.132	—1.152	—1.171	—1.191	—1.210	—1.230	—1.250	—1.270	**—20**
—10	—0.872	—0.893	—0.913	—0.933	—0.953	—0.973	—0.993	—1.013	—1.033	—1.053	—1.072	**—10**
(—)0	—0.670	—0.690	—0.710	—0.730	—0.751	—0.771	—0.792	—0.812	—0.832	—0.852	—0.872	**(—)0**
°F	0	1	2	3	4	5	6	7	8	9	10	**°F**

TABLE 50—*Type T thermocouples (continued).*
Temperature in Degrees Fahrenheit

EMF in Absolute Millivolts   Reference Junctions at 32 F

°F	0	1	2	3	4	5	6	7	8	9	10	°F
						Millivolts						
(+)0	−0. 670	−0. 649	−0. 629	−0. 608	−0. 588	−0. 567	−0. 546	−0. 526	−0. 505	−0. 484	−0. 463	(+)0
10	−0. 463	−0. 442	−0. 421	−0. 401	−0. 380	−0. 359	−0. 339	−0. 318	−0. 297	−0. 275	−0. 254	10
20	−0. 254	−0. 233	−0. 212	−0. 191	−0. 170	−0. 149	−0. 128	−0. 107	−0. 085	−0. 064	−0. 042	20
30	−0. 042	−0. 021	0. 000	0. 021	0. 042	0. 064	0. 086	0. 107	0. 129	0. 150	0. 171	30
40	0. 171	0. 193	0. 215	0. 236	0. 258	0. 280	0. 302	0. 324	0. 346	0. 367	0. 389	40
50	0. 389	0. 411	0. 433	0. 455	0. 477	0. 499	0. 521	0. 543	0. 565	0. 587	0. 609	50
60	0. 609	0. 631	0. 654	0. 676	0. 698	0. 720	0. 743	0. 765	0. 787	0. 809	0. 832	60
70	0. 832	0. 854	0. 877	0. 899	0. 922	0. 944	0. 967	0. 990	1. 012	1. 035	1. 057	70
80	1. 057	1. 080	1. 103	1. 126	1. 148	1. 171	1. 194	1. 217	1. 240	1. 263	1. 286	80
90	1. 286	1. 309	1. 332	1. 355	1. 378	1. 401	1. 424	1. 448	1. 471	1. 494	1. 517	90
100	1. 517	1. 540	1. 563	1. 587	1: 610	1. 633	1. 657	1. 680	1. 704	1. 727	1. 751	100
110	1. 751	1. 774	1. 798	1. 821	1. 845	1. 869	1. 893	1. 916	1. 940	1. 963	1. 987	110
120	1. 987	2. 011	2. 035	2. 059	2. 083	2. 107	2. 131	2. 154	2. 178	2. 202	2. 226	120
130	2. 226	2. 250	2. 274	2. 298	2. 322	2. 346	2. 370	2. 394	2. 418	2. 443	2. 467	130
140	2. 467	2. 491	2. 516	2. 540	2. 565	2.·589	2. 614	2. 638	2. 663	2. 687	2. 711	140
150	2. 711	2. 736	2. 760	2. 785	2. 810	2. 835	2. 859	2. 884	2. 908	2. 933	2. 958	150
160	2. 958	2. 982	3. 007	3. 032	3. 057	3. 082	3. 107	3. 132	3. 157	3. 182	3. 207	160
170	3. 207	3. 232	3. 257	3. 282	3. 307	3. 332	3. 357	3. 382	3. 407	3. 433	3. 458	170
180	3. 458	3. 483	3. 508	3. 534	3. 559	3. 584	3. 610	3. 635	3. 661	3. 686	3. 712	180
190	3. 712	3. 737	3. 762	3. 787	3. 813	3. 839	3. 864	3. 890	3. 915	3. 941	3. 967	190
200	3. 967	3. 993	4. 018	4. 044	4. 070	4. 096	4. 122	4. 148	4. 174	4. 199	4. 225	200
210	4. 225	4. 251	4. 277	4. 303	4. 329	4. 355	4. 381	4. 408	4. 434	4. 460	4. 486	210
220	4. 486	4. 512	4. 538	4. 564	4. 590	4. 617	4. 643	4. 670	4. 696	4. 722	4. 749	220
230	4. 749	4. 775	4. 801	4. 827	4. 854	4. 880	4. 907	4. 934	4. 960	4. 987	5. 014	230
240	5. 014	5. 040	5. 067	5. 094	5. 120	5. 147	5. 174	5. 200	5. 227	5. 254	5. 280	240
250	5. 280	5. 307	5. 334	5. 361	5. 388	5. 415	5. 442	5. 469	5. 496	5. 523	5. 550	250
260	5. 550	5. 577	5. 604	5. 631	5. 658	5. 685	5. 712	5. 739	5. 766	5. 794	5. 821	260
270	5. 821	5. 848	5. 875	5. 903	5. 930	5. 957	5. 985	6. 012	6. 040	6. 067	6. 094	270
280	6. 094	6. 122	6. 149	6. 177	6. 204	6. 232	6. 259	6. 287	6. 314	6. 342	6. 370	280
290	6. 370	6. 397	6. 425	6. 453	6. 481	6. 508	6. 536	6. 564	6. 592	6. 620	6. 647	290
300	6. 647	6. 675	6. 703	6. 731	6. 759	6. 786	6. 814	6. 842	6. 870	6. 898	6. 926	300
310	6. 926	6. 954	6. 982	7. 010	7. 038	7. 066	7. 095	7. 123	7. 151	7. 180	7. 208	310
320	7. 208	7. 236	7. 264	7. 292	7. 321	7. 349	7. 377	7. 405	7. 434	7. 462	7. 491	320
330	7. 491	7. 519	7. 548	7. 576	7. 605	7. 633	7. 661	7. 690	7. 719	7. 747	7. 776	330
340	7. 776	7. 805	7. 834	7. 862	7. 891	7. 920	7. 949	7. 978	8. 006	8. 035	8. 064	340
350	8. 064	8. 092	8. 120	8. 149	8. 178	8. 207	8. 236	8. 265	8. 294	8. 323	8. 352	350
360	8. 352	8. 381	8. 410	8. 439	8. 468	8. 497	8. 526	8. 555	8. 584	8. 613	8. 642	360
370	8. 642	8. 672	8. 701	8. 730	8. 759	8. 788	8. 818	8. 847	8. 876	8. 905	8. 935	370
380	8. 935	8. 964	8. 994	9. 023	9. 052	9. 082	9. 111	9. 141	9. 170	9. 200	9. 229	380
390	9. 229	9. 259	9. 288	9. 317	9. 347	9. 376	9. 406	9. 436	9. 466	9. 495	9. 525	390
400	9. 525	9. 555	9. 584	9. 614	9. 644	9. 674	9. 703	9. 733	9. 763	9. 793	9. 823	400
°F	0	1	2	3	4	5	6	7	8	9	10	°F

TABLE 50—*Type T thermocouples (continued).*
Temperature in Degrees Fahrenheit

EMF in Absolute Millivolts

Reference Junctions at 32 F

°F	0	1	2	3	4	5	6	7	8	9	10	°F
						Millivolts						
400	9. 525	9. 555	9. 584	9. 614	9. 644	9. 674	9. 703	9. 733	9. 763	9. 793	9. 823	400
410	9. 823	9. 853	9. 883	9. 913	9. 943	9. 973	10. 003	10. 033	10. 063	10. 093	10. 123	410
420	10. 123	10. 153	10. 183	10. 213	10. 243	10. 273	10. 303	10. 333	10. 363	10. 393	10. 423	420
430	10. 423	10. 453	10. 483	10. 514	10. 544	10. 574	10. 604	10. 635	10. 665	10. 695	10. 726	430
440	10. 726	10. 756	10. 787	10. 817	10. 848	10. 878	10. 909	10. 939	10. 969	11. 000	11. 030	440
450	11. 030	11. 061	11. 091	11. 122	11. 152	11. 183	11. 214	11. 244	11. 275	11. 305	11. 336	450
460	11. 336	11. 366	11. 397	11. 428	11. 459	11. 490	11. 520	11. 551	11. 581	11. 612	11. 643	460
470	11. 643	11. 674	11. 704	11. 735	11. 766	11. 797	11. 828	11. 859	11. 891	11. 922	11. 953	470
480	11. 953	11. 984	12. 015	12. 046	12. 077	12. 108	12. 138	12. 170	12. 201	12. 232	12. 263	480
490	12. 263	12. 294	12. 325	12. 356	12. 387	12. 418	12. 450	12. 481	12. 512	12. 543	12. 575	490
500	12. 575	12. 606	12. 637	12. 669	12. 700	12. 732	12. 763	12. 794	12. 825	12. 857	12. 888	500
510	12. 888	12. 919	12. 951	12. 983	13. 014	13. 046	13. 077	13. 108	13. 140	13. 172	13. 203	510
520	13. 203	13. 235	13. 267	13. 298	13. 330	13. 362	13. 393	13. 425	13. 457	13. 488	13. 520	520
530	13. 520	13. 552	13. 583	13. 615	13. 647	13. 678	13. 710	13. 742	13. 774	13. 806	13. 838	530
540	13. 838	13. 869	13. 901	13. 933	13. 965	13. 997	14. 029	14. 061	14. 093	14. 125	14. 157	540
550	14. 157	14. 189	14. 221	14. 253	14. 285	14. 317	14. 349	14. 381	14. 413	14. 445	14. 477	550
560	14. 477	14. 509	14. 541	14. 573	14. 605	14. 637	14. 670	14. 702	14. 734	14. 766	14. 799	560
570	14. 799	14. 831	14. 864	14. 896	14. 928	14. 961	14. 993	15. 025	15. 057	15. 090	15. 122	570
580	15. 122	15. 155	15. 187	15. 219	15. 252	15. 284	15. 317	15. 349	15. 382	15. 414	15. 447	580
590	15. 447	15. 480	15. 512	15. 545	15. 577	15. 610	15. 642	15. 675	15. 707	15. 740	15. 773	590
600	15. 773	15. 806	15. 838	15. 871	15. 904	15. 937	15. 969	16. 002	16. 035	16. 068	16. 101	600
610	16. 101	16. 133	16. 166	16. 199	16. 232	16. 264	16. 297	16. 330	16. 363	16. 396	16. 429	610
620	16. 429	16. 462	16. 495	16. 528	16. 560	16. 593	16. 626	16. 659	16. 692	16. 725	16. 758	620
630	16. 758	16. 791	16. 824	16. 857	16. 890	16. 924	16. 957	16. 990	17. 023	17. 056	17. 089	630
640	17. 089	17. 122	17. 155	17. 189	17. 222	17. 255	17. 288	17. 321	17. 354	17. 388	17. 421	640
650	17. 421	17. 454	17. 488	17. 521	17. 554	17. 588	17. 621	17. 654	17. 688	17. 721	17. 754	650
660	17. 754	17. 788	17. 821	17. 854	17. 888	17. 921	17. 955	17. 988	18. 022	18. 055	18. 089	660
670	18. 089	18. 123	18. 156	18. 190	18. 223	18. 257	18. 290	18. 324	18. 357	18. 391	18. 425	670
680	18. 425	18. 458	18. 492	18. 526	18. 560	18. 593	18. 627	18. 660	18. 694	18. 727	18. 761	680
690	18. 761	18. 795	18. 829	18. 863	18. 896	18. 930	18. 964	18. 998	19. 032	19. 066	19. 100	690
700	19. 100	19. 134	19. 168	19. 201	19. 235	19. 269	19. 303	19. 337	19. 371	19. 405	19. 439	700
710	19. 439	19. 473	19. 506	19. 540	19. 574	19. 608	19. 642	19. 676	19. 711	19. 745	19. 779	710
720	19. 779	19. 813	19. 847	19. 881	19. 915	19. 949	19. 983	20. 018	20. 052	20. 086	20. 120	720
730	20. 120	20. 154	20. 188	20. 223	20. 257	20. 291	20. 325	20. 359	20. 394	20. 428	20. 463	730
740	20. 463	20. 497	20. 531	20. 565	20. 599	20. 634	20. 668	20. 702	20. 736	20. 771	20. 805	740
750	20. 805	20. 840	20. 874	------	------	------	------	------	------	------	------	750
°F	0	1	2	3	4	5	6	7	8	9	10	°F

TABLE 51—*Type T thermocouples.*
Temperature in Degrees Celsius

EMF in Absolute Millivolts   Reference Junctions at 0 C

°C	0	1	2	3	4	5	6	7	8	9	10	°C
						Millivolts						
−190	−5.379	−5.395	−5.411	-----	-----	-----						−190
−180	−5.205	−5.223	−5.241	−5.258	−5.276	−5.294	−5.311	−5.328	−5.345	−5.362	−5.379	−180
−170	−5.018	−5.037	−5.056	−5.075	−5.094	−5.113	−5.132	−5.150	−5.169	−5.187	−5.205	−170
−160	−4.817	−4.838	−4.858	−4.878	−4.899	−4.919	−4.939	−4.959	−4.978	−4.998	−5.018	−160
−150	−4.603	−4.625	−4.647	−4.669	−4.690	−4.712	−4.733	−4.754	−4.775	−4.796	−4.817	−150
−140	−4.377	−4.400	−4.423	−4.446	−4.469	−4.492	−4.514	−4.537	−4.559	−4.581	−4.603	−140
−130	−4.138	−4.162	−4.187	−4.211	−4.235	−4.259	−4.283	−4.307	−4.330	−4.354	−4.377	−130
−120	−3.887	−3.912	−3.938	−3.964	−3.989	−4.014	−4.039	−4.064	−4.089	−4.114	−4.138	−120
−110	−3.624	−3.651	−3.678	−3.704	−3.730	−3.757	−3.783	−3.809	−3.835	−3.861	−3.887	−110
−100	−3.349	−3.377	−3.405	−3.432	−3.460	−3.488	−3.515	−3.542	−3.570	−3.597	−3.624	−100
−90	−3.062	−3.091	−3.120	−3.149	−3.178	−3.207	−3.235	−3.264	−3.292	−3.320	−3.349	−90
−80	−2.764	−2.794	−2.824	−2.854	−2.884	−2.914	−2.944	−2.974	−3.003	−3.033	−3.062	−80
−70	−2.455	−2.486	−2.518	−2.549	−2.580	−2.611	−2.642	−2.672	−2.703	−2.733	−2.764	−70
−60	−2.135	−2.167	−2.200	−2.232	−2.264	−2.296	−2.328	−2.360	−2.392	−2.423	−2.455	−60
−50	−1.804	−1.838	−1.871	−1.905	−1.938	−1.971	−2.004	−2.037	−2.070	−2.103	−2.135	−50
−40	−1.463	−1.498	−1.532	−1.567	−1.601	−1.635	−1.669	−1.703	−1.737	−1.771	−1.804	−40
−30	−1.112	−1.148	−1.183	−1.218	−1.254	−1.289	−1.324	−1.359	−1.394	−1.429	−1.463	−30
−20	−0.751	−0.788	−0.824	−0.860	−0.897	−0.933	−0.969	−1.005	−1.041	−1.076	−1.112	−20
−10	−0.380	−0.417	−0.455	−0.492	−0.530	−0.567	−0.604	−0.641	−0.678	−0.714	−0.751	−10
(−)0	0.000	−0.038	−0.077	−0.115	−0.153	−0.191	−0.229	−0.267	−0.305	−0.343	−0.380	(−)0
(+)0	0.000	0.038	0.077	0.116	0.154	0.193	0.232	0.271	0.311	0.350	0.389	(+)0
10	0.389	0.429	0.468	0.508	0.547	0.587	0.627	0.667	0.707	0.747	0.787	10
20	0.787	0.827	0.868	0.908	0.949	0.990	1.030	1.071	1.112	1.153	1.194	20
30	1.194	1.235	1.277	1.318	1.360	1.401	1.443	1.485	1.526	1.568	1.610	30
40	1.610	1.652	1.694	1.737	1.779	1.821	1.864	1.907	1.949	1.992	2.035	40
50	2.035	2.078	2.121	2.164	2.207	2.250	2.293	2.336	2.380	2.423	2.467	50
60	2.467	2.511	2.555	2.599	2.643	2.687	2.731	2.775	2.820	2.864	2.908	60
70	2.908	2.953	2.997	3.042	3.087	3.132	3.177	3.222	3.267	3.312	3.357	70
80	3.357	3.402	3.448	3.493	3.539	3.584	3.630	3.676	3.722	3.767	3.813	80
90	3.813	3.859	3.906	3.952	3.998	4.044	4.091	4.138	4.184	4.230	4.277	90
100	4.277	4.324	4.371	4.418	4.465	4.512	4.559	4.606	4.654	4.701	4.749	100
110	4.749	4.796	4.843	4.891	4.939	4.987	5.035	5.083	5.131	5.179	5.227	110
120	5.227	5.275	5.323	5.372	5.420	5.469	5.518	5.566	5.615	5.663	5.712	120
130	5.712	5.761	5.810	5.859	5.908	5.957	6.007	6.056	6.105	6.155	6.204	130
140	6.204	6.254	6.303	6.353	6.403	6.453	6.503	6.553	6.603	6.653	6.703	140
150	6.703	6.753	6.803	6.853	6.904	6.954	7.004	7.055	7.106	7.157	7.208	150
160	7.208	7.258	7.309	7.360	7.411	7.462	7.513	7.565	7.616	7.667	7.719	160
170	7.719	7.770	7.822	7.874	7.926	7.978	8.029	8.080	8.132	8.184	8.236	170
180	8.236	8.288	8.340	8.392	8.445	8.497	8.549	8.601	8.654	8.707	8.759	180
190	8.759	8.812	8.864	8.917	8.970	9.023	9.076	9.129	9.182	9.235	9.288	190
200	9.288	9.341	9.394	9.448	9.501	9.555	9.608	9.662	9.715	9.769	9.823	200
°C	0	1	2	3	4	5	6	7	8	9	10	°C

TABLE 51—*Type T thermocouples (continued)*.

Temperature in Degrees Celsius

EMF in Absolute Millivolts

Reference Junctions at 0 C

° C	0	1	2	3	4	5	6	7	8	9	10	° C
						Millivolts						
200	9. 288	9. 341	9. 394	9. 448	9. 501	9. 555	9. 608	9. 662	9. 715	9. 769	9. 823	200
210	9. 823	9. 877	9. 931	9. 985	10. 039	10. 093	10. 147	10. 201	10. 255	10. 309	10. 363	210
220	10. 363	10. 417	10. 471	10. 526	10. 580	10. 635	10. 689	10. 744	10. 799	10. 854	10. 909	220
230	10. 909	10. 963	11. 018	11. 073	11. 128	11. 183	11. 238	11. 293	11. 348	11. 403	11. 459	230
240	11. 459	11. 514	11. 569	11. 624	11. 680	11. 735	11. 791	11. 847	11. 903	11. 959	12. 015	240
250	12. 015	12. 071	12. 126	12. 182	12. 238	12. 294	12. 350	12. 406	12. 462	12. 518	12. 575	250
260	12. 575	12. 631	12. 688	12. 744	12. 800	12. 857	12. 913	12. 970	13. 027	13. 083	13. 140	260
270	13. 140	13. 197	13. 254	13. 311	13. 368	13. 425	13. 482	13. 539	13. 596	13. 653	13. 710	270
280	13. 710	13. 768	13. 825	13. 882	13. 939	13. 997	14. 055	14. 112	14. 170	14. 227	14. 285	280
290	14. 285	14. 343	14. 400	14. 458	14. 515	14. 573	14. 631	14. 689	14. 747	14. 805	14. 864	290
300	14. 864	14. 922	14. 980	15. 038	15. 096	15. 155	15. 213	15. 271	15. 330	15. 388	15. 447	300
310	15. 447	15. 506	15. 564	15. 623	15. 681	15. 740	15. 799	15. 858	15. 917	15. 976	16. 035	310
320	16. 035	16. 094	16. 153	16. 212	16. 271	16. 330	16. 389	16. 449	16. 508	16. 567	16. 626	320
330	16. 626	16. 685	16. 745	16. 804	16. 864	16. 924	16. 983	17. 043	17. 102	17. 162	17. 222	330
340	17. 222	17. 281	17. 341	17. 401	17. 461	17. 521	17. 581	17. 641	17. 701	17. 761	17. 821	340
350	17. 821	17. 881	17. 941	18. 002	18. 062	18. 123	18. 183	18. 243	18. 304	18. 364	18. 425	350
360	18. 425	18. 485	18. 546	18. 607	18. 667	18. 727	18. 788	18. 849	18. 910	18. 971	19. 032	360
370	19. 032	19. 093	19. 154	19. 215	19. 276	19. 337	19. 398	19. 459	19. 520	19. 581	19. 642	370
380	19. 642	19. 704	19. 765	19. 827	19. 888	19. 949	20. 011	20. 072	20. 134	20. 195	20. 257	380
390	20. 257	20. 318	20. 380	20. 442	20. 504	20. 565	20. 627	20. 688	20. 750	20. 812	20. 874	390
° C	0	1	2	3	4	5	6	7	8	9	10	° C

TABLE 52—*Type E thermocouples.*

Temperature in Degrees Fahrenheit

EMF in Absolute Millivolts

Reference Junctions at 32 F

°F	0	10	20	30	40	50	60	70	80	90	100	°F
					Millivolts							
−300	−8.30	−8.45	−8.60									−300
−200	−6.40	−6.62	−6.83	−7.04	−7.24	−7.44	−7.62	−7.80	−7.97	−8.14	−8.30	−200
−100	−3.94	−4.21	−4.47	−4.73	−4.98	−5.23	−5.48	−5.72	−5.95	−6.18	−6.40	−100
(−)0	−1.02	−1.33	−1.64	−1.94	−2.24	−2.54	−2.83	−3.11	−3.39	−3.67	−3.94	(−)0
(+)0	−1.02	−0.71	−0.39	−0.07	0.26	0.59	0.92	1.26	1.59	1.93	2.27	(+)0
100	2.27	2.62	2.97	3.32	3.68	4.04	4.40	4.77	5.13	5.50	5.87	100
200	5.87	6.25	6.62	7.00	7.38	7.76	8.15	8.54	8.93	9.32	9.71	200
300	9.71	10.11	10.51	10.91	11.31	11.71	12.11	12.52	12.93	13.34	13.75	300
400	13.75	14.17	14.59	15.00	15.42	15.84	16.26	16.68	17.10	17.52	17.95	400
500	17.95	18.38	18.81	19.23	19.66	20.09	20.52	20.95	21.39	21.82	22.25	500
600	22.25	22.69	23.13	23.57	24.00	24.44	24.88	25.32	25.76	26.20	26.65	600
700	26.65	27.09	27.53	27.97	28.42	28.86	29.31	29.75	30.19	30.64	31.09	700
800	31.09	31.54	31.98	32.43	32.87	33.32	33.77	34.22	34.67	35.12	35.57	800
900	35.57	36.02	36.47	36.92	37.37	37.82	38.26	38.71	39.16	39.61	40.06	900
1,000	40.06	40.51	40.96	41.41	41.86	42.31	42.76	43.21	43.66	44.11	44.56	1,000
1,100	44.56	45.01	45.46	45.91	46.36	46.81	47.26	47.71	48.15	48.60	49.04	1,100
1,200	49.04	49.49	49.93	50.37	50.82	51.27	51.72	52.16	52.61	53.05	53.50	1,200
1,300	53.50	53.94	54.38	54.83	55.27	55.71	56.15	56.59	57.03	57.48	57.92	1,300
1,400	57.92	58.36	58.80	59.24	59.68	60.11	60.55	60.99	61.43	61.86	62.30	1,400
1,500	62.30	62.74	63.17	63.60	64.04	64.47	64.90	65.34	65.77	66.20	66.63	1,500
1,600	66.63	67.05	67.48	67.91	68.34	68.76	69.19	69.62	70.05	70.47	70.90	1,600
1,700	70.90	71.32	71.75	72.17	72.60	73.02	73.44	73.86	74.28	74.70	75.12	1,700
1,800	75.12	75.53	75.95	76.37								1,800
°F	0	10	20	30	40	50	60	70	80	90	100	°F

TABLE 53—*Type E thermocouples.*

Temperature in Degrees Celsius

EMF in Absolute Millivolts   Reference Junctions at 0 C

°C	0	10	20	30	40	50	60	70	80	90	100	°C
						Millivolts						
−100	−5. 18	−5. 62	−6. 04	−6. 44	−6. 83	−7. 20	−7. 55	−7. 87	−8. 17	−8. 45	−8. 71	−100
(−) 0	0. 00	−0. 58	−1. 14	−1. 70	−2. 24	−2. 77	−3. 28	−3. 78	−4. 26	−4. 73	−5. 18	(−) 0
(+) 0	0. 00	0. 59	1. 19	1. 80	2. 41	3. 04	3. 68	4. 33	4. 99	5. 65	6. 32	(+) 0
100	6. 32	7. 00	7. 69	8. 38	9. 08	9. 79	10. 51	11. 23	11. 95	12. 68	13. 42	100
200	13. 42	14. 17	14. 92	15. 67	16. 42	17. 18	17. 95	18. 72	19. 49	20. 26	21. 1	200
300	21. 04	21. 82	22. 60	23. 39	24. 18	24. 97	25. 76	26. 56	27. 35	28. 15	28 95	300
400	28. 95	29. 75	30. 55	31. 36	32. 16	32. 96	33. 77	34. 58	35. 39	36. 20	37. 01	400
500	37. 01	37. 82	38. 62	39. 43	40. 24	41. 05	41. 86	42. 67	43. 48	44. 29	45. 10	500
600	45. 10	45. 91	46. 72	47. 53	48. 33	49. 13	49. 93	50. 73	51. 54	52. 34	53. 14	600
700	53. 14	53. 94	54. 74	55. 53	56. 33	57. 12	57. 92	58. 71	59. 50	60. 29	61. 08	700
800	61. 08	61. 86	62. 65	63. 43	64. 21	64. 99	65. 77	66. 54	67. 31	68. 08	68. 85	800
900	68. 85	69. 62	70. 39	71. 15	71. 92	72. 68	73. 44	74. 20	74. 95	75. 70	76. 45	900
°C	0	10	20	30	40	50	60	70	80	90	100	°C

(See Tables 54, 55, and 56 and Section 10.3, following)

TABLE 54—*Type B thermocouples.*

Temperature in Degrees Fahrenheit

EMF in Absolute Millivolts

Reference Junctions at 32 F

°F	0	1	2	3	4	5	6	7	8	9	10	°F
						Millivolts						
30			0.000	−0.000	−0.000	−0.000	−0.000	−0.001	−0.001	−0.001	−0.001	30
40	−0.001	−0.001	−0.001	−0.001	−0.001	−0.001	−0.001	−0.002	−0.002	−0.002	−0.002	40
50	−0.002	−0.002	−0.002	−0.002	−0.002	−0.002	−0.002	−0.002	−0.002	−0.002	−0.002	50
60	−0.002	−0.002	−0.002	−0.002	−0.002	−0.002	−0.002	−0.002	−0.002	−0.002	−0.002	60
70	−0.002	−0.002	−0.002	−0.002	−0.002	−0.002	−0.002	−0.002	−0.002	−0.002	−0.002	70
80	−0.002	−0.002	−0.002	−0.002	−0.002	−0.002	−0.002	−0.002	−0.002	−0.002	−0.002	80
90	−0.002	−0.002	−0.002	−0.001	−0.001	−0.001	−0.001	−0.001	−0.001	−0.001	−0.001	90
100	−0.001	−0.001	−0.001	−0.000	−0.000	−0.000	−0.000	0.000	0.000	0.000	0.001	100
110	0.001	0.001	0.001	0.001	0.001	0.001	0.001	0.002	0.002	0.002	0.002	110
120	0.002	0.002	0.002	0.003	0.003	0.003	0.003	0.003	0.004	0.004	0.004	120
130	0.004	0.004	0.005	0.005	0.005	0.005	0.005	0.006	0.006	0.006	0.006	130
140	0.006	0.007	0.007	0.007	0.007	0.008	0.008	0.008	0.009	0.009	0.009	140
150	0.009	0.009	0.010	0.010	0.010	0.011	0.011	0.011	0.011	0.012	0.012	150
160	0.012	0.012	0.013	0.013	0.013	0.014	0.014	0.014	0.015	0.015	0.015	160
170	0.015	0.016	0.016	0.017	0.017	0.017	0.018	0.018	0.018	0.019	0.019	170
180	0.019	0.020	0.020	0.020	0.021	0.021	0.022	0.022	0.022	0.023	0.023	180
190	0.023	0.024	0.024	0.024	0.025	0.025	0.026	0.026	0.027	0.027	0.028	190
200	0.028	0.028	0.029	0.029	0.029	0.030	0.030	0.031	0.031	0.032	0.032	200
210	0.032	0.033	0.033	0.034	0.034	0.035	0.035	0.036	0.036	0.037	0.037	210
220	0.037	0.038	0.038	0.039	0.040	0.040	0.041	0.041	0.042	0.042	0.043	220
230	0.043	0.043	0.044	0.045	0.045	0.046	0.046	0.047	0.047	0.048	0.049	230
240	0.049	0.049	0.050	0.050	0.051	0.052	0.052	0.053	0.054	0.054	0.055	240
250	0.055	0.055	0.056	0.057	0.057	0.058	0.059	0.059	0.060	0.061	0.061	250
260	0.061	0.062	0.063	0.063	0.064	0.065	0.065	0.066	0.067	0.067	0.068	260
270	0.068	0.069	0.069	0.070	0.071	0.072	0.072	0.073	0.074	0.074	0.075	270
280	0.075	0.076	0.077	0.077	0.078	0.079	0.080	0.080	0.081	0.082	0.083	280
290	0.083	0.083	0.084	0.085	0.086	0.087	0.087	0.088	0.089	0.090	0.090	290
300	0.090	0.091	0.092	0.093	0.094	0.094	0.095	0.096	0.097	0.098	0.099	300
310	0.099	0.099	0.100	0.101	0.102	0.103	0.104	0.105	0.105	0.106	0.107	310
320	0.107	0.108	0.109	0.110	0.111	0.111	0.112	0.113	0.114	0.115	0.116	320
330	0.116	0.117	0.118	0.119	0.120	0.120	0.121	0.122	0.123	0.124	0.125	330
340	0.125	0.126	0.127	0.128	0.129	0.130	0.131	0.132	0.133	0.134	0.135	340
350	0.135	0.136	0.137	0.138	0.139	0.139	0.140	0.141	0.142	0.143	0.144	350
360	0.144	0.145	0.146	0.147	0.148	0.149	0.151	0.152	0.153	0.154	0.155	360
370	0.155	0.156	0.157	0.158	0.159	0.160	0.161	0.162	0.163	0.164	0.165	370
380	0.165	0.166	0.167	0.168	0.169	0.170	0.172	0.173	0.174	0.175	0.176	380
390	0.176	0.177	0.178	0.179	0.180	0.182	0.183	0.184	0.185	0.186	0.187	390
400	0.187	0.188	0.189	0.191	0.192	0.193	0.194	0.195	0.196	0.197	0.199	400
°F	0	1	2	3	4	5	6	7	8	9	10	°F

TABLE 54—*Type B thermocouples (continued).*
Temperature in Degrees Fahrenheit

EMF in Absolute Millivolts   Reference Junctions at 32 F

°F	0	1	2	3	4	5	6	7	8	9	10	°F
						Millivolts						
410	0.199	0.200	0.201	0.202	0.203	0.205	0.206	0.207	0.208	0.209	0.210	410
420	0.210	0.212	0.213	0.214	0.215	0.217	0.218	0.219	0.220	0.221	0.223	420
430	0.223	0.224	0.225	0.226	0.228	0.229	·0.230	0.231	0.233	0.234	0.235	430
440	0.235	0.236	0.238	0.239	0.240	0.241	0.243	0.244	0.245	0.247	0.248	440
450	0.248	0.249	0.251	0.252	0.253	0.254	0.256	0.257	0.258	0.260	0.261	450
460	0.261	0.262	0.264	0.265	0.266	0.268	0.269	0.270	0.272	0.273	0.275	460
470	0.275	0.276	0.277	0.279	0.280	0.281	0.283	0.284	0.286	0.287	0.288	470
480	0.288	0.290	0.291	0.293	0.294	0.295	0.297	0.298	0.300	0.301	0.302	480
490	0.302	0.304	0.305	0.307	0.308	0.310	0.311	0.313	0.314	0.315	0.317	490
500	0.317	0.318	0.320	0.321	0.323	0.324	0.326	0.327	0.329	0.330	0.332	500
510	0.332	0.333	0.335	0.336	0.338	0.339	0.341	0.342	0.344	0.345	0.347	510
520	0.347	0.348	0.350	0.351	0.353	0.354	0.356	0.358	0.359	0.361	0.362	520
530	0.362	0.364	0.365	0.367	0.368	0.370	0.372	0.373	0.375	0.376	0.378	530
540	0.378	0.380	0.381	0.383	0.384	0.386	0.388	0.389	0.391	0.392	0.394	540
550	0.394	0.396	0.397	0.399	0.401	0.402	0.404	0.405	0.407	0.409	0.410	550
560	0.410	0.412	0.414	0.415	0.417	0.419	0.420	0.422	0.424	0.425	0.427	560
570	0.427	0.429	0.431	0.432	0.434	0.436	0.437	0.439	0.441	0.442	0.444	570
580	0.444	0.446	0.448	0.449	0.451	0.453	0.455	0.456	0.458	0.460	0.462	580
590	0.462	0.463	0.465	0.467	0.469	0.470	0.472	0.474	0.476	0.477	0.479	590
600	0.479	0.481	0.483	0.485	0.486	0.488	0.490	0.492	0.494	0.495	0.497	600
610	0.497	0.499	0.501	0.503	0.504	0.506	0.508	0.510	0.512	0.514	0.515	610
620	0.515	0.517	0.519	0.521	0.523	0.525	0.527	0.528	0.530	0.532	0.534	620
630	0.534	0.536	0.538	0.540	0.542	0.543	0.545	0.547	0.549	0.551	0.553	630
640	0.553	0.555	0.557	0.559	0.561	0.563	0.563	0.566	0.568	0.570	0.572	640
650	0.572	0.574	0.576	0.578	0.580	0.582	0.584	0.586	0.588	0.590	0.592	650
660	0.592	0.594	0.596	0.598	0.600	0.602	0.604	0.606	0.608	0.610	0.612	660
670	0.612	0.614	0.616	0.618	0.620	0.622	0.624	0.626	0.628	0.630	0.632	670
680	0.632	0.634	0.636	0.638	0.640	0.642	0.644	0.646	0.648	0.650	0.652	680
690	0.652	0.654	0.656	0.659	0.661	0.663	0.665	0.667	0.669	0.671	0.673	690
700	0.673	0.675	0.677	0.679	0.682	0.684	0.686	0.688	0.690	0.692	0.694	700
710	0.694	0.696	0.698	0.701	0.703	0.705	0.707	0.709	0.711	0.714	0.716	710
720	0.716	0.718	0.720	0.722	0.724	0.726	0.729	0.731	0.733	0.735	0.737	720
730	0.737	0.740	0.742	0.744	0.746	0.748	0.751	0.753	0.755	0.757	0.759	730
740	0.759	0.762	0.764	0.766	0.768	0.771	0.773	0.755	0.777	0.780	0.782	740
750	0.782	0.784	0.786	0.789	0.791	0.793	0.795	0.798	0.800	0.802	0.804	750
760	0.804	0.807	0.809	0.811	0.814	0.816	0.818	0.820	0.823	0.825	0.827	760
770	0.827	0.830	0.832	0.834	0.837	0.839	0.841	0.844	0.846	0.848	0.851	770
780	0.851	0.853	0.855	0.858	0.860	0.862	0.865	0.867	0.869	0.872	0.874	780
790	0.874	0.877	0.879	0.881	0.884	0.886	0.888	0.891	0.893	0.896	0.898	790
°F	0	1	2	3	4	5	6	7	8	9	10	°F

TABLE 54—*Type B thermocouples (continued).*
Temperature in Degrees Fahrenheit

EMF in Absolute Millivolts   Reference Junctions at 32 F

°F	0	1	2	3	4	5	6	7	8	9	10	°F
						Millivolts						
800	0.898	0.900	0.903	0.905	0.908	0.910	0.913	0.915	0.917	0.920	0.922	800
810	0.922	0.925	0.927	0.930	0.932	0.934	0.937	0.939	0.942	0.944	0.947	810
820	0.947	0.949	0.952	0.954	0.957	0.959	0.962	0.964	0.967	0.969	0.972	820
830	0.972	0.974	0.976	0.979	0.982	0.984	0.987	0.989	0.992	0.994	0.997	830
840	0.997	0.999	1.002	1.004	1.007	1.009	1.012	1.014	1.017	1.019	1.022	840
850	1.022	1.025	1.027	1.030	1.032	1.035	1.037	1.040	1.042	1.045	1.048	850
860	1.048	1.050	1.053	1.055	1.058	1.061	1.063	1.066	1.068	1.071	1.074	860
870	1.074	1.076	1.079	1.081	1.084	1.087	1.089	1.092	1.095	1.097	1.100	870
880	1.100	1.103	1.105	1.108	1.110	1.113	1.116	1.118	1.121	1.124	1.126	880
890	1.126	1.129	1.132	1.134	1.137	1.140	1.143	1.145	1.148	1.151	1.153	890
900	1.153	1.156	1.159	1.161	1.164	1.167	1.170	1.172	1.175	2.178	1.180	900
910	1.180	1.183	1.186	1.189	1.191	1.194	1.197	1.200	1.202	1.205	1.208	910
920	1.208	1.211	1.213	1.216	1.219	1.222	1.255	1.227	1.230	1.233	1.236	920
930	1.236	1.238	1.241	1.244	1.247	1.250	1.252	1.255	1.258	1.261	1.264	930
940	1.264	1.267	1.269	1.272	1.275	1.278	1.281	1.284	1.286	1.289	1.292	940
950	1.292	1.295	1.298	1.301	1.303	1.306	1.309	1.312	1.315	1.318	1.321	950
960	1.321	1.324	1.326	1.329	1.332	1.335	1.338	1.341	1.344	1.347	1.350	960
970	1.350	1.353	1.355	1.358	1.361	1.364	1.367	1.370	1.373	1.376	1.379	970
980	1.379	1.382	1.385	1.388	1.391	1.394	1.396	1.399	1.402	1.405	1.408	980
990	1.408	1.411	1.414	1.417	1.420	1.423	1.426	1.429	1.432	1.435	1.438	990
1000	1.438	1.441	1.444	1.447	1.450	1.453	1.456	1.459	1.462	1.465	1.468	1000
1010	1.468	1.471	1.474	1.477	1.480	1.483	1.486	1.489	1.492	1.495	1.499	1010
1020	1.499	1.502	1.505	1.508	1.511	1.514	1.517	1.520	1.523	1.526	1.529	1020
1030	1.529	1.532	1.535	1.538	1.541	1.545	1.548	1.551	1.554	1.557	1.560	1030
1040	1.560	1.563	1.566	1.569	1.573	1.576	1.579	1.582	1.585	1.588	1.591	1040
1050	1.591	1.594	1.598	1.601	1.604	1.607	1.610	1.613	1.616	1.620	1.623	1050
1060	1.623	1.626	1.629	1.632	1.635	1.639	1.642	1.645	1.648	1.651	1.655	1060
1070	1.655	1.658	1.661	1.664	1.667	1.671	1.674	1.677	1.680	1.683	1.687	1070
1080	1.687	1.690	1.693	1.696	1.700	1.703	1.706	1.709	1.712	1.716	1.719	1080
1090	1.719	1.722	1.725	1.729	1.732	1.735	1.739	1.742	1.745	1.748	1.752	1090
1100	1.752	1.755	1.758	1.761	1.765	1.768	1.771	1.775	1.778	1.781	1.785	1100
1110	1.785	1.788	1.791	1.794	1.798	1.801	1.804	1.808	1.811	1.814	1.818	1110
1120	1.818	1.821	1.824	1.828	1.831	1.834	1.838	1.841	1.844	1.848	1.851	1120
1130	1.851	1.855	1.858	1.861	1.865	1.868	1.871	1.875	1.878	1.882	1.885	1130
1140	1.885	1.888	1.892	1.895	1.898	1.902	1.905	1.909	1.912	1.916	1.919	1140
1150	1.919	1.922	1.926	1.929	1.933	1.936	1.939	1.943	1.946	1.950	1.953	1150
1160	1.953	1.957	1.960	1.964	1.967	1.970	1.974	1.977	1.981	1.984	1.988	1160
1170	1.988	1.991	1.995	1.998	2.002	2.005	2.009	2.012	2.016	2.019	2.023	1170
1180	2.023	2.026	2.030	2.033	2.037	2.040	2.044	2.047	2.051	2.054	2.058	1180
1190	2.058	2.061	2.065	2.068	2.072	2.075	2.079	2.083	2.086	2.090	2.093	1190
°F	0	1	2	3	4	5	6	7	8	9	10	°F

TABLE 54— *Type B thermocouples (continued).*
Temperature in Degrees Fahrenheit

EMF in Absolute Millivolts

Reference Junctions at 32 F

°F	0	1	2	3	4	5	6	7	8	9	10	°F
						Millivolts						
1200	2.093	2.097	2.100	2.104	2.107	2.111	2.115	2.118	2.122	2.125	2.129	1200
1210	2.129	2.132	2.136	2.140	2.143	2.147	2.150	2.154	2.158	2.161	2.165	1210
1220	2.165	2.168	2.172	2.176	2.179	2.183	2.186	2.190	2.194	2.197	2.201	1220
1230	2.201	2.205	2.208	2.212	2.216	2.219	2.223	2.226	2.230	2.234	2.237	1230
1240	2.237	2.241	2.245	2.248	2.252	2.256	2.259	2.263	2.267	2.271	2.274	1240
1250	2.274	2.278	2.282	2.285	2.289	2.293	2.296	2.300	2.304	2.308	2.311	1250
1260	2.311	2.315	2.319	2.322	2.326	2.330	2.334	2.337	2.341	2.345	2.349	1260
1270	2.349	2.352	2.356	2.360	2.363	2.367	2.371	2.375	2.379	2.382	2.386	1270
1280	2.386	2.390	2.394	2.397	2.401	2.405	2.409	2.412	2.416	2.420	2.424	1280
1290	2.424	2.428	2.431	2.435	2.439	2.443	2.447	2.450	2.454	2.458	2.462	1290
1300	2.462	2.466	2.470	2.473	2.477	2.481	2.485	2.489	2.493	2.496	2.500	1300
1310	2.500	2.504	2.508	2.512	2.516	2.520	2.523	2.527	2.531	2.535	2.539	1310
1320	2.539	2.543	2.547	2.551	2.554	2.558	2.562	2.566	2.570	2.574	2.578	1320
1330	2.578	2.582	2.536	2.589	2.593	2.597	2.601	2.605	2.609	2.613	2.617	1330
1340	2.617	2.621	2.625	2.629	2.633	2.637	2.641	2.644	2.648	2.652	2.656	1340
1350	2.656	2.660	2.664	2.663	2.672	2.676	2.680	2.684	2.688	2.692	2.696	1350
1360	2.696	2.700	2.704	2.703	2.712	2.716	2.720	2.724	2.728	2.732	2.736	1360
1370	2.736	2.740	2.744	2.748	2.752	2.756	2.760	2.764	2.768	2.772	2.776	1370
1380	2.776	2.780	2.784	2.788	2.792	2.796	2.800	2.804	2.808	2.812	2.816	1380
1390	2.816	2.821	2.825	2.829	2.833	2.837	2.841	2.845	2.849	2.853	2.857	1390
1400	2.857	2.861	2.865	2.869	2.874	2.878	2.882	2.886	2.890	2.894	2.898	1400
1410	2.898	2.902	2.906	2.910	2.915	2.919	2.923	2.927	2.931	2.935	2.939	1410
1420	2.939	2.943	2.948	2.952	2.956	2.960	2.964	2.968	2.972	2.977	2.981	1420
1430	2.981	2.985	2.989	2.993	2.997	3.002	3.006	3.010	3.014	3.018	3.023	1430
1440	3.023	3.027	3.031	3.035	3.039	3.043	3.048	3.052	3.056	3.060	3.064	1440
1450	3.064	3.069	3.073	3.077	3.081	3.086	3.090	3.094	3.098	3.102	3.107	1450
1460	3.107	3.111	3.115	3.119	3.124	3.128	3.132	3.136	3.141	3.145	3.149	1460
1470	3.149	3.153	3.158	3.162	3.166	3.170	3.175	3.179	3.183	3.188	3.192	1470
1480	3.192	3.196	3.200	3.205	3.209	3.213	3.218	3.222	3.226	3.231	3.235	1480
1490	3.235	3.239	3.243	3.248	3.252	3.256	3.261	3.265	3.269	3.274	3.278	1490
1500	3.278	3.282	3.287	3.291	3.295	3.300	3.304	3.308	3.313	3.317	3.321	1500
1510	3.321	3.326	3.330	3.335	3.339	3.343	3.348	3.352	3.356	3.361	3.365	1510
1520	3.365	3.369	3.374	3.378	3.383	3.387	3.391	3.396	3.400	3.405	3.409	1520
1530	3.409	3.413	3.418	3.422	3.427	3.431	3.435	3.440	3.444	3.449	3.453	1530
1540	3.453	3.458	3.462	3.466	3.471	3.475	3.480	3.484	3.489	3.493	3.497	1540
1560	3.497	3.502	3.506	3.511	3.515	3.520	3.524	3.529	3.533	3.538	3.542	1550
1560	3.542	3.547	3.551	3.556	3.560	3.564	3.569	3.573	3.578	3.582	3.587	1560
1570	3.587	3.591	3.596	3.600	3.605	3.609	3.614	3.618	3.623	3.627	3.632	1570
1580	3.632	3.637	3.641	3.646	3.650	3.655	3.659	3.664	3.668	3.673	3.677	1580
1590	3.677	3.682	3.686	3.691	3.696	3.700	3.705	3.709	3.714	3.718	3.723	1590
°F	0	1	2	3	4	5	6	7	8	9	10	°F

TABLE 54—*Type B thermocouples (continued).*

Temperature in Degrees Fahrenheit

EMF in Absoute Millivolts   Reference Junctons at 32 F

°F	0	1	2	3	4	5	6	7	8	9	10	°F
						Millivolts						
1600	3.723	3.727	3.732	3.737	3.741	3.746	3.750	3.755	3.759	3.764	3.769	1600
1610	3.769	3.773	3.778	3.782	3.787	3.792	3.796	3.801	3.805	3.810	3.815	1610
1620	3.815	3.819	3.824	3.829	3.833	3.838	3.842	3.847	3.852	3.856	3.861	1620
1630	3.861	3.866	3.870	3.875	3.879	3.884	3.889	3.893	3.898	3.903	3.907	1630
1640	3.907	3.912	3.917	3.921	3.926	3.931	3.935	3.940	3.945	3.949	3.954	1640
1650	3.954	3.959	3.963	3.968	3.973	3.978	3.982	3.987	3.992	3.996	4.001	1650
1660	4.001	4.006	4.010	4.015	4.020	4.025	4.029	4.034	4.039	4.044	4.048	1660
1670	4.048	4.053	4.058	4.062	4.067	4.072	4.077	4.081	4.086	4.091	4.096	1670
1680	4.096	4.100	4.105	4.110	4.115	4.119	4.124	4.129	4.134	4.139	4.143	1680
1690	4.143	4.148	4.153	4.158	4.162	4.167	4.172	4.177	4.182	4.186	4.191	1690
1700	4.191	4.196	4.201	4.206	4.210	4.215	4.220	4.225	4.230	4.234	4.239	1700
1710	4.239	4.244	4.249	4.254	4.259	4.263	4.268	4.273	4.278	4.283	4.288	1710
1720	4.288	4.292	4.297	4.302	4.307	4.312	4.317	4.322	4.326	4.331	4.336	1720
1730	4.336	4.341	4.346	4.351	4.356	4.360	4.365	4.370	4.375	4.380	4.385	1730
1740	4.385	4.390	4.395	4.400	4.404	4.409	4.414	4.419	4.424	4.429	4.434	1740
1750	4.434	4.439	4.444	4.449	4.454	4.458	4.463	4.468	4.473	4.478	4.483	1750
1760	4.483	4.488	4.493	4.498	4.503	4.508	4.513	4.518	4.523	4.528	4.533	1760
1770	4.533	4.538	4.543	4.547	4.552	4.557	4.562	4.567	4.572	4.577	4.582	1770
1780	4.582	4.587	4.592	4.597	4.602	4.607	4.612	4.617	4.622	4.627	4.632	1780
1790	4.632	4.637	4.642	4.647	4.652	4.657	4.662	4.667	4.672	4.677	4.682	1790
1800	4.682	4.687	4.692	4.697	4.702	4.707	4.712	4.717	4.722	4.728	4.733	1800
1810	4.733	4.738	4.743	4.748	4.753	4.758	4.763	4.768	4.773	4.778	4.783	1810
1820	4.783	4.788	4.793	4.798	4.803	4.808	4.814	4.819	4.824	4.829	4.834	1820
1830	4.834	4.839	4.844	4.849	4.854	4.859	4.864	4.870	4.875	4.880	4.885	1830
1840	4.885	4.890	4.895	4.900	4.905	4.910	4.916	4.921	4.926	4.931	4.936	1840
1850	4.936	4.941	4.946	4.951	4.957	4.962	4.967	4.972	4.977	4.982	4.987	1850
1860	4.987	4.993	4.998	5.003	5.008	5.013	5.018	5.024	5.029	5.034	5.039	1860
1870	5.039	5.044	5.049	5.055	5.060	5.065	5.070	5.075	5.080	5.086	5.091	1870
1880	5.091	5.096	5.101	5.106	5.112	5.117	5.122	5.127	5.132	5.138	5.143	1880
1890	5.143	5.148	5.153	5.158	5.164	5.169	5.174	5.179	5.185	5.190	5.195	1890
1900	5.195	5.200	5.206	5.211	5.216	5.221	5.226	5.232	5.237	5.242	5.247	1900
1910	5.247	5.253	5.258	5.263	5.268	5.274	5.279	5.284	5.290	5.295	5.300	1910
1920	5.300	5.305	5.311	5.316	5.321	5.326	5.332	5.337	5.342	5.348	5.353	1920
1930	5.353	5.358	5.363	5.369	5.374	5.379	5.385	5.390	5.395	5.401	5.406	1930
1940	5.406	5.411	5.417	5.422	5.427	5.432	5.438	5.443	5.448	5.454	5.459	1940
1950	5.459	5.464	5.470	5.475	5.480	5.486	5.491	5.496	5.502	5.507	5.512	1950
1960	5.512	5.518	5.523	5.529	5.534	5.539	5.545	5.550	5.555	5.561	5.566	1960
1970	5.566	5.571	5.577	5.582	5.588	5.593	5.598	5.604	5.609	5.614	5.620	1970
1980	5.620	5.625	5.631	5.636	5.641	5.647	5.652	5.658	5.663	5.668	5.674	1980
1990	5.674	5.679	5.685	5.690	5.695	5.701	5.706	5.712	5.717	5.723	5.728	1990
°F	0	1	2	3	4	5	6	7	8	9	10	°F

TABLE 54—*Type B thermocouples (continued)*.
Temperature in Degrees Fahrenheit

EMF in Absolute Millivolts   Reference Junctions at 32 F

°F	0	1	2	3	4	5	6	7	8	9	10	°F
						Millivolts						
2000	5.728	5.733	5.739	5.744	5.750	5.755	5.761	5.766	5.771	5.777	5.782	2000
2010	5.782	5.788	5.793	5.799	5.804	5.809	5.815	5.820	5.826	5.831	5.837	2010
2020	5.837	5.842	5.848	5.853	5.859	5.864	5.870	5.875	5.880	5.886	5.891	2020
2030	5.891	5.897	5.902	5.908	5.913	5.919	5.924	5.930	5.935	5.941	5.946	2030
2040	5.946	5.952	5.957	5.963	5.968	5.974	5.979	5.985	5.990	5.996	6.001	2040
2050	6.001	6.007	6.012	6.018	6.023	6.029	6.034	6.040	6.045	6.051	6.056	2050
2060	6.056	6.062	6.068	6.073	6.079	6.084	6.090	6.095	6.101	6.106	6.112	2060
2070	6.112	6.117	6.123	6.128	6.134	6.140	6.145	6.151	6.156	6.162	6.167	2070
2080	6.167	6.173	6.178	6.184	6.190	6.195	6.201	6.206	6.212	6.217	6.223	2080
2090	6.223	6.229	6.234	6.240	6.245	6.251	6.257	6.262	6.268	6.273	6.279	2090
2100	6.279	6.284	6.290	6.296	6.301	6.307	6.312	6.318	6.324	6.329	6.335	2100
2110	6.335	6.340	6.346	6.352	6.357	6.363	6.369	6.374	6.380	6.385	6.391	2110
2120	6.391	6.397	6.402	6.408	6.414	6.419	6.425	6.430	6.436	6.442	6.447	2120
2130	6.447	6.453	6.459	6.464	6.470	6.476	6.481	6.487	6.492	6.498	6.504	2130
2140	6.504	6.509	6.515	6.521	6.526	6.532	6.538	6.543	6.549	6.555	6.560	2140
2150	6.560	6.566	6.572	6.577	6.583	6.589	6.594	6.600	6.606	6.611	6.617	2150
2160	6.617	6.623	6.629	6.634	6.640	6.646	6.651	6.657	6.663	6.668	6.674	2160
2170	6.674	6.680	6.686	6.691	6.697	6.703	6.708	6.714	6.720	6.726	6.731	2170
2180	6.731	6.737	6.743	6.749	6.754	6.760	6.766	6.772	6.777	6.783	6.789	2180
2190	6.789	6.795	6.800	6.806	6.812	6.818	6.823	6.829	6.835	6.841	6.846	2190
2200	6.846	6.852	6.858	6.864	6.870	6.875	6.881	6.887	6.893	6.898	6.904	2200
2210	6.904	6.910	6.916	6.922	6.927	6.933	6.939	6.945	6.951	6.956	6.962	2210
2220	6.962	6.968	6.974	6.980	6.986	6.991	6.997	7.003	7.009	7.015	7.020	2220
2230	7.020	7.026	7.032	7.038	7.044	7.050	7.055	7.061	7.067	7.073	7.079	2230
2240	7.079	7.085	7.091	7.096	7.102	7.108	7.114	7.120	7.126	7.132	7.137	2240
2250	7.137	7.143	7.149	7.155	7.161	7.167	7.173	7.179	7.184	7.190	7.196	2250
2260	7.196	7.202	7.208	7.214	7.220	7.226	7.232	7.237	7.243	7.249	7.255	2260
2270	7.255	7.261	7.267	7.273	7.279	7.285	7.291	7.297	7.302	7.308	7.314	2270
2280	7.314	7.320	7.326	7.332	7.338	7.344	7.350	7.356	7.362	7.368	7.374	2280
2290	7.374	7.380	7.386	7.391	7.397	7.403	7.409	7.415	7.421	7.427	7.433	2290
2300	7.433	7.439	7.445	7.451	7.457	7.463	7.469	7.475	7.481	7.487	7.493	2300
2310	7.493	7.499	7.505	7.511	7.517	7.523	7.529	7.535	7.541	7.547	7.553	2310
2320	7.553	7.559	7.565	7.571	7.577	7.583	7.589	7.595	7.601	7.607	7.613	2320
2330	7.613	7.619	7.625	7.631	7.637	7.643	7.649	7.655	7.661	7.667	7.673	2330
2340	7.673	7.679	7.685	7.691	7.697	7.703	7.709	7.715	7.721	7.727	7.733	2340
2350	7.733	7.739	7.745	7.751	7.757	7.763	7.769	7.775	7.781	7.787	7.793	2350
2360	7.793	7.799	7.806	7.812	7.818	7.824	7.830	7.836	7.842	7.848	7.854	2360
2370	7.854	7.860	7.866	7.872	7.878	7.884	7.890	7.896	7.903	7.909	7.915	2370
2380	7.915	7.921	7.927	7.933	7.939	7.945	7.951	7.957	7.963	7.970	7.976	2380
2390	7.976	7.982	7.988	7.994	8.000	8.006	8.012	8.018	8.024	8.031	8.037	2390
F	0	1	2	3	4	5	6	7	8	9	10	°F

TABLE 54—*Type B thermocouples (continued)*.
Temperature in Degrees Fahrenheit

EMF in Absolute Millivolts   Reference Junctions at 32 F

°F	0	1	2	3	4	5	6	7	8	9	10	°F
					Millivolts							
2400	8.037	8.043	8.049	8.055	8.061	8.067	8.073	8.079	8.086	8.092	8.098	2400
2410	8.098	8.104	8.110	8.116	8.122	8.128	8.135	8.141	8.147	8.153	8.159	2410
2420	8.159	8.165	8.171	8.177	8.184	8.190	8.196	8.202	8.208	8.214	8.221	2420
2430	8.221	8.227	8.233	8.239	8.245	8.251	8.257	8.264	8.270	8.276	8.282	2430
2440	8.282	8.288	8.294	8.301	8.307	8.313	8.319	8.325	8.331	8.338	8.344	2440
2450	8.344	8.350	8.356	8.362	8.368	8.375	8.381	8.387	8.393	8.399	8.406	2450
2460	8.406	8.412	8.418	8.424	8.430	8.436	8.443	8.449	8.455	8.461	8.467	2460
2470	8.467	8.474	8.480	8.486	8.492	8.498	8.505	8.511	8.517	8.523	8.530	2470
2480	8.530	8.536	8.542	8.548	8.554	8.561	8.567	8.573	8.579	8.585	8.592	2480
2490	8.592	8.598	8.604	8.610	8.617	8.623	8.629	8.635	8.641	8.648	8.654	2490
2500	8.654	8.660	8.666	8.673	8.679	8.685	8.691	8.698	8.704	8.710	8.716	2500
2510	8.716	8.723	8.729	8.735	8.741	8.748	8.754	8.760	8.766	8.773	8.779	2510
2520	8.779	8.785	8.791	8.798	8.804	8.810	8.816	8.823	8.829	8.835	8.841	2520
2530	8.841	8.848	8.854	8.860	8.866	8.873	8.879	8.885	8.892	8.898	8.904	2530
2540	8.904	8.910	8.917	8.923	8.929	8.935	8.942	8.948	8.954	8.961	8.967	2540
2550	8.967	8.973	8.979	8.986	8.992	8.998	9.005	9.011	9.017	9.023	9.030	2550
2560	9.030	9.036	9.042	9.049	9.055	9.061	9.068	9.074	9.080	9.086	9.093	2560
2570	9.093	9.099	9.105	9.112	9.118	9.124	9.131	9.137	9.143	9.150	9.156	2570
2580	9.156	9.162	9.168	9.175	9.181	9.187	9.194	9.200	9.206	9.213	9.219	2580
2590	9.219	9.225	9.232	9.238	9.244	9.251	9.257	9.263	9.270	9.276	9.282	2590
2600	9.282	9.289	9.295	9.301	9.308	9.314	9.320	9.327	9.333	9.339	9.346	2600
2610	9.346	9.352	9.358	9.365	9.371	9.377	9.384	9.390	9.396	9.403	9.409	2610
2620	9.409	9.415	9.422	9.428	9.434	9.441	9.447	9.453	9.460	9.466	9.473	2620
2630	9.473	9.479	9.485	9.492	9.498	9.504	9.511	9.517	9.523	9.530	9.536	2630
2640	9.536	9.542	9.549	9.555	9.562	9.568	9.574	9.581	9.587	9.593	9.600	2640
2650	9.600	9.606	9.612	9.619	9.625	9.632	9.638	9.644	9.651	9.657	9.663	2650
2660	9.663	9.670	9.676	9.683	9.689	9.695	9.702	9.708	9.714	9.721	9.727	2660
2670	9.727	9.734	9.740	9.746	9.753	9.759	9.766	9.772	9.778	9.785	9.791	2670
2680	9.791	9.797	9.804	9.810	9.817	9.823	9.829	9.836	9.842	9.849	9.855	2680
2690	9.855	9.861	9.868	9.874	9.881	9.887	9.893	9.900	9.906	9.913	9.919	2690
2700	9.919	9.925	9.932	9.938	9.945	9.951	9.957	9.964	9.970	9.977	9.983	2700
2710	9.983	9.989	9.996	10.002	10.009	10.015	10.021	10.028	10.034	10.041	10.047	2710
2720	10.047	10.054	10.060	10.066	10.073	10.079	10.086	10.092	10.098	10.105	10.111	2720
2730	10.111	10.118	10.124	10.131	10.137	10.143	10.150	10.156	10.163	10.169	10.175	2730
2740	10.175	10.182	10.188	10.195	10.201	10.208	10.214	10.220	10.227	10.233	10.240	2740
2750	10.240	10.246	10.253	10.259	10.265	10.272	10.278	10.285	10.291	10.298	10.304	2750
2760	10.304	10.310	10.317	10.323	10.330	10.336	10.343	10.349	10.355	10.362	10.368	2760
2770	10.368	10.375	10.381	10.388	10.394	10.401	10.407	10.413	10.420	10.426	10.433	2770
2780	10.433	10.439	10.446	10.452	10.459	10.465	10.471	10.478	10.484	10.491	10.497	2780
2790	10.497	10.504	10.510	10.517	10.523	10.529	10.536	10.542	10.549	10.555	10.562	2790
°F	0	1	2	3	4	5	6	7	8	9	10	°F

TABLE 54—*Type B thermocouples (continued)*.
Temperature in Degrees Fahrenheit

EMF in Absolute Millivolts   Reference Junctions at 32 F

°F	0	1	2	3	4	5	6	7	8	9	10	°F
						Millivolts						
2800	10.562	10.568	10.575	10.581	10.587	10.594	10.600	10.607	10.613	10.620	10.626	2800
2810	10.626	10.633	10.639	10.646	10.652	10.658	10.665	10.671	10.678	10.684	10.691	2810
2820	10.691	10.697	10.704	10.710	10.717	10.723	10.729	10.736	10.742	10.749	10.755	2820
2830	10.755	10.762	10.768	10.775	10.781	10.788	10.794	10.801	10.807	10.813	10.820	2830
2840	10.820	10.826	10.833	10.839	10.846	10.852	10.859	10.865	10.872	10.878	10.885	2840
2850	10.885	10.891	10.898	10.904	10.910	10.917	10.923	10.930	10.936	10.943	10.949	2850
2860	10.949	10.956	10.962	10.969	10.975	10.982	10.988	10.995	11.001	11.007	11.014	2860
2870	11.014	11.020	11.027	11.033	11.040	11.046	11.053	11.059	11.066	11.072	11.079	2870
2880	11.079	11.085	11.092	11.098	11.105	11.111	11.118	11.124	11.130	11.137	11.143	2880
2890	11.143	11.150	11.156	11.163	11.169	11.176	11.182	11.189	11.195	11.202	11.203	2890
2900	11.203	11.215	11.221	11.226	11.234	11.241	11.247	11.254	11.260	11.266	11.273	2900
2910	11.273	11.279	11.286	11.292	11.299	11.305	11.312	11.318	11.325	11.331	11.338	2910
2920	11.338	11.344	11.351	11.357	11.364	11.370	11.377	11.383	11.390	11.396	11.403	2920
2930	11.403	11.409	11.416	11.422	11.429	11.435	11.441	11.448	11.454	11.461	11.467	2930
2940	11.467	11.474	11.480	11.487	11.493	11.500	11.506	11.513	11.519	11.526	11.532	2940
2950	11.532	11.539	11.545	11.552	11.558	11.565	11.571	11.578	11.584	11.591	11.597	2950
2960	11.597	11.604	11.610	11.617	11.623	11.630	11.636	11.642	11.649	11.655	11.662	2960
2970	11.662	11.668	11.675	11.681	11.688	11.694	11.701	11.707	11.714	11.720	11.727	2970
2980	11.727	11.733	11.740	11.746	11.753	11.759	11.766	11.772	11.779	11.785	11.792	2980
2990	11.792	11.798	11.805	11.811	11.818	11.824	11.831	11.837	11.844	11.850	11.857	2990
3000	11.857	11.863	11.869	11.876	11.882	11.889	11.895	11.902	11.903	11.915	11.921	3000
3010	11.921	11.928	11.934	11.941	11.947	11.954	11.960	11.967	11.973	11.980	11.986	3010
3020	11.986	11.993	11.999	12.006	12.012	12.019	12.025	12.032	12.038	12.045	12.051	3020
3030	12.051	12.058	12.064	12.071	12.077	12.084	12.090	12.096	12.103	12.109	12.116	3030
3040	12.116	12.122	12.129	12.135	12.142	12.148	12.155	12.161	12.168	12.174	12.181	3040
3050	12.181	12.187	12.194	12.200	12.207	12.213	12.220	12.226	12.233	12.239	12.246	3050
3060	12.246	12.252	12.259	12.265	12.272	12.278	12.285	12.291	12.297	12.304	12.310	3060
3070	12.310	12.317	12.323	12.330	12.336	12.343	12.349	12.356	12.362	12.369	12.375	3070
3080	12.375	12.382	12.388	12.395	12.401	12.408	12.414	12.421	12.427	12.434	12.440	3080
3090	12.440	12.447	12.453	12.459	12.466	12.472	12.479	12.485	12.492	12.498	12.505	3090
3100	12.505	12.511	12.518	12.524	12.531	12.537	12.544	12.550	12.557	12.563	12.570	3100
3110	12.570	12.576	12.583	12.589	12.596	12.602	12.608	12.615	12.621	12.628	12.634	3110
3120	12.634	12.641	12.647	12.654	12.660	12.667	12.673	12.680	12.686	12.693	12.699	3120
3130	12.699	12.706	12.712	12.719	12.725	12.731	12.738	12.744	12.751	12.757	12.764	3130
3140	12.764	12.770	12.777	12.783	12.790	12.796	12.803	12.809	12.816	12.822	12.829	3140
3150	12.829	12.835	12.841	12.848	12.854	12.861	12.867	12.874	12.880	12.887	12.893	3150
3160	12.893	12.900	12.906	12.913	12.919	12.926	12.932	12.938	12.945	12.951	12.958	3160
3170	12.958	12.964	12.971	12.977	12.984	12.990	12.997	13.003	13.010	13.016	13.022	3170
3180	13.022	13.029	13.035	13.042	13.048	13.055	13.061	13.068	13.074	13.081	13.087	3180
3190	13.087	13.093	13.100	13.106	13.113	13.119	13.126	13.132	13.139	13.145	13.152	3190
°F	0	1	2	3	4	5	6	7	8	9	10	°F

TABLE 54—*Type B thermocouples (continued)*.
Temperature in Degrees Fahrenheit

EMF in Absolute Millivolts   Reference Junctions at 32 F

°F	0	1	2	3	4	5	6	7	8	9	10	°F
						Millivolts						
3200	13.152	13.158	13.165	13.171	13.177	13.184	13.190	13.197	13.203	13.210	13.216	3200
3210	13.216	13.223	13.229	13.235	13.242	13.248	13.255	13.261	13.268	13.274	13.281	3210
3220	13.281	13.287	13.294	13.300	13.306	13.313	13.319	13.326	13.332	13.339	13.345	3220
3230	13.345	13.352	13.358	13.364	13.371	13.377	13.384	13.390	13.397	13.403	13.410	3230
3240	13.410	13.416	13.422	13.429	13.435	13.442	13.448	13.455	13.461	13.468	13.474	3240
3250	13.474	13.480	13.487	13.493	13.500	13.506	13.513	13.519	13.525	13.532	13.538	3250
3260	13.538	13.545	13.551	13.558	13.564	13.570	13.577	13.583	13.590	13.596	13.603	3260
3270	13.603	13.609	13.615	13.622	13.628	13.635	13.641	13.648	13.654	13.660	13.667	3270
3280	13.667	13.673	13.680	13.686	13.693	13.699	13.705	13.712	13.718	13.725	13.731	3280
3290	13.731	13.738	13.744	13.750	13.757	13.763	13.770	13.776	13.783	13.789	13.795	3290
3300	13.795	13.802	13.808	13.815	13.821	13.827	13.834	13.840	13.847			3300
°F	0	1	2	3	4	5	6	7	8	9	10	°F

TABLE 55—*Type B thermocouples.*

Temperature in Degrees Celsius

EMF in Absolute Millivolts   Reference Junctions at 0 C

°C	0	1	2	3	4	5	6	7	8	9	10	°C
						Millivolts						
0	-0.	-0.000	-0.000	-0.001	-0.001	-0.001	-0.001	-0.001	-0.002	-0.002	-0.002	0
10	-0.002	-0.002	-0.002	-0.002	-0.002	-0.002	-0.002	-0.002	-0.002	-0.002	-0.002	10
20	-0.002	-0.002	-0.002	-0.002	-0.002	-0.002	-0.002	-0.002	-0.002	-0.002	-0.002	20
30	-0.002	-0.002	-0.002	-0.002	-0.001	-0.001	-0.001	-0.001	-0.001	-0.001	-0.000	30
40	-0.000	-0.000	0.000	0.000	0.001	0.001	0.001	0.002	0.002	0.002	0.002	40
50	0.002	0.003	0.003	0.004	0.004	0.004	0.005	0.005	0.006	0.006	0.006	50
60	0.006	0.007	0.007	0.008	0.008	0.009	0.009	0.010	0.010	0.011	0.011	60
70	0.011	0.012	0.013	0.013	0.014	0.014	0.015	0.016	0.016	0.017	0.018	70
80	0.018	0.018	0.019	0.020	0.020	0.021	0.022	0.023	0.023	0.024	0.025	80
90	0.025	0.026	0.027	0.027	0.028	0.029	0.030	0.031	0.032	0.032	0.033	90
100	0.033	0.034	0.035	0.036	0.037	0.038	0.039	0.040	0.041	0.042	0.043	100
110	0.043	0.044	0.045	0.046	0.047	0.048	0.049	0.050	0.051	0.052	0.054	110
120	0.054	0.055	0.056	0.057	0.058	0.059	0.060	0.062	0.063	0.064	0.065	120
130	0.065	0.066	0.068	0.069	0.070	0.072	0.073	0.074	0.075	0.077	0.078	130
140	0.078	0.079	0.081	0.082	0.084	0.085	0.086	0.088	0.089	0.091	0.092	140
150	0.092	0.094	0.095	0.096	0.098	0.099	0.101	0.102	0.104	0.106	0.107	150
160	0.107	0.109	0.110	0.112	0.113	0.115	0.117	0.118	0.120	0.122	0.123	160
170	0.123	0.125	0.127	0.128	0.130	0.132	0.133	0.135	0.137	0.139	0.140	170
180	0.140	0.142	0.144	0.146	0.148	0.149	0.151	0.153	0.155	0.157	0.159	180
190	0.159	0.161	0.163	0.164	0.166	0.168	0.170	0.172	0.174	0.176	0.178	190
200	0.178	0.180	0.182	0.184	0.186	0.188	0.190	0.192	0.194	0.197	0.199	200
210	0.199	0.201	0.203	0.205	0.207	0.209	0.211	0.214	0.216	0.218	0.220	210
220	0.220	0.222	0.225	0.227	0.229	0.231	0.234	0.236	0.238	0.240	0.243	220
230	0.243	0.245	0.247	0.250	0.252	0.254	0.257	0.259	0.262	0.264	0.266	230
240	0.266	0.269	0.271	0.274	0.276	0.279	0.281	0.284	0.286	0.289	0.291	240
250	0.291	0.294	0.296	0.299	0.301	0.304	0.306	0.309	0.312	0.314	0.317	250
260	0.317	0.320	0.322	0.325	0.328	0.330	0.333	0.336	0.338	0.341	0.344	260
270	0.344	0.347	0.349	0.352	0.355	0.358	0.360	0.363	0.366	0.369	0.372	270
280	0.372	0.374	0.377	0.380	0.383	0.386	0.389	0.392	0.395	0.398	0.401	280
290	0.401	0.404	0.406	0.409	0.412	0.415	0.418	0.421	0.424	0.427	0.431	290
300	0.431	0.434	0.437	0.440	0.443	0.446	0.449	0.452	0.455	0.458	0.462	300
310	0.462	0.465	0.468	0.471	0.474	0.477	0.481	0.484	0.487	0.490	0.494	310
320	0.494	0.497	0.500	0.503	0.507	0.510	0.513	0.517	0.520	0.523	0.527	320
330	0.527	0.530	0.533	0.537	0.540	0.543	0.547	0.550	0.554	0.557	0.561	330
340	0.561	0.564	0.568	0.571	0.575	0.578	0.582	0.585	0.589	0.592	0.596	340
350	0.596	0.599	0.603	0.606	0.610	0.614	0.617	0.621	0.625	0.628	0.632	350
360	0.632	0.635	0.639	0.643	0.647	0.650	0.654	0.658	0.661	0.665	0.669	360
370	0.669	0.673	0.676	0.680	0.684	0.688	0.692	0.696	0.699	0.703	0.707	370
380	0.707	0.711	0.715	0.719	0.723	0.726	0.730	0.734	0.738	0.742	0.746	380
390	0.746	0.750	0.754	0.758	0.762	0.766	0.770	0.774	0.778	0.782	0.786	390
°C	0	1	2	3	4	5	6	7	8	9	10	°C

TABLE 55—*Type B thermocouples (continued).*

Temperature in Degrees Celsius

EMF in Absolute Millivolts   Reference Junctions at 0 C

°C	0	1	2	3	4	5	6	7	8	9	10	°C
						Millivolts						
400	0.786	0.790	0.794	0.799	0.803	0.807	0.811	0.815	0.819	0.823	0.827	400
410	0.827	0.832	0.836	0.840	0.844	0.848	0.853	0.857	0.861	0.865	0.869	410
420	0.869	0.874	0.878	0.882	0.887	0.891	0.895	0.900	0.904	0.908	0.913	420
430	0.913	0.917	0.921	0.926	0.930	0.934	0.939	0.943	0.948	0.952	0.957	430
440	0.957	0.961	0.966	0.970	0.975	0.979	0.984	0.988	0.993	0.997	1.002	440
450	1.002	1.006	1.011	1.015	1.020	1.025	1.029	1.034	1.038	1.043	1.048	450
460	1.048	1.052	1.057	1.062	1.066	1.071	1.076	1.080	1.085	1.090	1.095	460
470	1.095	1.099	1.104	1.109	1.114	1.118	1.123	1.128	1.133	1.138	1.143	470
480	1.143	1.147	1.152	1.157	1.162	1.167	1.172	1.177	1.182	1.186	1.191	480
490	1.191	1.196	1.201	1.206	1.211	1.216	1.221	1.226	1.231	1.236	1.241	490
500	1.241	1.246	1.251	1.256	1.261	1.267	1.272	1.277	1.282	1.287	1.292	500
510	1.292	1.297	1.302	1.307	1.313	1.318	1.323	1.328	1.333	1.339	1.344	510
520	1.344	1.349	1.354	1.359	1.365	1.370	1.375	1.381	1.386	1.391	1.396	520
530	1.396	1.402	1.407	1.412	1.418	1.423	1.429	1.434	1.439	1.445	1.450	530
540	1.450	1.455	1.461	1.466	1.472	1.477	1.483	1.488	1.494	1.499	1.505	540
550	1.505	1.510	1.516	1.521	1.527	1.532	1.538	1.543	1.549	1.554	1.560	550
560	1.560	1.566	1.571	1.577	1.583	1.588	1.594	1.599	1.605	1.611	1.616	560
570	1.616	1.622	1.628	1.634	1.639	1.645	1.651	1.656	1.662	1.668	1.674	570
580	1.674	1.680	1.685	1.691	1.697	1.703	1.709	1.714	1.720	1.726	1.732	580
590	1.732	1.738	1.744	1.750	1.756	1.761	1.767	1.773	1.779	1.785	1.791	590
600	1.791	1.797	1.803	1.809	1.815	1.821	1.827	1.833	1.839	1.845	1.851	600
610	1.851	1.857	1.863	1.869	1.875	1.882	1.888	1.894	1.900	1.906	1.912	610
620	1.912	1.918	1.924	1.931	1.937	1.943	1.949	1.955	1.962	1.968	1.974	620
630	1.974	1.980	1.986	1.993	1.999	2.005	2.011	2.018	2.024	2.030	2.037	630
640	2.037	2.043	2.049	2.056	2.062	2.068	2.075	2.081	2.087	2.094	2.100	640
650	2.100	2.107	2.113	2.120	2.126	2.132	2.139	2.145	2.152	2.158	2.165	650
660	2.165	2.171	2.178	2.184	2.191	2.197	2.204	2.210	2.217	2.224	2.230	660
670	2.230	2.237	2.243	2.250	2.257	2.263	2.270	2.276	2.283	2.290	2.296	670
680	2.296	2.303	2.310	2.316	2.323	2.330	2.337	2.343	2.350	2.357	2.363	680
690	2.363	2.370	2.377	2.384	2.391	2.397	2.404	2.411	2.418	2.425	2.431	690
700	2.431	2.438	2.445	2.452	2.459	2.466	2.473	2.480	2.486	2.493	2.500	700
710	2.500	2.507	2.514	2.521	2.528	2.535	2.542	2.549	2.556	2.563	2.570	710
720	2.570	2.577	2.584	2.591	2.598	2.605	2.612	2.619	2.626	2.633	2.641	720
730	2.641	2.648	2.655	2.662	2.669	2.676	2.683	2.690	2.698	2.705	2.712	730
740	2.712	2.719	2.726	2.733	2.741	2.748	2.755	2.762	2.770	2.777	2.784	740
750	2.784	2.791	2.799	2.806	2.813	2.821	2.828	2.835	2.843	2.850	2.857	750
760	2.857	2.865	2.872	2.879	2.887	2.894	2.901	2.909	2.916	2.924	2.931	760
770	2.931	2.939	2.946	2.953	2.961	2.968	2.976	2.983	2.991	2.993	3.006	770
780	3.006	3.013	3.021	3.028	3.036	3.043	3.051	3.059	3.066	3.074	3.081	780
790	3.081	3.089	3.097	3.104	3.112	3.119	3.127	3.135	3.142	3.150	3.158	790
°C	0	1	2	3	4	5	6	7	8	9	10	°C

TABLE 55—*Type B thermocouples (continued).*
Temperature in Degrees Celsius

EMF in Absolute Millivolt   Reference Junctions at 0 C

°C	0	1	2	3	4	5	6	7	8	9	10	°C
						Millivolts						
800	3.158	3.165	3.173	3.181	3.183	3.196	3.204	3.212	3.219	3.227	3.235	800
810	3.235	3.243	3.250	3.258	3.266	3.274	3.281	3.289	3.297	3.305	3.313	810
820	3.313	3.321	3.328	3.336	3.344	3.352	3.360	3.368	3.376	3.384	3.391	820
830	3.391	3.399	3.407	3.415	3.423	3.431	3.439	3.447	3.455	3.463	3.471	830
840	3.471	3.479	3.487	3.495	3.503	3.511	3.519	3.527	3.535	3.543	3.551	840
850	3.551	3.559	3.567	3.575	3.583	3.591	3.600	3.608	3.616	3.624	3.632	850
860	3.632	3.640	3.648	3.656	3.665	3.673	3.681	3.689	3.697	3.706	3.714	860
870	3.714	3.722	3.730	3.738	3.747	3.755	3.763	3.771	3.780	3.788	3.796	870
880	3.796	3.805	3.813	3.821	3.829	3.838	3.846	3.854	3.863	3.871	3.879	880
890	3.879	3.888	3.896	3.905	3.913	3.921	3.930	3.938	3.947	3.955	3.963	890
900	3.963	3.972	3.980	3.989	3.997	4.006	4.014	4.023	4.031	4.040	4.048	900
910	4.048	4.057	4.065	4.074	4.082	4.091	4.099	4.108	4.117	4.125	4.134	910
920	4.134	4.142	4.151	4.160	4.168	4.177	4.185	4.194	4.203	4.211	4.220	920
930	4.220	4.229	4.237	4.246	4.255	4.263	4.272	4.281	4.290	4.298	4.307	930
940	4.307	4.316	4.324	4.333	4.342	4.351	4.360	4.368	4.377	4.386	4.395	940
950	4.395	4.404	4.412	4.421	4.430	4.439	4.448	4.457	4.465	4.474	4.483	950
960	4.483	4.492	4.501	4.510	4.519	4.528	4.537	4.545	4.554	4.563	4.572	960
970	4.572	4.581	4.590	4.599	4.608	4.617	4.626	4.635	4.644	4.653	4.662	970
980	4.662	4.671	4.680	4.689	4.698	4.707	4.716	4.726	4.735	4.744	4.753	980
990	4.753	4.762	4.771	4.780	4.789	4.798	4.807	4.817	4.826	4.835	4.844	990
1000	4.844	4.853	4.862	4.872	4.881	4.890	4.899	4.908	4.918	4.927	4.936	1000
1010	4.936	4.945	4.954	4.964	4.973	4.982	4.992	5.001	5.010	5.019	5.029	1010
1020	5.029	5.038	5.047	5.057	5.066	5.075	5.085	5.094	5.103	5.113	5.122	1020
1030	5.122	5.131	5.141	5.150	5.160	5.169	5.178	5.188	5.197	5.207	5.216	1030
1040	5.216	5.225	5.235	5.244	5.254	5.263	5.273	5.282	5.292	5.301	5.311	1040
1050	5.311	5.320	5.330	5.339	5.349	5.358	5.368	5.377	5.387	5.396	5.406	1050
1060	5.406	5.415	5.425	5.435	5.444	5.454	5.463	5.473	5.483	5.492	5.502	1060
1070	5.502	5.511	5.521	5.531	5.540	5.550	5.560	5.569	5.579	5.589	5.598	1070
1080	5.598	5.608	5.618	5.627	5.637	5.647	5.657	5.666	5.676	5.686	5.695	1080
1090	5.695	5.705	5.715	5.725	5.734	5.744	5.754	5.764	5.774	5.783	5.793	1090
1100	5.793	5.803	5.813	5.823	5.832	5.842	5.852	5.862	5.872	5.882	5.891	1100
1110	5.891	5.901	5.911	5.921	5.931	5.941	5.951	5.961	5.970	5.980	5.990	1110
1120	5.990	6.000	6.010	6.020	6.030	6.040	6.050	6.060	6.070	6.080	6.090	1120
1130	6.090	6.100	6.110	6.120	6.130	6.140	6.150	6.160	6.170	6.180	6.190	1130
1140	6.190	6.200	6.210	6.220	6.230	6.240	6.250	6.260	6.270	6.280	6.290	1140
1150	6.290	6.300	6.310	6.320	6.330	6.340	6.351	6.361	6.371	6.381	6.391	1150
1160	6.391	6.401	6.411	6.421	6.432	6.442	6.452	6.462	6.472	6.482	6.492	1160
1170	6.492	6.503	6.513	6.523	6.533	6.543	6.554	6.564	6.574	6.584	6.594	1170
1180	6.594	6.605	6.615	6.625	6.635	6.646	6.656	6.666	6.676	6.687	6.697	1180
1190	6.697	6.707	6.718	6.728	6.738	6.749	6.759	6.769	6.780	6.790	6.800	1190
°C	0	1	2	3	4	5	6	7	8	9	10	°C

TABLE 55—*Type B thermocouples (continued).*
Temperature in Degrees Celsius

EMF in Absolute Millivolts   Reference Junctions at 0 C

°C	0	1	2	3	4	5	6	7	8	9	10	°C
					Millivolts							
1200	6.800	6.811	6.821	6.831	6.842	6.852	6.863	6.873	6.883	6.894	6.904	1200
1210	6.904	6.915	6.925	6.936	6.946	6.956	6.967	6.977	6.988	6.998	7.009	1210
1220	7.009	7.019	7.030	7.040	7.051	7.061	7.072	7.082	7.093	7.103	7.114	1220
1230	7.114	7.125	7.135	7.146	7.156	7.167	7.177	7.188	7.199	7.209	7.220	1230
1240	7.220	7.230	7.241	7.252	7.262	7.273	7.284	7.294	7.305	7.316	7.326	1240
1250	7.326	7.337	7.348	7.358	7.369	7.380	7.390	7.401	7.412	7.422	7.433	1250
1260	7.433	7.444	7.455	7.465	7.476	7.487	7.498	7.508	7.519	7.530	7.541	1260
1270	7.541	7.551	7.562	7.573	7.584	7.595	7.605	7.616	7.627	7.638	7.649	1270
1280	7.649	7.659	7.670	7.681	7.692	7.703	7.714	7.725	7.735	7.746	7.757	1280
1290	7.757	7.768	7.779	7.790	7.801	7.812	7.822	7.833	7.844	7.855	7.866	1290
1300	7.866	7.877	7.888	7.899	7.910	7.921	7.932	7.943	7.954	7.965	7.976	1300
1310	7.976	7.987	7.998	8.009	8.020	8.031	8.042	8.053	8.064	8.075	8.086	1310
1320	8.086	8.097	8.108	8.119	8.130	8.141	8.152	8.163	8.174	8.185	8.196	1320
1330	8.196	8.207	8.218	8.229	8.240	8.251	8.262	8.273	8.285	8.296	8.307	1330
1340	8.307	8.318	8.329	8.340	8.351	8.362	8.373	8.385	8.396	8.407	8.418	1340
1350	8.418	8.429	8.440	8.451	8.463	8.474	8.485	8.496	8.507	8.518	8.530	1350
1360	8.530	8.541	8.552	8.563	8.574	8.585	8.597	8.608	8.619	8.630	8.641	1360
1370	8.641	8.653	8.664	8.675	8.686	8.698	8.709	8.720	8.731	8.743	8.754	1370
1380	8.754	8.765	8.776	8.788	8.799	8.810	8.821	8.833	8.844	8.855	8.866	1380
1390	8.866	8.878	8.889	8.900	8.912	8.923	8.934	8.946	8.957	8.968	8.979	1390
1400	8.979	8.991	9.002	9.013	9.025	9.036	9.047	9.059	9.070	9.081	9.093	1400
1410	9.093	9.104	9.115	9.127	9.138	9.150	9.161	9.172	9.184	9.195	9.206	1410
1420	9.206	9.218	9.229	9.241	9.252	9.263	9.275	9.286	9.297	9.309	9.320	1420
1430	9.320	9.332	9.343	9.354	9.366	9.377	9.389	9.400	9.412	9.423	9.434	1430
1440	9.434	9.446	9.457	9.469	9.480	9.492	9.503	9.514	9.526	9.537	9.549	1440
1450	9.549	9.560	9.572	9.583	9.595	9.606	9.618	9.629	9.640	9.652	9.663	1450
1460	9.663	9.675	9.686	9.698	9.709	9.721	9.732	9.744	9.755	9.767	9.778	1460
1470	9.778	9.790	9.801	9.813	9.824	9.836	9.847	9.859	9.870	9.882	9.893	1470
1480	9.893	9.905	9.916	9.928	9.939	9.951	9.963	9.974	9.986	9.997	10.009	1480
1490	10.009	10.020	10.032	10.043	10.055	10.066	10.078	10.089	10.101	10.113	10.124	1490
1500	10.124	10.136	10.147	10.159	10.170	10.182	10.193	10.205	10.217	10.228	10.240	1500
1510	10.240	10.251	10.263	10.274	10.286	10.298	10.309	10.321	10.332	10.344	10.356	1510
1520	10.356	10.367	10.379	10.390	10.402	10.413	10.425	10.437	10.448	10.460	10.471	1520
1530	10.471	10.483	10.495	10.506	10.518	10.529	10.541	10.553	10.564	10.576	10.587	1530
1540	10.587	10.599	10.611	10.622	10.634	10.646	10.657	10.669	10.680	10.692	10.704	1540
1550	10.704	10.715	10.727	10.739	10.750	10.762	10.773	10.785	10.797	10.808	10.820	1550
1560	10.820	10.832	10.843	10.855	10.866	10.878	10.890	10.901	10.913	10.925	10.936	1560
1570	10.936	10.948	10.960	10.971	10.983	10.995	11.006	11.018	11.029	11.041	11.053	1570
1580	11.053	11.064	11.076	11.088	11.099	11.111	11.123	11.134	11.146	11.158	11.169	1580
1590	11.169	11.181	11.193	11.204	11.216	11.228	11.239	11.251	11.263	11.274	11.286	1590
°C	0	1	2	3	4	5	6	7	8	9	10	°C

TABLE 55—*Type B thermocouples (continued).*
Temperature in Degrees Celsius

EMF in Absolute Millivolts   Reference Junctions at 0 C

°C	0	1	2	3	4	5	6	7	8	9	10	°C
						Millivolts						
1600	11.286	11.298	11.309	11.321	11.333	11.344	11.356	11.368	11.379	11.391	11.403	1600
1610	11.403	11.414	11.426	11.438	11.449	11.461	11.473	11.484	11.496	11.508	11.519	1610
1620	11.519	11.531	11.543	11.554	11.566	11.578	11.589	11.601	11.613	11.624	11.636	1620
1630	11.636	11.648	11.659	11.671	11.683	11.694	11.706	11.718	11.729	11.741	11.753	1630
1640	11.753	11.764	11.776	11.788	11.799	11.811	11.823	11.834	11.846	11.858	11.869	1640
1650	11.869	11.881	11.893	11.905	11.916	11.928	11.940	11.951	11.963	11.975	11.986	1650
1660	11.986	11.998	12.010	12.021	12.033	12.045	12.056	12.068	12.080	12.091	12.103	1660
1670	12.103	12.115	12.126	12.138	12.150	12.161	12.173	12.185	12.196	12.208	12.220	1670
1680	12.220	12.231	12.243	12.255	12.266	12.278	12.290	12.301	12.313	12.325	12.336	1680
1690	12.336	12.348	12.360	12.371	12.383	12.395	12.406	12.418	12.430	12.441	12.453	1690
1700	12.453	12.465	12.476	12.488	12.500	12.511	12.523	12.535	12.546	12 558	12.570	1700
1710	12.570	12.581	12.593	12.605	12.616	12.628	12.640	12.651	12.663	12.675	12.686	1710
1720	12.686	12.698	12.709	12.721	12.733	12.744	12.756	12.768	12.779	12.791	12.803	1720
1730	12.803	12.814	12.826	12.838	12.849	12.861	12.872	12.884	12.896	12.907	12.919	1730
1740	12.919	12.931	12.942	12.954	12.966	12.977	12.989	13.000	13.012	13.024	13.035	1740
1750	13.035	13.047	13.059	13.070	13.082	13.093	13.105	13.117	13.128	13.140	13.152	1750
1760	13.152	13.163	13.175	13.186	13.198	13.210	13.221	13.233	13.245	13.256	13.268	1760
1770	13.268	13.279	13.291	13.303	13.314	13.326	13.337	13.349	13.361	13.372	13.384	1770
1780	13.384	13.395	13.407	13.419	13.430	13.442	13.453	13.465	13.477	13.488	13.500	1780
1790	13.500	13.511	13.523	13.534	13.546	13.558	13.569	13.581	13.592	13.604	13.616	1790
1800	13.616	13.627	13.639	13.650	13.662	13.673	13.685	13.696	13.708	13.720	13.731	1800
1810	13.731	13.743	13.754	13.766	13.777	13.789	13.801	13.812	13.824	13.835	13.847	1810
1820	13.847											1820
°C	0	1	2	3	4	5	6	7	8	9	10	°C

10.3 Generation of Smooth Temperature-Emf Relationships

10.3.1 *Need for Smooth Temperature-Emf Relationship*

A table of reference values for use with thermocouples should be capable of easy and unique generation to facilitate its use in computer and similar applications. Furthermore, the reference values should agree closely with the characteristics of the thermocouple type under study, so that differences will change only slowly and smoothly with temperature level.

Reference Tables 42 to 55 (Sections 10.2) give values to two or three decimal places only, and this roundoff in table values results in inherent discontinuities in the temperature-emf relation as represented by the tables. Such discontinuities prevent unique interpolation between the table values, and introduce abrupt changes in the slopes of difference curves, in contrast to the requirement of slow, smooth changes in differences with level.

This section provides means for generating smooth, continuous temperature-emf relationships, to as many decimal places as have any useful significance, for the thermocouple types included in this standard. These data will be useful in all applications where discontinuities in the established reference values are objectionable, or where storage of complete tables is impractical.

10.3.2 *Methods of Interpolation*

The method of interpolation employed in Illustration A consists of applying the second degree Lagrange interpolation equation to the key values given in Table 56, and is based on Ref *144*. The key values are given to four decimal places at temperature intervals of approximately 50 F, and are compatible with the two- and three-place values given in the reference tables of this standard.

Temperature-emf relationships developed by this method are generated easily and stored for convenience in digital computer applications. They represent a smooth, continuous relationship, and the values will round off, with a few exceptions, to the established temperature-emf values given in Tables 42 to 55. However, because various interpolation methods were used in generating the reference tables, it is not to be expected that values generated by the Lagrange equation will round off in every instance to those given in the reference tables. The relationships so generated, however, are as representative of the characteristics of the thermocouple as the values given in Tables 42 to 55.

An alternate method of generating the temperature-emf equivalents for the Type S thermocouple is shown in Illustration B. It makes use of the equation

$$e = a + bt + ct^2 + dt^3$$

ILLUSTRATION A—*Method of interpolation.*

Example of second-degree Lagrange interpolation equation applied to key values of Table 56.

To find

E Corresponding to $T = 72$ F (for a Type J Thermocouple):

Step 1

$$T_0 < T_1 < T_2 \text{; also, } T_0 < T < T_2$$

Step 2

The key values that apply are (Table 56):

$$
\begin{array}{ll}
T_0 = \quad 0 \text{ F} & E_0 = -0.8865 \text{ mV} \\
T_1 = \quad 50 \text{ F} & E_1 = \quad 0.5043 \text{ mV} \\
T_2 = 100 \text{ F} & E_2 = \quad 1.9412 \text{ mV}
\end{array}
$$

Step 3

The second-degree Lagrange interpolation equation is:

$$E = \frac{(T - T_1)(T - T_2)}{(T_0 - T_1)(T_0 - T_2)} E_0 + \frac{(T - T_0)(T - T_2)}{(T_1 - T_0)(T_1 - T_2)} E_1 + \frac{(T - T_0)(T - T_1)}{(T_2 - T_0)(T_2 - T_1)} E_2$$

Step 4

Substituting key values into Lagrange equation:

$$E = \frac{(72 - 50)(72 - 100)}{(0 - 50)(0 - 100)}(-0.8865) + \frac{(72 - 0)(72 - 100)}{(50 - 0)(50 - 100)}(0.5043)$$

$$+ \frac{(72 - 0)(72 - 50)}{(100 - 0)(100 - 50)}(1.9412)$$

$$E = 0.1092168 + 0.40666752 + 0.61497216$$

Step 5

Thus, at 72 F, $E = 1.130856$.

This value, rounded off to two decimal places, agrees with the value of 1.13 mV given in Table 42.

ILLUSTRATION B—*Alternate method of interpolation for type S thermocouples.*

Interpolation equation:

$$e = a + bt + ct^2 + dt^3$$

where:

c = emf in absolute millivolts,
t = temperature in degrees Fahrenheit, and
$a, b, c; d$ = Constants (see below).

Example:
To find e corresponding to 1000 F

$$e = \frac{-4.5634297}{10} + \frac{4.5817882}{10^3}(1000) + \frac{4.4221848}{10^7}(1000)^2 + \frac{2.8026406}{10^{11}}(1000)^3$$

$e = -0.45634297 + 4.5817882 + 0.44221848 + 0.028026406$
$e = 4.59569$ millivolts
Value of 4.59569 millivolts rounds off to the three decimal place value of 4.596 from Table 48.

Temperature Range, deg F	Constants			
	a	b	c	d
32.0 to 449.4	$-9.3396807 \times 10^{-2}$	2.8063975×10^{-3}	3.5712104×10^{-6}	$-1.9785322 \times 10^{-9}$
449.4 to 621.3	$-1.4121145 \times 10^{-1}$	3.2447172×10^{-3}	2.3308013×10^{-6}	$-8.6189301 \times 10^{-10}$
621.3 to 787.0	$-1.6364488 \times 10^{-1}$	3.3668170×10^{-3}	2.1120570×10^{-6}	$-7.3261489 \times 10^{-10}$
787.0 to 1166.9	$-4.5634297 \times 10^{-1}$	4.5817882×10^{-3}	4.4221848×10^{-7}	$+2.8026406 \times 10^{-11}$
1166.9 to 1945.4[a]	$-4.0943621 \times 10^{-1}$	4.4786356×10^{-3}	5.2859573×10^{-7}	0
1945.4 to 2285.0	$+1.0223098$	1.9677953×10^{-3}	1.9752832×10^{-6}	$-2.7459019 \times 10^{-10}$
2285.0 to 2650.0	-3.5117422	7.5033780×10^{-3}	$-2.6419300 \times 10^{-7}$	$+2.5323215 \times 10^{-11}$

* International Practical Temperature Scale Range.

This procedure is used in the International Practical Temperature Scale to interpolate temperatures between 630.5 and 1063 C. The method is included here for the benefit of those who may prefer it to the Lagrange equation.

10.4 References

[144] Benedict, R. P. and Ashby, H. F., "Improved Reference Tables for Thermocouples," *Temperature, Its Measurement and Control in Science and Industry,* Part 2, Vol. 3, 1962, p. 51.

TABLE 56—*Key values for generation of thermocouple reference tables*
(based on second degree Lagrange interpolation).

Emf—Millivolts (Reference Junction 32 F)

Temperature, deg F	Type E	Type J	Type K	Type T	Type R	Type S	Type B
−300	...	−7.5245	...	−5.2840	...	...	...
−250	...	−6.7135	...	−4.7470	...	...	...
−200	...	−5.7568	...	−4.1107	...	...	...
−150	...	−4.6750	...	−3.3800	...	...	...
−100	...	−3.4887	...	−2.5590	...	...	...
−50	...	−2.2230	...	−1.6540	...	...	...
0	−1.0250	−0.8865	−0.6850	−0.6695	−0.0920	−0.0929	+0.0060
25	...	...	−0.1517	...	...	...	...
30	−0.0651	...	...	...	...	...	...
50	...	+0.5043	+0.3956	+0.3890	+0.0550	+0.0556	−0.0018
70	+1.2550	...	...	...	...	...	...
100	2.2725	1.9412	1.5163	1.5170	0.2199	0.2210	−0.0008
150	4.0394	3.4104	2.6621	2.7114	0.4002	0.4012	+0.0091
200	5.8736	4.9090	3.8194	3.9673	0.5959	0.5949	0.0276
225	...	...	4.3990	...	...	...	...
250	7.7650	6.4170	4.9704	5.2805	0.8066	0.8005	0.0548
300	9.7130	7.9431	6.0904	6.6468	1.0296	1.0166	0.0905
325	...	...	6.6459	...	...	...	...
350	11.7098	9.4824	7.1984	8.0635	1.2614	1.2415	0.1346
400	13.7548	11.0278	8.3117	9.5252	1.5038	1.4740	0.1871
420	14.5850	...	...	...	...	...	...
450	...	12.5738	9.4311	11.0303	1.7541	1.7124	0.2479
460	16.2550	...	...	...	...	...	...
490	17.5250	...	...	...	...	...	...
500	17.9540	14.1153	10.5650	12.5750	2.0122	1.9560	0.3169
550	20.0900	15.6510	11.7068	14.1569	2.2768	2.2050	0.3940
600	22.2540	17.1836	12.8566	15.7731	2.5474	2.4584	0.4792
650	24.4425	18.7206	14.0153	17.4211	2.8232	2.7159	0.5722
700	26.6460	20.2569	15.1793	19.0995	3.1033	2.9767	0.6731
750	28.8608	21.7887	16.3514	20.8055	3.3875	3.2404	0.7818
780	30.1949	...	...		...	...	...
800	31.0888	23.3198	17.5262		3.6765	3.5064	0.8981
850	33.3220	24.8531	18.7022		3.9699	3.7749	1.0220
900	35.5690	26.3952	19.8879		4.2644	4.0459	1.1533
929	...	...	20.5754		...	...	...
950	37.8150	27.9517	21.0718		4.5632	4.3194	1.2921
1000	40.0600	29.5240	22.2552		4.8682	4.5956	1.4381
1050	42.3100	31.1151	23.4402		5.1765	4.8744	1.5913
1100	44.5630	32.7248	24.6261		5.4881	5.1559	1.7516

TABLE 56—*Key values for generation of thermocouple reference tables (based on second degree Lagrange interpolation) (continued).*

Emf—Millivolts (Reference Junction 32 F)

Temperature, deg F	Type E	Type J	Type K	Type T	Type R	Type S	Type B
1150	46.8110	34.3550	25.8110		5.8045	5.4400	1.9189
1194	...	...	26.8450		...	...	...
1200	49.0410	36.0120	26.9850		6.1245	5.7265	2.0932
1230	50.3749	...	...		...	...	...
1250	...	37.7069	28.1547		6.4461	6.0152	2.2742
1260	51.7150	...	...		...	...	...
1300	...	39.4324	29.3231		6.7726	6.3065	2.4620
1310	53.9401	...	...		...	...	...
1330	54.8251	...	...		...	...	...
1350	55.7066	41.1870	30.4867		7.1025	6.6005	2.6563
1400	57.9172	42.9605	31.6462		7.4361	6.8972	2.8572
1450	60.1148	...	32.7961		7.7742	7.1964	3.0645
1500	62.2998	...	33.9350		8.1156	7.4983	3.2780
1550	64.4710	...	35.0690		8.4603	7.8029	3.4975
1600	66.6262	...	36.1950		8.8088	8.1101	3.7229
1650	68.7622	...	37.3144		9.1605	8.4199	3.9541
1700	...	...	38.4257		9.5162	8.7325	4.1912
1710	71.3250	...	...		...	...	...
1750	73.0170		39.5263		9.8738	9.0475	4.4339
1800	75.1160		40.6192		10.2373	9.3653	4.6822
1850	...		41.7038		10.6027	9.6857	4.9360
1900			42.7800		10.9732	10.0087	5.1950
1950			43.8480		11.3480	10.3345	5.4591
2000			44.9084		11.7265	10.6623	5.7279
2050			45.9587		12.1054	10.9919	6.0013
2100			46.9990		12.4876	11.3230	6.2789
2150			48.0288		12.8712	11.6554	6.5604
2200			49.0490		13.2549	11.9885	6.8464
2250			50.0561		13.6406	12.3225	7.1375
2300			51.0509		14.0265	12.6568	7.4331
2350			52.0344		14.4125	12.9908	7.7330
2400			53.0074		14.7984	13.3246	8.0366
2450			53.9685		15.1838	13.6580	8.3437
2500			54.9189		15.5685	13.9911	8.6539
2550					15.9539	14.3237	8.9669
2600					16.3399	14.6559	9.2823
2650					16.7254	14.9876	9.5997
2700					17.1098	15.3187	9.9190
2750					17.4931	15.6492	10.2397
2800					17.8752	15.9792	10.5617
2850					18.2555	16.3083	10.8846
2900					18.6360	16.6368	11.2082

TABLE 56—*Key values for generation of thermocouple reference tables*
(based on second degree Lagrange interpolation) (continued).

Emf—Millivolts (Reference Junction 32 F)

Temperature, deg F	Type E	Type J	Type K	Type T	Type R	Type S	Type B
2950					19.0161	16.9646	11.5323
3000					19.3936	17.2915	11.8565
3050							12.1808
3100							12.5049
3150							12.8285
3200							13.1516
3250							13.4739
3300							13.7954

11. CRYOGENICS

11.1 General Remarks

Although there is some variation in the defined temperature range involved, cryogenics (kri-o-jen-iks) usually indicates concern with temperatures in the liquid oxygen range (about 90 K or −183 C) or lower. This temperature range will be discussed primarily in this chapter. Since a triple point (of water) or ice bath reference junction often is used, additional comments and values will be given for the entire subzero (0 C) range.

Most aspects of cryogenic thermometry are similar to those applicable at room or high temperatures. In particular, the measurement systems and thermoelectric theory are nearly identical. However, there are significant differences with respect to some materials, techniques or assembly and fabrication, calibration schemes, and methods of practical usage. Fortunately the added difficulties with some details are offset by the removal of a few problems peculiar to high-temperature thermometry: chemical transformations are insignificant; oxidation, reduction, and impurity migration do not occur because of the low temperatures. Annealing of physical imperfections is also absent for the same reasons. Maintenance of fixed points and techniques of calibration are usually considerably easier and sometimes much more accurate. Thermal radiation is usually not important, at least if simple precautions are taken to account for it.

Several books have been written on thermometry and on the experimental techniques necessary for cryogenic research by Scott [145] on cryogenic engineering, and by White [146] and by Rose-Innes [147] on smaller, scientific systems.

11.2 **Materials**

Many thermocouple materials developed for high-temperature usage have too low a sensitivity at low temperatures to be practical; others have a reasonable sensitivity, but are very erratic with large inhomogeneities and lot-to-lot variations. Only three common commercial thermocouple combinations have proven themselves for cryogenic use: (1) Type E, Chromel versus constantan; (2) Type T, copper versus constantan; and (3) Type K, Chromel versus Alumel. The first, Type E, is definitely recommended as the most valuable for general low temperature use. It is better because of its higher sensitivity (coupled with only average inhomogeneity voltages) and lower thermal conductivity of both thermocouple materials. The latter property is particularly important for obtaining good thermal tempering of measuring junctions. Tables for the total voltage and Seebeck coefficient are given in Section 11.3 for each of the above combinations.

There are other, but uncommon, materials that have a higher sensitivity at very low temperatures, below 50 K: *gold*-cobalt and *gold*-iron. Both are negative thermoelectric materials and should be matched with Chromel, "normal" silver, or less preferably, copper, to obtain a thermocouple pair. The older material, *gold*-cobalt (2.1 atomic percent cobalt) has been found to have significant instabilities caused by room temperature annealing of the metastable solution of cobalt in gold. Because of those problems *gold*-cobalt should not be used any more in the future. The alloy *gold*-0.07 atomic percent iron has replaced *gold*-cobalt: it does not have alloying instabilities, it is more homogeneous, and it has a larger Seebeck coefficient. Tables for the total voltage and Seebeck coefficient of Chromel versus *gold*-0.07 atomic percent iron are given also in Section 11.3. The values given for this combination are interim: the *gold*-iron material is not yet standardized. Figure 49 compares various Seebeck coefficients for materials used in the cryogenic range. Type K is not shown, but would be slightly above Type T.

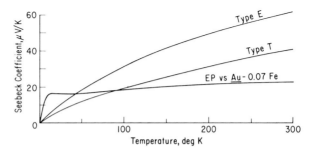

FIG. 49—*Seebeck coefficients for Types E, T, and EP versus Au-0.07Fe.*

11.3 **Reference Tables (for use in the cryogenic range)**

TABLE 57—*Type E thermocouples (Chromel versus constantan).*

T, deg K	E, μV	dE/dT, μV/K	T, deg K	E, μV	dE/dT, μV/K	T, deg K	E, μV	dE/dT, μV/K
1	0.41	0.660	51	522.68	19.001	101	1806.73	31.648
2	1.31	1.140	52	541.83	19.303	102	1838.49	31.865
3	2.70	1.623	53	561.28	19.603	103	1870.46	32.081
4	4.56	2.085	54	581.04	19.900	104	1902.65	32.295
5	6.87	2.535	55	601.08	20.194	105	1935.05	32.509
6	9.62	2.975	56	621.42	20.486	106	1967.67	32.722
7	12.81	3.406	57	642.05	20.776	107	2000.50	32.934
8	16.43	3.829	58	662.97	21.063	108	2033.54	33.145
9	20.47	4.244	59	684.18	21.348	109	2066.79	33.355
10	24.92	4.653	60	705.67	21.630	110	2100.25	33.565
11	29.77	5.057	61	727.44	21.911	111	2133.92	33.773
12	35.03	5.455	62	749.49	22.188	112	2167.79	33.981
13	40.68	5.848	63	771.82	22.464	113	2201.88	34.187
14	46.72	6.238	64	794.42	22.737	114	2236.17	54.393
15	53.15	6.623	65	817.29	23.008	115	2270.66	34.598
16	59.97	7.005	66	840.43	23.277	116	2305.36	34.802
17	67.16	7.385	67	863.84	23.544	117	2340.27	35.005
18	74.74	7.761	68	887.52	23.809	118	2375.37	35.207
19	82.69	8.135	69	911.46	24.072	119	2410.68	35.408
20	91.01	8.506	70	935.66	24.333	120	2446.19	35.609
21	99.70	8.876	71	960.13	24.592	121	2481.90	35.809
22	108.76	9.243	72	984.85	24.849	122	2517.81	36.007
23	118.18	9.608	73	1009.82	25.104	123	2553.91	36.205
24	127.97	9.971	74	1035.05	25.357	124	2590.22	36.402
25	138.12	10.332	75	1060.54	25.608	125	2626.72	26.599
26	148.64	10.691	76	1086.27	25.858	126	2663.41	36.794
27	159.51	11.049	77	1112.25	26.106	127	2700.30	36.989
28	170.73	11.404	78	1138.48	26.353	128	2737.39	37.182
29	182.31	11.758	79	1164.96	26.598	129	2774.67	37.375
30	194.25	12.110	80	1191.68	26.841	130	2812.14	37.567
31	206.53	12.460	81	1218.64	27.083	131	2849.80	37.759
32	219.17	12.808	82	1245.84	27.323	132	2887.66	37.949
33	232.15	13.154	83	1273.28	27.562	133	2925.70	38.139
34	245.47	13.498	84	1300.96	27.799	134	2963.94	38.328
35	259.14	13.840	85	1328.88	28.035	135	3002.36	38.516
36	273.15	14.179	86	1357.03	28.270	136	3040.97	38.703
37	287.50	14.517	87	1385.42	28.505	137	3079.76	38.890
38	302.19	14.853	88	1414.04	28.735	138	3118.75	39.075
39	317.20	15.186	89	1442.89	28.966	139	3157.91	39.260
40	332.56	15.517	90	1471.97	29.196	140	3197.27	39.445
41	348.24	15.846	91	1501.28	29.424	141	3236.80	39.628
42	364.25	16.172	92	1530.82	29.652	142	3276.52	39.811
43	380.58	16.496	93	1560.59	29.878	143	3316.42	39.993
44	397.24	16.818	94	1590.58	30.103	144	3356.51	40.174
45	414.22	17.137	95	1620.79	30.327	145	3396.77	40.355
46	431.51	17.454	96	1651.23	30.549	146	3437.22	40.534
47	449.13	17.769	97	1681.89	30.771	147	3477.84	40.715
48	467.05	18.081	98	1712.77	30.992	148	3518.64	40.892
49	485.29	18.390	99	1743.87	31.212	149	3559.62	41.070
50	503.83	18.697	100	1775.19	31.430	150	3600.78	41.247

TABLE 57—*Type E thermocouples (Chromel versus constantan) (continued).*

T, deg K	E, μV	dE/dT, μV/K	T, deg K	E, μV	dE/dT, μV/K	T, deg K	E, μV	dE/dT, μV/K
151	3642.12	41.425	201	5920.81	49.483	251	8566.06	56.133
152	3683.63	41.599	202	5970.36	49.629	252	8622.25	56.255
153	3725.31	41.774	203	6020.06	49.775	253	8678.57	56.377
154	3767.18	41.948	204	6069.91	49.919	254	8735.01	56.498
155	3809.21	42.122	205	6119.90	50.064	255	8791.56	56.619
156	3851.42	42.295	206	6170.04	50.207	256	8848.24	56.739
157	3893.80	42.467	207	6220.52	40.350	257	8905.04	56.858
158	3936.35	42.639	208	6270.74	50.492	258	8961.96	56.977
159	3979.08	42.811	209	6321.30	50.634	259	9019.00	57.095
160	4021.97	42.981	210	6372.01	50.775	260	9076.15	57.212
161	4065.04	43.151	211	6422.85	50.915	261	9133.42	57.329
162	4108.28	43.321	212	6475.84	51.055	262	9190.81	57.445
163	4151.68	43.490	213	6524.96	51.194	263	9248.31	57.559
164	4195.26	43.658	214	6576.22	51.333	264	9305.93	57.673
165	4239.00	43.825	215	6627.63	51.471	265	9363.66	57.786
166	4282.91	43.993	216	6679.17	51.608	266	9421.50	57.898
167	4326.98	44.159	217	6730.84	51.745	267	9479.45	58.009
168	4371.22	44.325	218	6782.66	51.882	268	9537.51	58.119
169	4415.63	44.491	219	6834.61	52.017	269	9595.69	58.227
170	4460.21	44.655	220	6886.69	52.152	270	9653.97	58.335
171	4504.94	44.820	221	6938.91	52.287	271	9712.36	58.442
172	4549.84	44.984	222	6991.27	52.421	272	9770.85	58.547
173	4594.91	45.147	223	7043.75	52.555	273	9829.45	58.651
174	4640.14	45.309	224	7096.38	52.688	274	9888.15	58.755
175	4685.53	45.471	225	7149.13	52.821	275	9946.96	58.857
176	4731.08	45.633	226	7202.02	52.953	276	10005.87	58.958
177	4776.79	45.794	227	7255.04	53.085	277	10064.88	59.059
178	4822.67	45.954	228	7308.19	53.216	278	10123.98	59.159
179	4868.70	46.114	229	7361.47	53.347	279	10183.19	59.258
180	4914.90	46.274	230	7414.88	53.477	280	10242.50	59.356
181	4961.25	46.432	231	7468.42	53.607			
182	5007.76	46.590	232	7522.09	53.737			
183	5054.43	46.748	233	7575.89	53.866			
184	5101.26	46.905	234	7629.83	53.995			
185	5148.24	47.061	235	7683.88	54.124			
186	5195.38	47.217	236	7738.07	54.252			
187	5242.67	47.373	237	7792.39	54.380			
188	5290.12	47.527	238	7846.83	54.507			
189	5337.73	47.681	239	7901.40	54.634			
190	5385.49	47.835	240	7956.10	54.761			
191	5433.40	47.988	241	8010.92	54.888			
192	5481.46	48.140	242	8065.87	55.014			
193	5529.68	48.292	243	8120.95	55.140			
194	5578.05	48.443	244	8176.15	55.265			
195	5626.56	48.593	245	8231.48	55.390			
196	5875.23	48.743	246	8286.93	55.515			
197	5724.05	48.893	247	8342.51	55.639			
198	5773.02	49.041	248	8398.21	55.763			
199	5822.13	49.189	249	8454.04	55.887			
200	5871.40	49.336	250	8509.99	56.010			

TABLE 58—*Type T thermocouples (copper versus constantan).*

T, deg K	E, μV	dE/dT, μV/K	T, deg K	E, μV	dE/dT, μV/K	T, deg K	E, μV	dE/dT, μV/K
1	−0.09	0.147	51	343.30	12.345	101	1147.25	19.498
2	0.28	0.586	52	355.73	12.519	102	1166.81	19.629
3	1.07	0.985	53	368.33	12.690	103	1186.51	19.758
4	2.24	1.351	54	381.11	12.059	104	1206.33	19.888
5	3.76	1.690	55	394.05	13.025	105	1226.28	20.017
6	5.61	2.006	56	407.16	13.189	106	1246.37	20.147
7	7.77	2.304	57	420.43	13.350	107	1266.58	20.275
8	10.21	2.587	58	433.86	13.510	108	1286.92	20.404
9	12.94	2.859	59	447.44	13.668	109	1307.38	20.532
10	15.93	3.121	60	461.19	13.824	110	1327.98	20.660
11	19.18	3.377	61	475.09	13.978	111	1348.71	20.788
12	22.68	3.628	62	489.15	14.130	112	1369.56	20.916
13	26.43	3.876	63	503.35	14.281	113	1390.54	21.043
14	30.43	4.120	64	517.71	14.431	114	1411.64	21.170
15	34.67	4.364	65	532.21	14.579	115	1432.88	21.297
16	39.16	4.606	66	546.86	14.726	116	1454.24	21.424
17	43.89	4.848	67	561.66	14.872	117	1475.75	21.551
18	48.85	5.091	68	576.61	15.017	118	1497.34	21.677
19	54.07	5.333	69	591.70	15.160	119	1519.08	21.803
20	59.52	5.576	70	606.93	15.303	120	1540.95	21.929
21	65.22	5.818	71	622.30	15.445	121	1562.94	22.055
22	71.16	6.062	72	637.82	15.587	122	1585.06	22.181
23	77.34	6.305	73	653.48	15.727	123	1607.30	22.306
24	83.77	6.548	74	669.27	15.868	124	1629.67	22.431
25	90.44	6.791	75	685.21	16.007	125	1652.16	22.557
26	97.35	7.033	76	701.29	16.146	126	1674.78	22.682
27	104.50	7.274	77	717.50	16.284	127	1697.53	22.807
28	111.90	7.515	78	733.86	16.422	128	1720.39	22.932
29	119.53	7.754	79	750.35	16.559	129	1743.39	23.056
30	127.40	7.991	80	766.97	16.696	130	1766.51	23.181
31	135.51	8.227	81	783.74	16.833	131	1789.75	23.305
32	143.86	8.461	82	800.64	16.969	132	1813.12	23.430
33	152.43	8.692	83	817.68	17.105	133	1836.61	23.554
34	161.24	8.921	84	834.85	17.241	134	1860.23	23.678
35	170.27	9.147	85	852.16	17.376	135	1883.97	25.802
36	179.53	9.371	86	869.60	17.511	136	1907.83	23.926
37	189.01	9.591	87	887.18	17.645	137	1931.82	24.050
38	198.72	9.809	88	904.89	17.779	138	1955.93	24.174
39	208.63	10.023	89	922.74	17.913	139	1930.17	24.297
40	218.76	10.234	90	940.72	18.047	140	2004.52	24.420
41	229.10	10.442	91	958.83	18.180	141	2029.01	24.544
42	239.64	10.647	92	977.08	18.314	142	2053.61	24.667
43	250.39	10.848	93	995.46	18.446	143	2078.34	24.790
44	261.34	11.046	94	1013.97	18.579	144	2103.19	24.913
45	272.48	11.241	95	1032.62	18.711	145	2128.17	25.035
46	283.82	11.433	96	1051.39	18.843	146	2153.26	25.158
47	295.35	11.621	97	1070.30	18.975	147	2178.48	25.280
48	307.06	11.807	98	1089.34	19.106	148	2203.82	25.402
49	318.96	11.999	99	1108.51	19.237	149	2229.28	25.524
50	331.04	12.168	100	1127.82	19.368	150	2254.87	25.646

TABLE 58—*Type J thermocouples (copper versus constantan) (continued).*

T, deg K	E, μV	dE/dT, μV/K	T, deg K	E, μV	dE/dT, μV/K	T, deg K	E, μV	dE/dT, μV/K
151	2280.58	25.767	201	3715.39	31.527	251	5423.15	36.677
152	2306.40	25.889	202	3746.97	31.637	252	5459.87	36.776
153	2332.35	26.010	203	5778.66	31.746	253	5496.69	36.874
154	2358.42	26.130	204	3810.46	31.856	254	5533.62	36.972
155	2384.61	26.251	205	3842.58	31.965	255	5570.64	37.070
156	2410.93	26.371	206	3874.39	32.074	256	5607.75	37.167
157	2437.36	26.491	207	3906.52	32.182	257	5644.97	37.264
158	2463.91	26.611	208	3938.76	32.291	258	5682.28	37.360
159	2490.58	26.731	209	3971.10	32.399	259	5719.69	37.456
160	2517.37	26.850	210	4003.56	32.506	260	5757.20	37.551
161	2544.28	26.969	211	4036.12	32.614	261	5794.79	37.645
162	2571.31	27.088	212	4068.78	32.721	262	5832.49	37.739
163	2598.46	27.206	213	4101.56	32.828	263	5870.27	37.831
164	2625.72	27.325	214	4134.44	32.934	264	5908.15	37.923
165	2653.10	27.442	215	4167.43	33.040	265	5946.12	38.014
166	2680.61	27.560	216	4200.52	33.146	266	5984.17	38.103
167	2708.22	27.677	217	4233.72	33.252	267	6022.32	38.193
168	2735.96	27.795	218	4267.02	33.357	268	6060.56	38.281
169	2763.81	27.911	219	4300.43	33.462	269	6098.88	38.368
170	2791.78	28.028	220	4333.95	33.566	270	6137.30	38.455
171	2819.87	28.144	221	4367.56	33.670	271	6175.79	38.541
172	2848.07	28.260	222	4401.29	33.774	272	6214.37	38.627
173	2876.39	28.376	223	4435.11	33.877	273	6253.05	38.714
174	2904.82	28.491	224	4469.04	33.980	274	6291.80	38.802
175	2933.37	28.606	225	4503.07	34.083	275	6330.65	38.891
176	2962.04	28.721	226	4537.21	34.185	276	6369.59	38.983
177	2990.81	28.856	227	4571.44	34.287	277	6408.62	39.078
178	3019.71	28.950	228	4605.78	34.388	278	6447.74	39.178
179	3048.71	29.065	229	4640.22	34.490	279	6486.97	39.283
180	3077.84	29.179	230	4674.76	34.591	280	6526.31	39.397
181	3107.07	29.292	231	4709.40	34.691			
182	3136.42	29.406	232	4744.14	34.792			
183	3165.88	29.519	233	4778.98	34.892			
184	3195.46	29.632	234	4813.93	34.992			
185	3225.15	29.745	235	4848.97	35.092			
186	3254.95	29.858	236	4884.11	35.192			
187	3284.86	29.971	237	4919.35	35.291			
188	3314.89	30.083	238	4954.69	35.391			
189	3345.03	30.195	239	4990.13	35.490			
190	3375.28	30.307	240	5025.67	35.589			
191	3405.64	30.419	241	5061.31	35.688			
192	3456.12	30.531	242	5097.05	35.787			
193	3466.71	30.642	243	5132.89	35.886			
194	3497.40	30.753	244	5168.82	· 35.986			
195	3528.21	30.864	245	5204.86	36.084			
196	3559.13	30.975	246	5240.99	36.183			
197	3590.16	31.086	247	5277.22	36.282			
198	3621.30	31.197	248	5313.55	36.381			
199	3652.56	31.307	249	5349.99	36.480			
200	3693.92	31.417	250	5386.51	36.579			

TABLE 59—*Type K thermocouples (Chromel versus Alumel).*

T, deg K	E, μV	dE/dT, μV/K	T, deg K	E, μV	dE/dT, μV/K	T, deg K	E, μV	dE/dT, μV/K
1	0.30	0.382	51	276.73	10.940	101	1061.84	20.118
2	0.77	0.541	52	287.77	11.146	102	1082.04	20.282
3	1.39	0.707	53	299.02	11.351	103	1102.41	20.446
4	2.18	0.879	54	310.47	11.555	104	1122.93	20.610
5	3.15	1.056	55	322.13	11.758	105	1143.62	20.772
6	4.30	1.239	56	333.99	11.960	106	1164.48	20.935
7	5.63	1.426	57	346.05	12.161	107	1185.49	21.096
8	7.15	1.618	58	358.31	12.360	108	1206.67	21.257
9	8.86	1.814	59	370.77	12.559	109	1228.01	21.418
10	10.78	2.014	60	383.43	12.757	110	1249.51	21.578
11	12.89	2.217	61	396.28	12.953	111	1271.16	21.737
12	15.21	2.423	62	409.33	13.149	112	1292.98	21.896
13	17.74	2.631	63	422.58	13.344	113	1314.95	22.054
14	20.48	2.843	64	436.02	13.537	114	1337.09	22.211
15	23.42	3.056	65	449.65	13.730	115	1359.38	22.368
16	26.59	3.272	66	463.48	13.922	116	1381.82	22.525
17	29.97	3.489	67	477.50	14.113	117	1404.43	22.681
18	33.57	3.708	68	491.70	14.302	118	1427.18	22.836
19	37.38	3.928	69	506.10	14.491	119	1450.10	22.991
20	41.42	4.149	70	520.69	14.679	120	1473.16	23.145
21	45.68	4.371	71	535.46	14.866	121	1496.39	23.298
22	50.17	4.594	72	550.42	15.053	122	1519.76	23.451
23	54.87	4.818	73	565.56	15.238	123	1543.29	23.604
24	59.80	5.042	74	580.89	15.422	124	1566.97	23.756
25	64.96	5.267	75	596.41	15.606	125	1590.80	23.907
26	70.34	5.491	76	612.11	15.789	126	1614.78	24.058
27	75.94	5.716	77	627.99	15.971	127	1638.91	24.208
28	81.77	5.941	78	644.05	16.152	128	1663.20	24.357
29	87.82	6.165	79	660.29	16.332	129	1687.63	24.506
30	94.10	6.390	80	676.71	16.512	150	1712.21	24.655
31	100.60	6.614	81	693.31	16.690	131	1736.94	24.802
32	107.33	6.837	82	710.09	16.868	132	1761.81	24.930
33	114.28	7.061	83	727.05	17.046	133	1786.84	25.096
34	121.45	7.283	84	744.18	17.222	134	1812.01	25.242
35	128.84	7.505	85	761.49	17.398	135	1837.32	25.388
36	136.46	7.727	86	778.98	17.575	136	1862.78	25.533
37	144.29	7.947	87	796.64	17.747	137	1888.39	25.677
38	152.35	8.167	88	814.47	17.921	138	1914.14	25.821
39	160.63	8.386	89	832.48	18.094	139	1940.03	25.964
40	169.12	8.604	90	850.66	18.266	140	1966.06	26.107
41	177.84	8.821	91	869.01	18.438	141	1992.24	26.249
42	186.77	9.038	92	887.53	18.608	142	2018.56	26.390
43	195.91	9.253	93	906.23	18.779	143	2045.02	26.531
44	205.27	9.468	94	925.09	18.948	144	2071.62	26.671
45	214.85	9.681	95	944.13	19.117	145	2098.37	26.811
46	224.63	9.893	96	963.33	19.286	146	2125.25	26.950
47	234.63	10.105	97	982.70	19.453	147	2152.26	27.088
48	244.84	10.315	98	1002.23	19.620	148	2179.42	27.226
49	255.26	10.524	99	1021.94	19.787	149	2206.72	27.363
50	265.89	10.733	100	1041.81	19.952	150	2234.15	27.500

TABLE 59—*Type K thermocouples (Chromel versus Alumel) (continued).*

T, deg K	E, μV	dE/dT, μV/K	T, deg K	E, μV	dE/dT, μV/K	T, deg K	E, μV	dE/dT, μV/K
151	2261.72	27.636	201	3800.43	33.652	251	5600.01	38.058
152	2289.42	27.772	202	3834.13	33.756	252	5630.10	38.129
153	2317.26	27.906	203	3867.94	33.860	253	5676.26	38.200
154	2345.23	28.041	204	3901.85	33.962	254	5714.50	38.270
155	2373.34	28.174	205	3935.86	34.065	255	5752.80	38.339
156	2401.58	28.507	206	3969.98	34.166	256	5791.18	38.407
157	2429.96	28.440	207	4004.20	34.267	257	5029.62	38.475
158	2458.46	28.572	208	4038.51	34.367	258	5868.13	38.543
159	2487.10	28.703	209	4072.93	34.467	259	5906.70	38.609
160	2515.87	28.833	210	4107.45	34.566	260	5945.34	38.675
161	2544.77	28.963	211	4142.06	34.664	261	5984.05	38.740
162	2573.79	29.093	212	4176.78	34.762	262	6022.82	38.804
163	2602.95	29.222	213	4211.59	34.859	263	6061.66	38.868
164	2632.24	29.350	214	4246.49	34.955	264	6100.56	38.931
165	2661.65	29.477	215	4281.50	35.051	265	6139.52	38.993
166	2691.19	29.604	216	4316.60	35.146	266	6178.54	39.054
167	2720.86	29.730	217	4351.79	35.240	267	6217.63	39.114
168	2750.65	29.856	218	4387.08	35.334	268	6256.77	39.174
169	2780.57	29.981	219	4422.46	35.427	269	6295.97	39.232
170	2810.61	30.106	220	4457.93	35.520	270	6335.24	39.290
171	2840.78	30.229	221	4493.50	35.611	271	6374.55	39.347
172	2871.07	30.353	222	4529.15	35.702	272	6413.93	39.403
173	2901.49	30.475	223	4564.90	35.793	273	6453.36	39.458
174	2932.02	30.597	224	4600.74	35.883	274	6492.84	39.512
175	2962.68	30.719	225	4636.67	35.972	275	6532.38	39.565
176	2993.46	30.839	226	4672.68	36.060	276	6571.97	39.616
177	3024.36	30.959	227	4708.79	36.148	277	6611.62	39.667
178	3055.38	31.079	228	4744.98	36.235	278	6651.31	39.716
179	3086.52	31.198	229	4781.25	36.321	279	6691.05	39.765
180	3117.77	31.316	230	4817.62	36.407	280	6730.84	39.812
181	3149.15	31.433	231	4854.07	36.492			
182	3180.64	31.550	232	4890.60	36.577			
183	3212.25	31.667	233	4927.22	36.661			
184	3243.97	31.782	234	4963.93	36.744			
185	3275.81	31.898	235	5000.71	36.826			
186	3307.77	32.012	236	5037.58	36.908			
187	3339.84	32.126	237	5074.53	36.989			
188	3372.02	32.239	238	5111.56	37.070			
189	3404.32	32.351	239	5148.67	37.150			
190	3436.72	32.463	240	5185.86	37.229			
191	3469.24	32.575	241	5223.12	37.308			
192	3501.87	32.685	242	5260.47	37.386			
193	3534.61	32.795	243	5297.90	37.463			
194	3567.46	32.905	244	5335.40	37.540			
195	3600.42	33.013	245	5372.97	37.616			
196	3633.49	33.121	246	5410.63	37.691			
197	3666.66	33.229	247	5448.36	37.766			
198	3699.95	33.335	248	5486.16	37.840			
199	3733.33	33.442	249	5524.04	37.913			
200	3766.83	33.547	250	5561.98	37.986			

TABLE 60—*Type EP versus gold-0.07 percent iron (Chromel versus gold-0.07 atomic percent iron) interim values.*

T, deg K	E, μV	dE/dT, μV/K	T, deg K	E, μV	dE/dT, μV/K	T, deg K	E, μV	dE/dT, μV/K
1	7.86	8.645	51	785.00	16.402	101	1665.72	18.751
2	17.21	10.035	52	801.42	16.450	102	1684.47	18.770
3	27.86	11.220	53	817.89	16.498	103	1703.26	18.809
4	39.59	12.226	54	834.42	16.548	104	1722.09	18.848
5	52.26	13.075	55	850.99	16.598	105	1740.95	18.886
6	65.69	13.782	56	867.61	16.648	106	1759.86	18.924
7	79.78	14.369	57	884.29	16.699	107	1778.80	18.961
8	94.40	14.852	58	901.01	16.750	108	1797.78	18.998
9	109.45	15.243	59	917.79	16.801	109	1816.80	19.035
10	124.86	15.555	60	934.61	16.852	110	1855.85	19.071
11	140.54	15.801	61	951.49	16.903	111	1854.94	19.107
12	156.44	15.989	62	968.42	16.953	112	1874.06	19.143
13	172.50	16.128	63	985.40	17.004	113	1893.22	19.178
14	188.68	16.228	64	1002.43	17.055	114	1912.42	19.213
15	204.94	16.293	65	1019.51	17.105	115	1931.65	19.247
16	221.26	16.331	66	1036.64	17.155	116	1950.91	19.281
17	237.60	16.347	67	1053.82	17.205	117	1970.21	19.315
18	253.95	16.346	68	1071.05	17.255	118	1989.54	19.349
19	270.29	16.330	69	1088.33	17.304	119	2008.91	19.382
20	286.60	16.305	70	1105.66	17.354	120	2028.31	19.415
21	302.89	16.272	71	1123.03	17.402	121	2047.74	19.447
22	319.15	16.235	72	1140.46	17.451	122	2067.20	19.480
23	335.36	16.195	73	1157.94	17.499	123	2086.70	19.512
24	351.54	16.154	74	1175.46	17.547	124	2106.23	19.543
25	367.67	16.114	75	1193.03	17.595	125	2125.78	19.575
26	383.77	16.076	76	1210.65	17.642	126	2145.37	19.606
27	399.82	16.040	77	1228.31	17.689	127	2165.00	19.637
28	415.85	16.008	78	1246.03	17.736	128	2184.65	19.668
29	431.84	15.980	79	1263.79	17.783	129	2204.33	19.698
30	447.81	15.957	80	1281.59	17.829	130	2224.04	19.728
31	463.76	15.938	81	1299.44	17.875	131	2243.79	19.758
32	479.69	15.924	82	1317.34	17.921	132	2263.56	19.788
33	495.61	15.915	83	1335.29	17.966	133	2283.36	19.818
34	511.52	15.911	84	1353.28	18.011	134	2303.20	19.847
35	527.43	15.912	85	1371.31	18.056	135	2323.06	19.876
36	543.35	15.917	86	1389.39	18.101	136	2342.95	19.905
37	559.27	15.927	87	1407.51	18.145	137	2362.87	19.934
38	575.20	15.941	88	1425.68	18.189	138	2382.82	19.963
39	591.15	15.960	89	1443.89	18.233	139	2402.80	19.991
40	607.12	15.982	90	1462.14	18.276	140	2422.80	20.020
41	623.12	16.007	91	1480.44	18.319	141	2442.83	20.048
42	639.14	16.036	92	1498.78	18.362	142	2462.90	20.076
43	655.19	16.068	93	1517.16	18.404	143	2482.99	20.104
44	671.28	16.103	94	1535.59	18.446	144	2503.10	20.131
45	687.40	16.140	95	1554.06	18.488	145	2523.25	20.159
46	703.56	16.180	96	1572.56	18.529	146	2543.42	20.186
47	719.76	16.221	97	1591.11	18.570	147	2563.62	20.213
48	736.00	16.264	98	1609.70	18.611	148	2583.85	20.240
49	752.29	16.309	99	1628.34	18.651	149	2604.10	20.267
50	768.62	16.355	100	1647.01	18.691	150	2624.38	20.293

TABLE 60—*Type EP versus gold-0.07 percent iron (Chromel versus gold-0.07 atomic percent iron) interim values (continued).*

T, deg K	E, μV	dE/dT, μV/K	T, deg K	E, μV	dE/dT, μV/K	T, deg K	E, μV	dE/dT, μV/K
151	2644.69	20.320	201	3689.08	21.374	251	4775.99	22.040
152	2665.02	20.346	202	3710.46	21.391	252	4798.03	22.050
153	2685.38	20.372	203	3731.86	21.408	253	4820.09	22.060
154	2705.76	20.398	204	3755.27	21.424	254	4842.15	22.071
155	2726.18	20.425	205	3774.71	21.441	255	4864.25	22.081
156	2746.61	20.449	206	3796.16	21.457	256	4886.31	22.090
157	2767.07	20.474	207	3817.62	21.474	257	4908.41	22.100
158	2787.56	20.499	208	3839.10	21.490	258	4930.51	22.109
159	2808.07	20.524	209	3860.60	21.506	259	4952.63	22.118
160	2828.61	20.548	210	3882.12	21.522	260	4974.75	22.127
161	2849.17	20.573	211	3903.65	21.538	261	4996.88	22.135
162	2869.75	20.597	212	3925.19	21.554	262	5019.03	22.143
163	2890.36	20.621	213	3946.75	21.569	263	5041.17	22.150
164	2910.99	20.644	214	3968.33	21.585	264	5063.32	22.157
165	2931.65	20.668	215	3989.92	21.600	265	5085.40	22.164
166	2952.33	20.691	216	4011.53	21.615	266	5107.64	22.170
167	2973.03	20.714	217	4033.15	21.630	267	5129.82	22.175
168	2993.76	20.736	218	4054.79	21.645	268	5152.00	22.180
169	3014.50	20.759	219	4076.44	21.659	269	5174.18	22.185
170	3035.27	20.781	220	4098.11	21.674	270	5196.37	22.190
171	3056.06	20.803	221	4119.79	21.688	271	5218.56	22.194
172	3076.88	20.825	222	4141.49	21.702	272	5240.75	22.198
173	3097.71	20.846	223	4163.19	21.716	273	5262.96	22.203
174	3118.57	20.867	224	4184.92	21.729	274	5285.16	22.209
175	3139.45	20.888	225	4206.65	21.743	275	5307.37	22.216
176	3160.35	20.909	226	4228.40	21.756	276	5329.59	22.224
177	3181.27	20.930	227	4250.17	21.769	277	5351.83	22.235
178	3202.21	20.950	228	4271.94	21.782	278	5374.06	22.248
179	3223.17	20.970	229	4293.73	21.794	279	5396.32	22.266
180	3244.15	20.990	230	4315.53	21.807	280	5418.60	22.290
181	3265.15	21.010	231	4337.34	21.819			
182	3286.17	21.030	232	4359.17	21.831			
183	3307.21	21.049	233	4381.00	21.843			
184	3328.27	21.068	234	4402.65	21.855			
185	3349.34	21.088	235	4424.71	21.866			
186	3370.44	21.106	236	4446.59	21.878			
187	3391.56	21.125	237	4468.47	21.889			
188	3412.69	21.144	238	4490.36	21.900			
189	3433.85	21.162	239	4512.27	21.911			
190	3455.02	21.180	240	4534.19	21.922			
191	3476.21	21.199	241	4556.11	21.933			
192	3497.41	21.217	242	4578.05	21.944			
193	3518.64	21.235	243	4600.00	21.955			
194	3539.88	21.252	244	4621.96	21.966			
195	3561.14	21.270	245	4643.94	21.977			
196	3582.42	21.288	246	4865.92	21.987			
197	3603.72	21.305	247	4687.91	21.998			
198	3625.03	21.322	248	4709.91	22.009			
199	3646.36	21.340	249	4731.93	22.019			
200	3667.71	21.357	250	4753.95	22.030			

11.4 References

[145] Scott, R. B., *Cryogenic Engineering,* Van Nostrand, Princeton, N. J., 1959.
[146] White, G. K., *Experimental Techniques in Low Temperature Physics,* Oxford University Press, London, 1959.
[147] Rose-Innes, A. C., *Low Temperature Techniques,* Van Nostrand, Princeton, N. J., 1964.

12. BIBLIOGRAPHY

12.1 Introduction

The material contained herein was collected from two general sources: (1) scientific, technical, and trade journals and (2) reports of investigations sponsored or conducted by various governmental agencies. The material is an addition to the references and bibliographies appended to the other sections of the manual. The entries are generally for the period 1963 to 1967, with a few earlier and later items that were brought to our attention. Because of the rapid expansion of the technology, the list does not claim to be exhaustive. Not all of the items are necessarily of current interest, since a few years have made obsolete many that can now be considered only source material of historic interest.

While reasonable coverage was intended, it is inevitable that oversights and other unintentional omissions have occurred; however, references to the various volumes and parts of the series "Temperature, Its Measurement and Control in Science and Industry" were deliberately omitted because the many references to the series throughout the manual are sufficient to assure reference to the series.

The method used to identify the periodicals follows that used in ASTM *Special Technical Publication 329,* 1963 and *STP 329-51, 1964 Coden for Periodical Titles.* For the convenience of the reader who may not have access to these publications, the Coden used in this bibliography are identified in Section 12.3. Numbers prefixed by the letters AD refer to report numbers in the U. S. Government Technical Abstract Bulletin; those by N63, etc., are identifying numbers in the Scientific and Technical Aerospace Reports.

12.2 Bibliography

Anon., "Automatic Recording and Control Used in Thermocouple Comparator," DSER, July-Aug. 1963.

Anon., "Automatic Thermocouple Comparator," INCS, Vol. 37, No. 5, May 1964.

Anon., "Cable Application Information Data Sheet," Thermo-Electric Co. Inc. EDS-72, Oct. 1963.

Anon., "Calibrating Platinum Thermocouples," PTMR, Vol. 8, No. 4, Oct. 1964.

Anon., "Calibrating and Checking Thermocouples in Plants and Laboratories," Leeds and Northrup Co., Folder A1-1121-1965, 1965.

Anon., "Continuous Immersion Thermocouple Aids Blast Furnace Control," IRSE, Vol. 41, No. 7, July 1964.

Anon., "High Temperature Thermometry at Pratt and Whitney Aircraft CANEL," Pratt and Whitney Aircraft, CANEL, PWAC-462, June 1965.

Anon., "Improved Thermocouple Techniques for Safe Operation of Modern Boilers," POWE, Jan. 1963.

Anon., "Metallic Thermocouples for Measurement of Temperature above 1600 C," METL, Vol. 61, No. 365, March 1960.

Anon., "Methods for Testing Used Thermocouples and for Calibrating Thermocouple Materials," Honeywell, Technical Bulletin B15-15A, 1951.

Anon., "New Thermocouple Tables," NBST, May 1966.

Anon., "Pallador II—A New High-Output Noble Metal Thermocouple," PTMR, Vol. 9, No. 5, July 1965.

Anon., "The Physical Properties of the Noble Metals," EITB, Vol. 6, No. 3, Dec. 1965.

Anon., "A Rhodium-Platinum Thermocouple for High Temperatures," PTMR, Vol. 9, No. 1, Jan. 1965.

Anon., "Rugged Reliable Space Sensors," MEEN, Vol. 85, No. 9, Sept. 1963.

Anon., "Soviet Infrared Sensors, I—Thermal Sensors," L. C. ATD Report P-65-44, July 1965, AD-619813.

Anon., "Standards and Calibration—NBS Thermocouple Research," NBST, Vol. 48, No. 12, Dec. 1964.

Anon., "Steam, Its Generation and Use," The Babcock and Wilcox Co., 1965.

Anon., "Surface Temperature Measurements with Thermoelectric Materials," Wright-Patterson Air Force Base, Ohio, Technical Document Report No. ASD-TR-61-373, Aug. 1962.

Anon., "Telemetry Transducer Handbook," Vol. 1, Revision 1, Radiation Inc., WPAFB WADD-TR-61-67, Sept. 1963.

Anon., "Thin Film Thermocouple," IRES, Sept. 1968.

Anon., "Transducer Measures Low-Heat Flux Values," PREN, Vol. 35, No. 26, 21 Dec. 1964.

Anon., "Unusual Thermocouples and Accessories," INCS, Vol. 36, No. 5, 6, 7, 8, May, June, July, Aug. 1963.

Abrahamsen, R. F., "Thermocouple Installations for Tube Surface Temperature Measurement," ISAP, Paper No. 44-TC-61, June 1961.

Accinno, D. J. and Schneider, J. F., "Platinel—A Noble Metal Thermocouple to Replace Chromel-Alumel," EITB, Vol. 1, No. 2, Sept. 1960.

Accinno, D. J., "A Vacuum Furnace for High Temperature Thermocouple Calibration," EITB, Vol. 2, No. 1, June 1961.

Allen, J. G. and Pears, C. D., "Evaluation of the Performance of Tungsten-Tungsten 26 Rhenium Thermocouples to about 5000 F," SRI, STAR N64-16269.

Almond, R. J., "Errors in Thermocouple Circuits," INCS, Vol. 33, No. 1, Jan. 1960.

Anderson, A. R. and MacKenzie, D. J., "Materials for High (2500–4000 F) Gas Turbine Temperature Measurements," SAE, Paper No. 158B, April 1960.

Angerhofer, A. W., "Cryogenic Instrumentation—I," CEGN, Vol. 12, No. 10, Oct. 1965.

Bailleal-Langlais, J., "Les Thermocouples," VIDE, Vol. 14, No. 84, Nov.–Dec. 1959.

Beakley, W. R., "Method of Modifying the Thermal EMF Temperature Characteristics of Constantan Thermocouple Wire," JSIN, Vol. 31, No. 7, July 1954.

Beck, J. V., "Thermoelectric Temperature Disturbances in Low Conductivity Materials," JHTR, May 1962.

Beck, J. V. and Hurwicz, H., "Effect of Thermocouple Cavity on Heat Sink Temperature," JHTR, Vol. 82, No. 1, Feb. 1960.

Beck, J. V., "Study of Thermal Discontinuities and Associated Temperature Disturbances in a Solid Subject to a Surface Heat Flux," "Effects of Sensors in Low Conductivity Materials upon Temperature Distribution and its Measurements," AVCO Corp., RAD-TR-9(7)-59-26, Feb. 1960, AD-461644.

Bedford, R. E., "Reference Tables for Platinum 20% Rhodium-Platinum 5% Rhodium Thermocouples," RSIN, Vol. 35, No. 9, Sept. 1964.

Bedford, R. E., "Reference Tables for Platinum 40% Rhodium-Platinum 20% Rhodium Thermocouples," RSIN, Vol. 36, No. 11, Nov. 1965.

Belling, P. R., "High Pressure Thermocouple Feed Through," INCS, Vol., No. 11, Nov. 1962.

Benedict, R. P. and Ashby, H. F., "Empirical Determination of Thermocouple Characteristics," JEPO (A), Vol. 85, No. 1, Jan. 1963.

Benedict, R. P., "The Calibration of Thermocouples by the Freezing-Point Baths and Empirical Equations," JEPO, April 1959.

Benedict, R. P., "International Practical Temperature Scale of 1968," LNTJ, No. 6, Spring 1969.

Benedict, R. P., "Review of Practical Thermometry," ASMS, Paper No. 57-A-203, Dec. 1957.

Benedict, R. P., "Temperature and Its Measurement," ELTE, Vol. 72, No. 7, July 1963.

Benedict, R. P., "Temperature Measurement in Moving Fluids," ASMS, Paper No. 59-A-257, Dec. 1959.

Benedict, R. P., "Thermoelectric Effects," ELMA, Feb. 1960.

Bennett, H. E., "Noble Metal Thermocouples," Johnson, Matthey and Co., Ltd., Feb. 1956.

Bennett, H. E., "Pallador Thermocouple," PTMR, Vol. 4, No. 2, April 1960.

Bennett, H. E., "The Care and Maintenance of Thermocouples," PTMR, Vol. 9, No. 2, April 1965.

Bergund, C. N. and Beairsto, R. C., "An Automatic Technique for Accurate Measurement of Seebeck Coefficient," RSIN, Vol. 38, No. 1, Jan. 1967.

Berry, J. M. and Martine, D. L., "Thermocouple Immersion Errors," ASTT, No. 178, 1955.

Bertodo, R. J., "Thermocouple for the Measurement of Gas Temperatures up to 2000 C," PTML, Vol. 177, No. 22, 1963.

Blackburn, G. F. and Caldwell, F. R., "Reference Tables for Thermocouples of Iridium-Rhodium Alloys Versus Iridium," JNBC (C), Vol. 68, No. 1, Jan. 1964.

Boerdijk, A. H., "Diagrams Representing States of Operation of General Thermocouple," JAPI, Vol. 31, No. 7, July 1960.

Boerdijk, A. H., "Contributions to General Theory of Thermocouples," JAPI, Vol. 30, No. 7, July 1959.

Boon, G. and Van-Der Woulde, F., "Very Stable and Simple Compensator for Thermocouples," RSIN, Vol. 36, No. 6, June 1965.

Bose, B. N., "Thermocouple Well Design," ISAJ, Vol. 9, No. 9, Sept. 1962.

Boudreaux, P. J., "Providing Thermocouple Compensation at Low Cost," ELEC, Vol. 35, No. 11, Nov. 1962.

Brooks, E. J. and O'Neal, E. W., "Keep Your Thermocouple Leads Dry," JISA, Vol. 12, No. 3, March 1965.

Brown, D. A. H. et al, "Construction of Radiation Thermocouples Using Semi-Conducting Thermoelectric Materials," JSIN, Vol. 30, No. 6, June 1953.

Brownell, H. R., "Thermoelement Transfer," INCS, Vol. 36, No. 1, Jan. 1963.

Bullis, L. H., "Vacuum-Deposited Thin-Film Thermocouples for Accurate Measurement of Substrate Surface Temperature," JSIN, Vol. 40, No. 12, Dec. 1963.

Burns, G. W. and Gallagher, J. S., "Reference Tables for the Pt-30 Percent Rh Versus Pt-6 Percent Rh Thermocouple," JNBS(C), Vol. 700, No. 2, April–June 1966.

Burns, G. W., "Studies at NBS of Platinum-6% Rhodium Versus Platinum-30% Rhodium Thermocouple—Preliminary Report," SEPP, Paper 750B, 23–27 Sept. 1963.

Caldwell, F. R. et al, "Intercomparison of Thermocouple Response Data," SAE, Paper 158F, National Aeronautic Meeting, April 1960.

Caldwell, R., "Temperature of Thermocouple Reference Junctions in an Ice Bath," JNBS(C), Vol. 69C, No. 2, April–June 1965.

Cameron, G. and Blanchard, R. L., "Performance and Use of Metal Freezing Point Cells Which Generate Precise Temperature," ISAJ, Vol. 2, No. 3, 1963.

Carlson, A. V., "An Evaluation of Radiation Compensating Thermocouple Temperature Sensors," U.S. Army Electronics Research and Development Activity, Arizona, USAERDAA-MET-10-64, Sept. 1964, AD449874.

Chambers, J. T. et al, "Experimental Evaluation of a Dual Element Transducer for High Temperature Gas Measurements," American Radiation and Standard Sanitary Corp., Mountain View, Calif., ARL 63-58, March 1963, AD-409-816.

Charet, I., "High Detectivity Microcalorimeters," JSIN, Vol. 40, No. 8, Aug. 1963.

Chenoweth, J. M., "Gun-Barrel Measurements Involve Rapidly Fluctuating Temperatures," INST, Vol. 26, No. 11, Nov. 1953.

Clagett, T. J., "External Thermocouple Compensation," INSR, Vol. 18, No. 2, 1965.

Cook, G. E. and McCampbell, W. M., "Thermocouple Probe Controls Welding Torch," CENG, Vol. 14, No. 11, Nov. 1967.

Cox, J. E., "High Vacuum Thermocouple," RSIN, Vol. 34, No. 8, Aug. 1963.

Culpin, M. F. and Martin, K. A., "Instruments for Measuring the Temperature of a Running Thread-Line and a Jet of Viscous Liquid," JSIN, July 1960.

Cutt, "Proportional Control Steadies Cryosurgical Probe Temperature," CEGN, Vol. 12, No. 3, March 1965.

Dallman, A. C., "Mechanical Reliability and Thermoelectrical Stability of Noble Metal Thermocouples at 2000 F Temperature and Dose Rates up to 10 to 2 Power Nvt.," WPAFB, SEG. TDR-64 7, April 1964, AD-601-267.

Danberg, J. E., "The Equilibrium Temperature Probe, A Device for Measuring Temperature in Hypersonic Boundary Layers," Naval Ordnance Laboratory, White Oak, Md., NOL-TR-61-2, Dec. 1961, AD-439-624.

Daneshesku, S. K., "Selection and Calibration of Tungsten and Molybdenum Wire for Thermocouples," MSTC (IZTE), No. 5, May 1959.

Dauphinee, T. M., "An Apparatus for Comparison of Thermocouples," CJPH, Vol. 33, 1955.

Dauphinee, T. M., "Use Thermocouples for Measuring Temperatures Below 70 K," JSIN, Vol. 30, No. 11, Nov. 1953.

Davis, D. A., "Two Thermocouples Suitable for Measurements of Temperature up to 2800 C," JSIN, Vol. 1, Jan. 1960.

Decker, W. H., "Bellows Seals Furnace Couples," PEPR, Vol. 8, Aug. 1953.

DeLee, R. V. and Seari, A. E., "Study to Determine Suitable High Temperature, High Altitude, Total Temperature Sensors," Rosemount Engineering Co., Report No. 11652B, Nov. 1965, NASA No. N66-13968.

Denison, J. N., "Technique to Construct Fine Wire Thermocouples," RSIN, Vol. 33, No. 8, Aug. 1962.

Devita, V. A., "Predicting Cryogenic Thermocouple Time Constant," CEGN, Vol. 12, No. 7, July 1965.

Diven, J. L., "Temperature Deficiency of a Heat Transfer Calorimeter," Brown Engineering Co., Huntsville, Ala., Technical No. R-44, April 1963, AD 415146.

Donne, M. D., "Tests and Data Concerning Platinel—A New High Temperature Thermocouple," EITB, Vol. 5, No. 1, June 1964.

Dow, M. B., "Comparison of Measurement of Internal Temperatures in Ablation Material by Various Thermocouple Configurations," National Aeronautics and Space Administration, NASA-TND-2165, Nov. 1964.

Dutton, R. and Lee, E. C., "Surface Temperature Measurement of Current Carrying Objects," ISAJ, Vol. 6, No. 12, Dec. 1959.

Eisenman, W. L. et al, "Properties of Photodetectors. Photodetector Series, 58th Report," Naval Ordnance Laboratory, Corona, Calif., Report No. NOLC-588, July 1963, AD-413-131.

Eldridge, D. A. G. and Mansell, D. H. L., "An Automatic Temperature Measuring System for Propellants Undergoing Stability Trials," Ministry of Aviation, Explosives Research and Development Establishment, (GB), ERDE, Sept. 1965.

Elliott, R. D., "Dynamic Behavior of Missile Skin Temperature Transducers," ISAP, Paper No. 160-LA-61, Sept. 1961.

Evans, D. J. and Strnat, K., "An Apparatus for Direct and Differential Thermal Analysis at Temperature up to 1500 C," WPAFB-ASD, ASD-TDR-63504, July, 1963, AD-418-225.

Falweiler, R. C., "Investigation of High Temperature Thermal Expansion Devices and Design of a Unit for Use with Thermal Image Heating," Lexington Labs. Inc., Cambridge, Mass., AFCRL 62-952, Dec. 1962, AD-296433.

Farrow, R. L. and Levitt, A. P., "Tungsten/Tungsten-Rhenium Thermocouples in a Carbon

Atmosphere," U.S. Army Materials Research Agency, Watertown, Mass., AMRA-TR-64-12, July 1964, AD-606-565.

Faul, J. C., "Thermocouple Performance in Gas Streams," INCS, Vol. 35, No. 12, Dec. 1962.

Fenton, A. W., "The Theory of Thermoelectric Thermometers," United Kingdom Atomic Energy Authority, unpublished, 1969.

Fenton, A. W., "Errors in Thermoelectric Thermometers," *Proceedings,* Institution of Electrical Engineers, Vol. 116, No. 7, July 1969.

Fenton, A. W. et al, "Thermocouples: Instabilities of Seebeck Coefficients," United Kingdom Atomic Energy Authority, Report No. TRG 1447(R), Jan. 1967.

Finch, D. I., "General Principles of Thermoelectric Thermometry," Leeds and Northrup Co., Technical Publication ENS2 (1), 1961.

Finnemore, D. K. et al, "Secondary Thermometer for the 4 to 20 K Range," RSIN, Vol. 36, No. 9, Sept. 1965.

Fluke, G. L., "Tungsten-Rhenium Thermocouple Performance in STR-13 Test," Aerojet-General Corp., 7461:0094, July 1964, AD-453293.

Foote, P. D. et al, "Pyrometric Practice," NBSE, No. 170, 1921.

Francinllo, S., "Thermocouple Development, Lithium-Cooled Reactor Experiment," Pratt-Whitney Aircraft Co., Report No. 422, N-64-16000.

Freedman, J. W., "Panel Discussion on Pyrometric Practice in Elevated Temperature Testing," ASTT, No. 178, 1955.

Freeze, P. D. et al, "Reference Table for the Palladium Versus Platinum-15% Iridium Thermocouple," NBS, Report No. ASD-TDR-62-523, Dec. 1962, AD-295-607.

Fuschilla, N., "Inhomogeneity Emfs in Thermoelectric Thermometers," JSIN, Vol. 31, No. 4, April 1954.

Garsuch, R. G., "Thermal Transfer Voltmeter," INCS, Vol. 36, No. 1, Jan. 1963.

Gelb, G. H. et al, "Manufacture on Fine Wire Thermocouple Probes," RSNI, Vol. 35, No. 1, Jan. 1964.

George, A. L., "Conversion Formula for Thermocouples," INCS, Vol. 36, No. 7, July 1963.

Glawe, G. E. et al, "Radiation and Recovery Corrections and Time Constants of Several Chromel-Alumel Thermocouple Probes in High Temperature, High Velocity Gas Streams," NACA, TN-3766, Oct. 1956.

Goldberg, J. L. and Ling, H. M., "Automatic Switching of Sensitive Thermocouples," ELEG, Vol. 36, No. 9, Sept. 1964.

Graham, J. W., "Project Asset Instrumentation Test of Columbium Sheathed Platinum Rhodium Thermocouples," McDonnell Aircraft Corp., St. Louis, Mo., Report No. 051-052, Aug. 1962, AD-298-369.

Greenberg, H. J. and Sysk, E. D., "Applied Research, Fabrication and Testing of 2300F Thermocouples for Air-Breathing Propulsion Systems," Engelhard Industries, Inc., East Newark, N.J., Report ASD-TDR-C2-891, Jan. 1963, AD-400-470.

Greenwood, T. L., "Thermocouple Continuity Monitor," Marshall Space Flight Center, NASA, NASA Technical Brief 63-10567.

Griffin, J. E. and Hermach, F. L., "Differential Thermocouple Voltmeter," CELE, Nov. 1962.

Gross, J. and Griffith, C. B., "Thermocouple for Molten Steel," MEPD, Vol. 83, No. 6, June 1963.

Gross, P. M. et al, "An Advanced High Temperature Thermocouple for Use on Aerospace Vehicles," Weston Instruments Co., A-FFDL-TR-66-24, May 1966, AD-487-788.

Gross, P. M. et al, "A Re-Entry Thermocouple for Use up to 4500 F," ISAT, Vol. 3, No. 4, Oct. 1964.

Gucker, F. T., Jr., and Peterson, A. H., "Simple Circuit for Adapting Thermocouple Recorders to Measure Voltage in High Resistance Circuits," ANCH, Vol. 25, No. 10, Oct. 1953.

Gumprecht, D. L., "Cycle Counter for Zone-Refining Systems," CHEG, Vol. 69, Nov. 26, 1963.

Gutt, W., "Microfurnace for High Temperature Microscopy and X-Ray Analysis up to 2150 C," JSIN, Vol. 41, No. 6, June 1964.

Hall, B. F., Jr., and Spooner, N. F., "Application and Performance Data for Tungsten-Rhenium Alloy Thermocouples," SEPP, Paper No. 750C, Sept. 1963.

Hartz, R. A., "Temperature Transducers—A Guide for Selection," MADE, Vol. 38, No. 21, 15 Sept. 1966.

Harvey, N. D., "Instrumenting Models for Aerodynamic Heat Transfer Studies Involving Transient Heating Rates," ISAT, Vol. 6, No. 1, Jan. 1967.

Heitzman, W. P., "Dual Thermocouple Temperature Control," INCS, Vol. 37, No. 5, May 1964.

Hendricks, J. W. and McElroy, D. L., "High Temperature-High Vacuum Thermocouple Drift Tests," ASTM Paper, unpublished, Annual Meeting, June 1964.

Henning, C. D. and Parker, R., "Transient Response of an Intrinsic Thermocouple," JHTR, Vol. 89, No. 2, May 1967.

Hicks, E. W., Jr., "The Requirements for a Direct Reading 5000 F Thermocouple," ISPA, Paper No. 29.2.63, Chicago, Ill., Sept. 1963.

Hicksman, M. E. and Ivory, J. E., "Basic Studies of the Metallic State (Thermoelectric Properties of Tungsten and Kovar-A)," RNRL, April 1965.

Hill, J. S., "The Use of Fibro Platinum in Thermocouple Elements," EITB, Vol. 11, No. 3, Dec. 1961.

Hoig, L. B., "Design Procedure for Thermocouple Probes," SAE, Paper No. 158C, National Aeronautic Meeting, April 1960.

Houke, H. F., "Study of High-Current Resistance Testing of Grounded Shealded Chromel-Alumel Thermocouples," Aerojet Liquid Rocket Plant, Report No. 347,60.00.00-A6-02, Aug. 1961, AD-45984.

Houska, C. R. and Keplin, E. J., "A High Temperature Furnace for Quantitative X-Ray Intensity Measurements," Union Carbide Research Department, Tarrytown, N. Y., TRC 14, April 1963, AD-405-495.

Howley, S. A. et al, "Fabrication of Miniature Thermocouple for UHF Acoustic Detectors," RSIN, Vol. 33, No. 10, Oct. 1962.

Hughes, P. C. and Berley, N. A., "Metallurgical Factors Affecting Stability of Nickel-Base Thermocouples," JIME, Vol. 91, 1962–1963.

Huhes, W. F. and Gaylord, E. W., "On the Theoretical Analysis of a Dynamic Thermocouple, 2—The Continuous Area Interface," JAMC, Vol. 27, No. 2, June 1960.

Hurst, N. J., "Effect of Thermocouple Size and Installation on Thermocouple Readings," U.S. Army Missile Command, Redstone Arsenal, Ala., RS-TM-64-1, April 1964, AD-612493.

Irvine, F. H. et al, "Measurements of the Response of Various Thermocouple Arrangements," Royal Aircraft Establishment, AERO-2959, April, 1964, AD-449469.

Janeschitz-Kriegel et al, "A Temperature Probe for Flowing Polymer Melts," JSIN, Vol. 40, No. 8, Aug. 1963.

Jensan, J. T. et al, "Thermocouple Errors Using Pt-Pt(Rh) Thermocouples on Nickel Surfaces," RSIN, Vol. 32, No. 12, Dec. 1964.

Jones, E. W., "Facile Thermocouple Calibrations," INCS, Vol. 40, No. 5, May 1967.

Jordan, H. C., "Welded Thermocouple Junctions," INCS, Vol. 33, No. 6, June 1960.

Jury, S. H., "Mechanically Refrigerated Ice Bath," INCS, Vol. 40, No. 5, May 1967.

Kanter, J. J., "Panel Discussion on Pyrometric Practices," ASTT, No. 178, 1955.

Kashmiry, M. A., "Hermach-Engelhard Transfer Standard," INCS, Vol. 36, No. 1, Jan. 1963.

Kaufman, A. B., "Cryogenic Characteristics of Alloy Wires," INCS, Vol. 37, No. 3, March 1964.

Kelly, D., "Thermocouple Temperature Measurement without Special Instruments," INCS, Vol. 33, No. 1, Jan. 1960.

Kendall, D. N. et al, "SemiConductor Surface Thermocouples for Determining Heat Transfer Rates," ITAE, July 1967.

Kepner, A. A., "Thermocouple Temperature Transducer," North American Aviation Inc., Downey, Calif., Report No. MC 449 0006, Nov. 1962, AD-415-586.

Kilkin, G. M. et al, "Lithium-Boiling Potassium Test Loop—Interim Report," Jet Propulsion Laboratory, TR 32-1083, Sept. 1966, N67-26629.

King, J., "The Design of a Temperature Compensated Thermocouple Reference Circuit," Naval Ordnance Test Station, China Lake, Calif., NOTS TR-3462, NAVWEPS 8492, May 1964, AD-600675.

Kisely, V. M., "Thermal Probes for Measuring the Density of Heat Flows," Foreign Translation Division, WPAFB, Ohio, FTD-TT-64-1151(1+2), April 1965, AD-614949.

Klein, C. A. et al, "Analysis and Experimental of an Ultra-High Temperature Pyrolytic Graphite Thermocouple," Raytheon Co. Research Division, AFFDL-TR-6627, June 1966, AD-802214.

Klein, C. A. et al, "Development of an Ultra-High Temperature Pyrolytic Graphite Thermo-couple," Raytheon Co., Waltham, Mass., AD-431-203.

Knowlson, P. M. and Bartie, P. M., "Sealing a Stainless Steel Sheathed Thermocouple into a Magnox End-Cap," JIME, Vol. 93, 1964–1965.

Koselapor, V. I. and Skvortsov, Y. M., "Thermocouple Needle," INDL (ZVDL), July 1963.

Kostkowski, H. J. and Burns, G. W., "Thermocouple and Radiation Thermometry above 900 K," NASA, Symposium, Measurement of Thermal Radiation Properties of Solids, Sept. 1963.

Kovacs, A. and Mesler, R. B., "Making and Testing a Small Surface Thermocouple for Fast Response," RSIN, Vol. 35, No. 4, April 1964.

Krag, J., "Improved Needle Thermocouple for Subcutaneous and Intramuscular Tempera-ture Measurements in Animals and Men," RSIN, Vol. 25, No. 8, Aug. 1954.

Krivka, M. A., "Make Your Own Thermocouples," MADE, 9 June 1960.

Krysiak, K. F. et al, "A Thermocouple Suitable for Use from 2500 to 4500 F on Aerospace Vehicles," Weston Instrument Division, Daystrom Inc., Newark, N. J., X10232, June 1963, AD-414-562.

Kuhlan, W. C., "Status Report of Investigation of Thermocouple Materials for Use at Temp-eratures above 4500 F," SEPP, Paper No. 750D, Sept. 1963.

Kuhlman, W. C., "Research and Evaluation of Materials for Thermoelectric Application Suitable for Temperature Measurement up to 4500 F on the Surface of Glide Re-Entry Vehicles," General Electric Co., Cincinatti, Ohio, TDR-63-233, March 1962 and 1963, AD-404-577.

Lachman, J. C., "Calibration of Rhenium-Molybdenum and Rhenium-Tungsten Thermo-couple to 4000 F," XAER, APEX 365, April 1958, AD-159-884.

Lachman, J. C., "New Development in Tungsten/Tungsten-Rhenium Thermocouples," ISAC, Paper No. 150-LA-61, Sept. 1961.

Lachman, J. C., "Refractory Metal Thermocouples," ASME, Paper No. 59-HT-21, Aug. 1959.

Lanza, G. et al, "Studies on Plasmas and Pulsed Magnetic Field Effects (Plasma Thermo-couples)," N 64-15493, 1963.

Lee, J. D., "The Measurement of Velocity and Temperature in a Hypersonic Laminar Bound-ary Layer," Ohio State University, Columbus, Ohio, ASD TRD-62-914, Dec. 1962, AD-295-466.

Levinstein, H., "Infrared Detectors," Syracuse University Research Institutes, N. Y., Physics 105-1, Feb. 1963, AD-296-831.

Lewis, E. L., ed., "Instrument Standards," Naval Ship Engineering Center, Philadelphia Division, Philadelphia, Pa., 1966.

Lewis, H. W. and Reitz, T. R., "Efficiency of the Plasma Thermocouple," JAPI, Vol. 31, April 1960.

Leyda, W. E., "Summary of Pyrometric Procedure Employed by One Company in Creep-Rupture Testing," ASTT, No. 178, 1955.

Lipson, H. G. and Bouthillette, L. Q., "Epoxy Vacuum Seal for Introduction of Leads into Cryogenic Equipment," Air Force Cambridge Research Laboratory, Bedford, Mass., AFCRL 63-78, March 1963, AD-410018.

Loeffler, R. F., "Thermocouples, Resistance Temperature Detectors, Thermistors–Which," INCS, Vol. 40, No. 5, May 1967.

Lyell, J. A., "Designing a Modern Metrology Laboratory," JIST, Vol. 11, No. 6, June 1964.

Maevaskaya, L. A. and Morozor, A. D., "Thermoelectric Anemometer for Measuring Air Flow Speeds," MSTC (IZTE), No. 7, July 1962.

Marglis, O. M. et al, "Tips for Immersion Type Thermocouples Made from Zirconium Dioxide of Increased Resistance to Heat," OGNP, No. 12, 1962, AD-411-058.

Matthew, H. L., "Thermocouple Calibration in a Subsonic Air Flow," University of Wyoming Master Thesis, May 1965, AD-617-305.

McCoy, H. E., Jr., "Influence of CO–CO$_2$ Environments on the Calibration of Chromel-P Alumel Thermocouples," Oak Ridge National Laboratory, Report No. ORNL-P-1070, March 1965.

McElroy, D. L., "Progress Report of Thermocouple Research—Report for Period Nov. 1, 1956 to Oct. 31, 1957, Oak Ridge National Laboratory, Report No. ORNL-2467, 1968.

McSherry, P. B., "Metal Ceramic Wall Lengthens Thermocouple Life," IRAG, Vol. 172, 29 Oct. 1953.

Meador, J. D., "Dynamic Testing of Gas Sample Thermocouples," SAE, Paper No. 158G, April 1960.

Mesalovich, G. I., "Local Temperature Measurements in Forging and Stamping," MSTC (IZRE), Nov. 1962.

Miksch, E. S., "Equilibration of the Ice-Water Temperature Standard," RSIN, Vol, 36, No. 6, June 1965.

Miller, C. E. and Flynn, T. M., "On the Problems of Measuring Transient Temperature in Cryogenic Fluids," ISAT, Vol. 6, No. 2, April 1967.

Miller, J. T., "The Revised Cource in Industrial Technology," INPA, Vol. 18, No. 1, Jan. 1964.

Moeller, C. E., "Special Surface Thermocouples," INCS, Vol. 36, No. 5, May 1963.

Moen, N. K., "Use Thermocouples Right and Melt Away Weight," SAEJ, Vol. 71, No. 11, Nov. 1963.

Moffat, R. J., "Stable High Temperature Thermometry Rig," SAE, Paper No. 158E, April 1960.

Monroc, J., "Automatic Thermal Transfer Meter," INCS, Vol. 36, No. 1, Jan. 1963.

Morrison, R. D. and Lachenmayer, R. R., "Thin Film Thermocouples for Substrate Temperature Measurement," RSIN, Vol. 34, No. 1, Jan. 1963.

Morrison, R. G., "Application of Miniature Intrinsic Thermocouples for Reactor Transient Diagnostics," Los Alomos Scientific Laboratory, LA-DC-7004, March 1965.

Muth, S., Jr., "Reference Junctions," INCS, Vol. 40, No. 5, May 1967.

Nanigian, J. et al, "Instantaneous Heat Transfer, Pressure and Surface Temperature Characteristics of Solid Propellant Rocket Igniters," U.S. Naval Propellant Plant, Indian Head, Md., Report No. 178, April 1960.

Nanigian, J., "Matching Thermocouple Well Material to Wall Improves Accuracy," ISAJ, Vol. 10, No. 2, Feb. 1963.

Nanigian, J., "Ribbon Thermocouples in the 3000 to 5000 F Range," INCS, Vol. 39, No. 5, May 1966.

Nanigian, J., "Temperature Measurements and Heat Transfer Calculation in Rocket Nozzle Throats and Exit Cones," ISAC, Paper No. 29.3.63, Sept. 1963.

Nanigian, J., "Thermal Properties of Thermocouples," INCS, Vol. 36, No. 10, Oct. 1963.

Nilson, J. R., "Elimination of Electrical Interference In High Temperature Thermocouple Installations," ISAC, Paper No. 36 SF60, May 1960.

Nydick, S. E., "Thermocouple Errors in Ablation Materials," ISAC, Paper No. 16.12.-3-66, Oct. 1966.

Olsen, K. N., "Temperature System Standardizer," Rocketdyne, Canoga Park, Calif., TR-6113, AD-409-338.

Olsen, L. O. and Freeze, P. D., "Reference Tables for the Platinel II Thermocouples," JRNB, Vol. 68C, No. 4, Oct.–Dec. 1964.

Olsen, L. O., "Catalytic Effects of Thermocouple Materials," MEPD, Vol. 85, No. 1, Jan. 1964.

Olsen, L. O., "Guideposts to Keep Thermocouple Users Out of Hot Water," SAEJ, Vol. 72, No. 8, Aug. 1964.

Olsen, L. O., "Some Recent Developments in Noble Metal Thermocouples," SEPP, Paper No. 750 A, 23–27 Sept. 1963.

Pak, V., "Calibration of Thermocouples in a Dynamic Condition," MSTC(IZTE), Vol. 6, No. 6, June 1963.

Pak, V., "A Thermocouple for Measuring Temperature Drops in Electrically Conductive Media," INDL (ZVDL), July 1963.

Pak, V. et al, "Simplified Equipment for Calibration of Precious Metal Thermocouples in a Dynamic Condition," MSTC (IZTE), June 1963.

Palmer, E. P. and Turner, G. H., "Response of Thermocouple Junctions to Shock Waves," Utah Research and Development Co., Inc., Report No. ATL-TR-66-36, April 1966, AD-484507.

Palmer, R. B. P., "Thermoelectric Fields in Liquids," JSIN, Vol. 30, No. 6, June 1953.

Parker, R., "Method for Determining the Response of Temperature Sensors to a Rapid Temperature Rise," ISAC, Paper No. 16.10-1-66, Oct. 1966.

Patter, R. D., "Open Circuit Thermocouple," MEPD, Vol. 64, No. 11, Nov. 1953.

Paul, M. C. and Obery, P. E., "Modified Pyrometer Temperature Measurement Technique," RSIN, Vol. 35, No. 1, Aug. 1964.

Plyukhin, V. S. and Kologrivov, V. N., "Emf of Thermocouples Compressed by a Shock Wave," ZPMF, 1962, WPAFB-FTD, Report No. FTD-TT-63-1057, AD-426901.

Pokhodnya, I. K. and Fruman, I. I., "The Temperature of a Weld Pool," AUSV, No. 5, 1955, AD-400830.

Potts, J. F., Jr., and McElroy, D. L., "Basic Studies on Base Metal Thermocouples," SAE, Paper No. 158 A, April 1960.

Potts, J. F. and McElroy, D. L., "Thermocouple Research to 1000 C Final Report, Nov. 1, 1957 to June 30, 1959," Oak Ridge National Laboratory, Report No. ORNL 2773, 1959.

Powell, R. L. et al, "Low Temperature Thermocouples 1. Gold-Cobalt or Constantan Versus Copper or 'Normal Silver'," CRYO, No. 5, March 1961.

Powell, W. B. and Price, E. W., "A Method for Determination of Level Heat Flux from Transient Temperature Measurements," ISAC, Paper No. 8.2.63, Sept. 1963.

Prentice, J. L., "Flashdown in Solid Propellants," Naval Ordnance Test Station, China Lake, Calif., NOTS TP 3009, Dec. 1962, AD-295979.

Rabinowitz, E., "Equations Give Quick Estimates of Friction Temperature," PREN, Vol. 35, No. 4, March 1964.

Rainey, W. T., Jr., and Bennett, R. L., "Stability of Base Metal Thermocouples in Helium Atmosphere," ISAT, Vol. 2, No. 1, Jan. 1963.

Ramachandran, S. and Acre, T. R., "Measuring Molten Steel Temperature," ISAJ, Vol. 11, No. 3, March 1964.

Rauch, W. G., "Design and Construction of Needle Thermocouples," MEDD, Vol. 65, No. 3, March 1954.

Redr, M. et al, "Thermocouples for Prolonged and Deep Temperature Measurement of Liquids in Moulds," HUTL, Vol. 18, No. 7, 1963.

Robertson, D., "A New Sensitive Temperature Detector for Use in High Pressure Fluid Piping," ASMS, Paper No. 59-A-201, Dec. 1959.

Roeser, W. F. and Lonberger, S. L., "Methods of Testing Thermocouples and Thermocouple Material," NBSC, No. 590, Feb. 1958.

Rosenbaum, "Some Low Temperature Thermometry Observations," RSIN, Vol. 40, No. 4, April 1969.

Rozenblit, G. B., "Measurement of Small Temperature Variations on the Surface of Solid Bodies," MSTC, Vol. 2, No. 2, Feb. 1962.

Rubin, L. G., "Measuring Temperature," ISCT, No. 25, Jan. 1964.

Rubin, L. G., "Temperature Concepts, Seals and Measurement Techniques," Raytheon Co. Technical Memorandum T-538.

Ruff, A. W., Jr., "Open-Probe Thermocouple Control of Radio-Frequency Heating," RSIN, Vol. 35, No. 6, June 1964.

Russell, A., "Inexpensive Thermocouple-Scanner Using Gold-Plated Relays," CNTL, Vol. 7, No. 63, Sept. 1963.

Sannes, T., "Tiny Thermocouples—Advances in Design Extended Applications," INTH, Vol. 14, No. 3, March 1967.

Savage, B., "Multipoint Thermocouple Reference Junction," JSIN, Vol. 40, No. 1, Jan. 1963.

Scadron, L., "Ceramic Insulated Thermocouple," INCS, May 1961.

Scadron, M. D., "Time Response Characteristics of Temperature Sensors," SAE, Paper No. 158H, April 1960.

Schinizlein, J. G. et al, "Temperature Transducer for Thermogravimetric Studies," RSIN, Vol. 36, No. 5, May 1965.

Schriempf, J. T., "Basic Studies of the Metallic State (Thermoelectric Power of 0.02 AT-% Iron in Gold)," RNRL, April 1965.

Sheldon, L. E., "Multiplexing Thermocouple Signal to a Computer-Controlled Data System," ISAC, Paper No. 53.3.63, Sept. 1963.

Shepard, C. E. and Warshawsky, I., "Electrical Techniques for Time Lag Compensation of Thermocouples Used on Gas Turbine Temperature Measurements," INST, Vol. 26, No. 11, Nov. 1953.

Sherman, A., "Thermocouple Circuit Restorer," INAU, Vol. 27, No. 1, Jan. 1954.

Shmonin, A. A., "Measurement and Control of Temperature in Furnaces," MSTC (IZTE), Vol. 2, No. 2, Feb. 1962.

Shortland, M. and Standly, D. J., "Thermocouple Switch Works inside Slab Furnace," INPA, Vol. 20, No. 1, Jan. 1966.

Shu, H. H. H., "Relation Between the Rubbing Interface Temperature Distribution and Dynamic Thermocouple Temperature," ASMS(D), Vol. 86, No. 3, Sept. 1964.

Sibley, F. S., "A New Oxidation and Corrosion Resistant Nickel Alloy Thermocouple Matching Type J Iron-Constantan in Calibration," ISAC, Paper No. 21.2.62, Oct. 1962.

Siede, O. A. and Edison, L. R., "Mechanics of Drawing and Swaging Ceramic Filled Tubes," ASTM, Paper, unpublished, Annual Meeting, June 1966.

Siede, O. A., "Manufacturing Procedures and Techniques for Metal Sheathed Ceramic Insulated Thermocouples," ASTM, Paper, unpublished, Annual Meeting, June 1966.

Silverman, L., "Reference Junctions," INCS, Vol. 36, No. 6, June 1963.

Simmons, F. S. and Glawe, G. E., "Theory and Design of a Pneumatic Temperature Probe and Experimental Results Obtained in a High Temperature Gas Stream," NACN, ON-3893, Jan. 1957, AD-118-995.

Simmons, J. D., "Tungsten-Molybdenum Thermocouple for Immersion Pyrometer," JISI, Vol. 175, No. 12, Dec. 1953.

Slater, H. W., "Improvements in Fine Wire Thermocouples," JSIN, Vol. 30, No. 8, Aug. 1953.

Slaughter, J. I. and Margrave, J. L., "Temperature Measurement, U.S. Air Force, Space Systems Division, Report No. TDR-169 (3240-20) TN-1, Oct, 1962, AD-298-142.

Sparks, L. L. and Hall, W. J., "Cryogenic Thermocouple Tables–Part II, Reference Materials Versus Thermocouple Alloys," NBS, Report No. 9719, Dec. 1968.

Sparks, L. L. and Hall, W. J., "Cryogenic Thermocouple Tables–Part III, Miscellaneous and Comparative Material Combinations," NBS, Report No. 9721, Jan. 1969.

Sparks, L. L. and Powell, R. L., "Available Low Temperature Thermocouple Information and Services," NBS, Report No. 875, Feb. 1965.

Sparks, L. L. and Powell, R. L., "Final Report on Thermometry Project," NBS, Report No. 9249, June 1966.

Sparks, L. L. and Powell, R. L., "Cryogenic Thermocouple Thermometry," MESD, Vol. 1, No. 2, March–April 1967.

Sparks, L. L. et al, "Cryogenic Thermocouple Tables," NBS, Report No. 9721, July 1968.

Stamper, J. A., "Differential Sensing Controlled Thermocouple," RSIN, Vol. 34, No. 4, April 1963.

Starr, C. D. and Wang, T. P., "Effect of Oxidation on Stability of Thermocouples," ASTE, Vol. 63, 1963.

Stepka, F. S. and Hickel, R. O., "Methods for Measuring Temperatures of Thin-Walled Gas Turbine Blades," NACR, E56G17, Nov. 1956, AD-112-316.

Stevenson, J. A., "Metal-Clad Thermocouples," PTMR, Vol. 4, No. 4, Oct. 1960.

Stimson, H. F., "International Temperature Scale of 1948, Test Revision of 1960," NBSM, No. 37, Sept. 1961.

Stover, C. M., "Method of Butt Welding Small Thermocouples 0.001 to 0.010 Inches in Diameter," RSIN, Vol. 31, No. 6, June 1960.

Strittmaster, R. C. et al, "Measurement of Temperature Profiles in Burning Solid Propellant," U.S. Army Material Command, Ballistic Research Laboratory, Report No. BRL-MR-17-37, March 1966, AD-635188.

Strub, J. W., "The Use of Chord Thermocouples for Monitoring the Thermal Resistance of Boiler Waterside Deposits," PAPW, Vol. 23, 1961.

Sturm, W. J. and Jones, R. J., "Applications of Thermocouples to Target Temperature Measurement in the Internal Beam of a Cyclotron," RSIN, Vol. 25, No. 4, April 1954.

Swaney, F. S., "Open Hearth Bath Temperature Measurement," INST, Vol. 26, No. 2, Feb. 1953.

Swindells, J. F., ed., "Precision Measurement and Calibration–Temperature," NBS, Special Publication No. 300, Vol. 2, Aug. 1968.

Taligren, V. and Kracht, V., "Reliable D.C. Amplifier for Thermocouple Vacuum Gauges," ELEG, Vol. 36, No. 10, Oct. 1964.

Thielke, M. R., "Thermoelectric Materials," National Carbon Co., Jan. 1959, AD-220537.

Thomas, D. B., "Furnace for Thermocouple Calibrations to 2200 C," JNBC(C), Vol. 66C, No. 7, July 1962.

Thomas, D. B., "Studies on the Tungsten-Rhenium Thermocouple to 2000 C," JNBC(C), Vol. 67, No. 4, Oct.–Dec. 1963.

Tillinger, M. H., "Noise Thermometry at High Pressure," Polytechnic Institute of Brooklyn, TR-1, June 1965, AD-467-990.

Toenshuff, D. A., "Automatic Calibration of Thermocouples," EITB, Vol. 2, No. 3, Dec. 1961.

Tschang, P. S., "Temperature Determination in Moderately Dense High Temperature Gases by Transient Thermocouple Probes," Wright-Patterson Air Force Base, Ohio, ARL-65-95, May 1965, AD-617-702.

Tseng, Y. et al, "Platinum-Molybdenum Thermocouple," EITB, Vol. 9, No. 3, Dec. 1968.

Tsyplyatnikov, G. P. and Aleskorskiy, V. B., "A Thermochemical Gas Analyzer for Continuous Detection of Oxygen," IVVK, Vol. 3, No. 3, 1960, AD-295-607.

Turner, G. H., "Response of a Thermocouple Junction to Shock Waves in Copper," JAPI, Vol. 35, No. 10, Oct. 1964.

Turner, R. C. and Gordon, G. D., "Thermocouple for Vacuum Tests Minimizes Errors," SPAE, Vol. 41, No. 1, Jan. 1964.

Vrolyk, J. J. and Kinzie, P. A., "Feasibility Investigation of Heat-Flow Rate Measuring Techniques," Wright-Patterson Air Force Base, Ohio, RTD-TDR-63-4077, Jan. 1964, AD-431189.

Walker, B. E. et al, "High Temperature Materials Research," RNRL, April 1965.

Walter, B. E. et al, "Instability of Refractory Metal Thermocouples," RSIN, Vol. 36, No. 6, June 1965, AD-621-484.

Walker, B. E. et al, "Study of the Instability of Noble Metal Thermocouples in Vacuum," RSIN, Vol. 36, No. 5, May 1965.

Walker, B. E. et al, "Thermoelectric Instability of Some Noble Metal Thermocouples at High Temperature," RSIN, Vol. 33, No. 9, Sept. 1962.

Wang, T. P., "Temperature Sensors," INCS, Vol. 40, No. 5, May 1967.

Watson, G. G., "Techniques for Measuring Surface Temperatures," INPA, Vol. 20, Nos. 3–8, March–Aug. 1966.

Wechter, G. H., "The Accuracy of Thermocouples in Measuring Surface Temperatures of Stainless Steel Specimens Subject to Radiant Heat-Phase II," Aeronautical Structures Laboratory, Naval Air Development Center, NAEC, ASL 13 R 360FR101, July 1963, AD-412-317.

Weiner, S. and Schwartz, F., "Thermopile IR Detectors," SPAE, Vol. 40, No. 8, Aug. 1963.

White, F. J., "Accuracy of Thermocouples in Radiant Heat Testing," EXMC, Vol. 2, July 1962.

Whitow, L. and Porter, M. J., "Method of Reducing Time Lag of Transducers Which Have Exponential Response," ELOE, Vol. 31, No. 379, Sept. 1959.

Wilks, C. R., "Creep and Rupture Test Pyrometry," ASTT, No. 178, 1955.

Williams, R. J. and Kaiura, E. H., "Time Constant Study of Some Commercial Temperature Sensors at Cryogenic Temperatures," The Boeing Co., Report No. D2-23831-1, July 1965, AD-485191L.

Wormser, A. F., "Experimental Determination of Thermocouple Time Constants with Use of a Variable Turbulence, Variable Density Wind Tunnel and the Analytic Evaluation of Conduction Radiation and Other Secondary Effects," SAE, Paper No. 158 D, April 1960.

Wormser, A. F. and Pfuntner, R. A., "Pulse Technique Extends Range of Chromel Alumel to 7000 F," INCS, Vol. 37, No. 5, May 1964.

Wormser, A. F. and Pfuntner, R. A., "Pulse Thermocouple Measures at 5700 F," SPAE, Vol. 40, No. 7, July 1963.

Wright, D. F. and Acheson, G. F., "Effect on the Static Strength of Aluminum Alloy Test Specimens of the Attachment of Thermocouples by a Welding Technique," Royal Aircraft Establishment, Technical Note, Structures 345, Jan. 1964, AD-440894.

Young, R., "Thermoelectric Thermometry," CEEN, Vol. 33, No. 14, Oct. 1960.

Zabawsky, Z. and Gavan, F. M., "Thermocouples and Their Usage in ASTM Standards," MTRS, Vol. 5, No. 2, Feb. 1965.

Zopf, W. D., "Thermal-Electric Type Instrumentation for the Detection and Relative Measurement of Infrared Radiation in a Selected Wave Length Bound of Essentially Black Body Radiation," Arkansas University Masters Thesis, 1963, AD-298162.

Zysk, E. D. and Toenshoff, D. A., "Calibration of Refractory Metal Thermocouples," ISAP, Paper No. 12.11.4.66, Oct. 1966.

Zysk, E. D., "Noble Metals in Thermometry—Recent Developments," EITB, Vol. 5, No. 3, Dec. 1964.

Zysk, E. D., "A Review of Recent Work with the Platinel Thermocouple," EITB, Vol. 4, No. 1, June 1963.

Zysk, E. D. et al, "Tungsten 3 Rhenium Versus Tungsten 25 Rhenium—A New High Temperature Thermocouple," EITB, Vol. 3, No. 4, March 1963.

12.3 Coden for Periodical Titles

CODEN	Title
ANCH	Analytical Chemistry
ASMS	American Society of Mechanical Engineers, Paper
ASTE	American Society for Testing and Materials, Proceedings, Committee Reports, Technical Papers
ASTT	American Society for Testing and Materials, Special Technical Publication
ATEL	Avto Matika I Telemekhanika (USSR)
AUSV	Automaticheskaya Svarka (USSR) Translated by Administrative Center for Scientific Information and Liaison (UK)
CEEN	Certificated Engineer (UK)
CELE	Communication and Electronics
CENG	Control Engineering
CHEG	Chemical Engineer, The
CJPH	Canadian Journal of Physics
CNTL	Control
CRYO	Cryogenics
DSER	Data Systems Engineering
EITB	Engelhard Industries, Inc., Technical Bulletin
ELEC	Electronics
ELEG	Electronic Engineering
ELOE	Electronic Engineer
ELTE	Electro-Technology
EXMC	Experimental Mechanics
HUTL	Hutnicke Listy (Czech)
INAU	Instruments and Automation
INCS	Instruments and Control Systems
INDL	Industrial Laboratory (USSR)
INET	Instruments and Experimental Techniques (USSR)
INLA	Industrial Laboratories
INPA	Instrument Practice
INSR	Instrumentation
INST	Instruments
INTH	Instrumentation Technology
IRAG	Iron Age, The
IRES	Industrial Research
IRSE	Iron and Steel Engineer
ISAC	Instrument-Automation Conference and Exhibit Proceedings, Annual, Instrument Society of America
ISAJ	Instrument Society of America Journal

CODEN	*Title*
ISAP	Instrument Society of America Preprint (Paper)
ISAT	Instrument Society of America, Transactions
ISCT	International Science and Technology
ITAE	Institute of Electrical and Electronic Engineers, Transactions on Aerospace and Electronic Systems
IVUK	IZ Vestiya Vysshikh Uchebnykh Zavedeniy, Khimiya I Khimicheskaya Tekhnologiya (USSR)
IZTE	Izmeritel'naia Tekhnika (USSR)
JAMC	Journal of Applied Mechanics (ASME)
JAPI	Journal of Applied Physics
JBAE	Journal of Basic Engineering (ASME)
JEPO	Journal of Engineering for Power (ASME)
JHTR	Journal of Heat Transfer (ASME)
JIME	Journal of the Institute of Metals (UK)
JISI	Journal of Iron and Steel Institute (UK)
JNBC(C)	Journal of Research, National Bureau of Standards, C. Engineering and Instrumentation
JSIN	Journal of Scientific Instruments
LNTJ	Leeds and Northrup Co. Technical Journal
MADE	Machine Design
MEEN	Mechanical Engineering
MEPO	Metal Progress
MESD	Measurements and Data
METL	Metallurgia
MSTC	Measurement Techniques (USSR)
NAAR	National Advisory Committee for Aeronautics, Research Abstracts
NACN	National Advisory Committee for Aeronautics, Technical Notes
NBSE	National Bureau of Standards Technological Papers
NBSM	National Bureau of Standards Monograph
NBST	National Bureau of Standards, Technical News Bulletin
NSSP	National Aeronautics and Space Administration, Special Publication
OGNP	Ogneapory (USSR)
PAPW	Proceedings of the American Power Conference
PEPR	Petroleum Processing
PIML	Proceedings of the Institution of Mechanical Engineers (UK)
POWE	Power
PREN	Product Engineering
PRTE	Pribory I Teknika Eksperimenta (USSR)
PTMR	Platinum Metals Review
RNRL	Report of Naval Research Laboratory Progress
RSIN	Review of Scientific Instruments
SAEJ	SAE Journal, Society of Automotive Engineers
SEPP	SAE, Preprint, Society of Automotive Engineers
SPAE	Space Aeronautics
STAE	Scientific and Technical Aerospace Reports (STAR), National Aeronautic and Space Administration
VIDE	Vide (France)
XAER	United States Atomic Energy Commission
ZPMF	Zhurnal Prikladnei Mekhaniki I Tekhnicheskoi Fiziki (USSR)
ZVDL	Zavodskaya Laboratoriya (USSR)

13. DEFINITIONS

adjusting device (liquid-in-glass thermometer) *n.*—a device to adjust the liquid in the bulb and main capillary to that needed for the intended temperature interval.

bulb (liquid-in-glass thermometer) *n.*—the reservoir for the thermometer liquid.

bulb length (liquid-in-glass thermometer) *n.*—the distance from the bottom of the bulb to the point where the internal bulb diameter begins to decrease as the bulb merges into the stem.

calibrate, *v.*: 1. *general*—to determine the indication or output of a measuring device with respect to that of a standard.

2. *liquid-in-glass thermometer*—to determine the indication of the thermometer with respect to temperature established by a standard.

3. *thermocouple*—to determine the emf developed by a thermocouple with respect to temperature established by a standard.

calibration point, *n.*: 1. *general*—a specific value, established by a standard, at which the indication or output of a measuring device is determined.

2. *liquid-in-glass thermometer*—a temperature, established by a standard, at which the indication of the thermometer is determined.

3. *thermocouple*—a temperature, established by a standard, at which the emf developed by a thermocouple is determined.

Celsius, *n.*—the designation of the degree on the International Practical Temperature Scale of 1948. Also used for the name of the Scale, as "Celsius temperature scale". Formerly (prior to 1948) called "Centigrade."

centigrade, *n.*—the designation of the degree on the International Temperature Scale prior to 1948. (See *Celsius*.)

coaxial thermocouple element, *n.*—a thermocouple element consisting of a thermoelement in wire form, within a thermoelement in tube form and insulated from the tube except at the measuring junction.

complete immersion thermometer, *n.*—a liquid-in-glass thermometer designed to indicate temperatures correctly when the entire thermometer is exposed to the temperature being measured. (Compare *total immersion thermometer*.)

connecting wire (metal-sheathed heater), *n.*—a conductor used to connect the heater resistance wire to the power supply terminals.

connection head, *n.*—a housing enclosing a terminal block for an electrical temperature-sensing device and usually provided with threaded openings for attachment to a protecting tube and for attachment of conduit.

contraction chamber (liquid-in-glass thermometer), *n.*—an enlargement of the bore of the stem which serves to reduce its length, or to prevent contraction of the liquid column into the bulb.

defining fixed points, *n.*—the reproducible temperatures upon which the International Practical Temperature Scale is based.

degree, *n.*—the unit of a temperature scale.

diameter, (liquid-in-glass thermometer), *n.*—the diameter as measured with a ring gage.

electromotive force (emf) *n.*—the electrical potential difference which produces or tends to produce an electric current.

expansion chamber (liquid-in-glass thermometer),*n.*—an enlargement at the top of the capillary to provide protection in case of overheating.

extension wire, *n.*—a pair of wires having such temperature-emf characteristics relative to the thermocouple with which the wires are intended to be used that, when properly connected to the thermocouple, the reference junction is transferred to the other end of the wires.

Fahrenheit, *n.*—the designation of the degree and the temperature scale used commonly in public life and engineering circles in English-speaking countries. Related to the International Practical Temperature Scale by means of the equation:

$$t_F = 9/5\, t_C + 32$$

fixed point, *n.*—a reproducible temperature of equilibrium between different phases of a material. (See *Defining Fixed Points* and *Secondary Reference Points*.)

freezing point, *n.*—the fixed point between the solid and liquid phases of a material when approached from the liquid phase under a pressure of 1 standard atm (101325 N/m^2). For a pure material this is also the melting point.

heater, metal sheathed, electrical resistance, *n.*—one consisting of resistance wire or wires, with or without connecting wires, embedded in ceramic insulation compacted within a metal protecting tube.

ice point, *n.*—the fixed point between ice and air-saturated water under a pressure of 1 standard atm (101325 N/m^2). This temperature is 0.000 C on the International Practical Temperature Scale of 1948.

International Practical Temperature Scale of 1948 (IPTS), *n.*—the temperature scale, which through adoption by the 11th General Conference on Weights and Measures in 1960, is defined in terms of fixed and reproducible equilibrium temperatures (defining fixed points) to which numerical values have been assigned, and equations establishing the relation between temperature and the indications of sensing instruments calibrated by means of the values assigned to the defining fixed points.

Kelvin, *n.*—the designation of the thermodynamic temperature scale and the degree on this scale. This Kelvin scale was defined by the Tenth General Conference on Weights and Measures in 1954 by assigning the temperature of 273.16 K to the triple point of water. Also the degree on the International Practical Kelvin Temperature Scale.

liquid-in-glass thermometer, *n.*—a temperature-measuring instrument whose indications are based on the temperature coefficient of expansion of a liquid relative to that of its containing glass bulb.

lower range value, *n.*—the lowest quantity that an instrument is adjusted to measure.

measuring junction, *n.*—that junction of a thermocouple which is subjected to the temperature to be measured.

melting point, *n.*—the fixed point between the solid and liquid phases of a material when approached from the solid phase under a pressure of 1 standard atm (101325 N/m^2). For a pure material this is also the freezing point.

partial immersion thermometer, *n.*—a liquid-in-glass thermometer designed to indicate temperatures correctly when the bulb and a specified part of the stem are exposed to the temperature being measured.

Peltier coefficient, *n.*—the reversible heat which is absorbed or evolved at a thermocouple junction when unit current passes in unit time. Synonymous with *Peltier emf.*

Peltier emf, *n.*—Synonymous with *Peltier coefficient.*

platinum 27, *n.*—the platinum standard to which the National Bureau of Standards refers thermoelectric measurements.

potentiometer, Group A, *n.*—a laboratory high-precision type potentiometer having limits of error of approximately 0.2 μV at 1000 μV, and 5 μV or less at 50,000 μV.

potentiometer, Group B, *n.*—a laboratory precision-type potentiometer having limits of error of approximately 1 μV at 1000 μV and 12 μV or less at 50,000 μV.

primary standard resistance thermometer, *n.*—a resistance thermometer that has had its temperature-resistance relationship determined in accordance with methods described in the text establishing the International Practical Temperature Scale of 1948.

primary standard thermocouple, *n.*—a thermocouple that has had its temperature-emf relationship determined in accordance with methods described in the text establishing the International Practical Temperature Scale of 1948.

protecting tube, *n.*—a tube designed to enclose a temperature-sensing device and protect it from the deleterious effects of the environment. It may provide for attachment to a connection head but is not primarily designed for pressure-tight attachment to a vessel.

range, *n.*—the region between the limits within which a quantity is measured. It is expressed by stating the lower and upper range-values.

reference junction, *n.*—that junction of a thermocouple which is at a known temperature.

reference point (liquid-in-glass thermometer), *n.*—a temperature at which a thermometer is checked for changes in bulb volume.

refractory metal thermocouple, *n.*—a thermocouple whose thermoelements have melting points above that of 60 percent platinum, 40 percent rhodium, 1935 C (3515 F).

saddle (liquid-in-glass thermometer), *n.*—the bottom support of the enclosed scale of an enclosed-scale thermometer.

secondary reference points, *n.*—reproducible temperatures (other than the *defining fixed points*) listed in the text establishing the International Practical Temperature Scale as being useful for calibration purposes.

secondary standard thermocouple, *n.*—a thermocouple that has had its temperature-emf relationship determined by reference to a primary standard of temperature.

Seebeck coefficient, *n.*—the rate of change of thermal emf with temperature at a given temperature. Normally expressed as emf per unit of temperature. Synonymous with *thermoelectric power.*

Seebeck emf, *n.*—the net emf set up in a thermocouple under condition of zero current. It represents the algebraic sum of the Peltier and Thomson emf. Synonymous with *thermal emf.*

setting temperature (liquid-in-glass thermometer), *n.*—the temperature which causes a reading of zero on the main scale of an adjustable-range thermometer.

sheath (enclosed-scale thermometer), *n.*—the cylindrical glass envelope which encloses the scale and capillary tube.

sheathed thermocouple, *n.*—a thermocouple having its thermoelements, and sometimes its measuring junction, embedded in ceramic insulation compacted within a metal protecting tube.

sheathed thermocouple wire, *n.*—one or more pairs of thermoelements (without measuring junction(s)) embedded in ceramic insulation compacted within a metal protecting tube.

sheathed thermoelement, *n.*—a thermoelement embedded in ceramic insulation compacted within a metal protecting tube.

span, *n.*—the algebraic difference between the upper and lower range-values.

standard thermoelement, *n.*—a thermoelement that has been calibrated with reference to platinum 27.

stem (liquid-in-glass thermometer), *n.*—the capillary tube through which the meniscus of the liquid moves with change of temperature.

temperature interval (liquid-in-glass thermometer), *n.*—a specified portion of the range of a thermometer.

test thermocouple, *n.*—a thermocouple that is to have its temperature-emf relationship determined by reference to a temperature standard.

test thermoelement, *n.*—a thermoelement that is to be calibrated with reference to platinum 27 by comparing its thermal emf with that of a standard thermoelement.

thermal electromotive force (thermal emf), *n.*—the net emf set up in a thermocouple under conditions of zero current. Synonymous with *Seebeck emf.*

thermocouple, *n.*—two dissimilar thermoelements so joined as to produce a thermal emf when the junctions are at different temperatures.

thermocouple assembly, *n.*—an assembly consisting of a thermocouple element and one or more associated parts such as terminal block, connection head, and protecting tube.

thermocouple element, *n.*—a pair of bare or insulated thermoelements joined at one end to form a measuring junction and intended for use as a thermocouple or as part of a thermocouple assembly.

thermocouple, Type E, B, J, K, R, S, or T, *n.*—a thermocouple having an emf-temperature relationship corresponding to the appropriate letter-designated table in ASTM Standard E 230, Temperature Electromotive Force (EMF) Tables for Thermocouples, within the limits of error specified in that Standard.

thermoelectric power, *n.*—the rate of change of thermal emf with temperature at a given temperature. Synonymous with *Seebeck coefficient.* Normally expressed as emf per unit of temperature.

thermoelectric pyrometer, *n.*—an instrument that senses the output of a thermocouple and converts it to equivalent temperature units.

thermoelement, *n.*—one of the two dissimilar electrical conductors comprising a thermocouple.

thermopile, *n.*—a number of thermocouples connected in series, arranged so that alternate junctions are at the reference temperature and at the measured temperature, to increase the output for a given temperature difference between reference and measuring junctions.

thermowell, *n.*—a closed end reentrant tube designed for the insertion of a temperature-sensing element, and provided with means for pressure-tight attachment to a vessel.

omson coefficient, *n.*—the rate at which heat is absorbed or evolved reversibly in a thermo-element per unit temperature difference per unit current.

omson emf, *n.*—the product of the Thomson coefficient and the temperature difference across a thermoelement.

al immersion thermometer, *n.*—a liquid-in-glass thermometer designed to indicate temperatures correctly when just that portion of the thermometer containing the liquid is exposed to the temperature being measured. (Compare *complete immersion thermometer*).

al length (liquid-in-glass thermometer), *n.*—the over-all length including any special finish at the top.

ple point (water), *n.*—the temperature of equilibrium between ice, water, and water vapor. This temperature is +0.01 C on the International Practical Temperature Scale of 1948.

per range-value, *n.*—the highest quantity that an instrument is adjusted to measure.

ification (liquid-in-glass thermometer), *n.*—the process of testing a thermometer for compliance with specifications.

ification temperatures (liquid-in-glass thermometer), *n.*—the specified temperatures at which thermometers are tested for compliance with scale error limits.

rking standard thermocouple, *n.*—a thermocouple that has had its temperature-emf relationship determined by reference to a secondary standard of temperature.